American Mathematical Society
COLLOQUIUM PUBLICATIONS

Volume 24

Structure of Algebras

A. Adrian Albert

American Mathematical Society
Providence, Rhode Island

1991 *Mathematics Subject Classification.* Primary 16–01; Secondary 16K20.

Library of Congress Catalog Card Number 41-9
International Standard Book Number 0-8218-1024-3
International Standard Serial Number 0065-9258

Revised edition, 1961

Printed in the United States of America

♾ The paper used in this book is acid-free and falls within the guidelines established to ensure permanence and durability.

♻ Printed on recycled paper.

12 11 10 9 8 7   99 98 97 96 95 94

TO F. D. A.

PREFACE

The theory of linear associative algebras probably reached its zenith when the solution was found for the problem of determining all rational division algebras. Since that time it has been my hope that I might develop a reasonably self-contained exposition of that solution as well as of the theory of algebras upon which it depends and which contains the major portion of my own discoveries. The first step in carrying out this desire was necessarily that of writing a text with contents selected so as to provide a foundation adequate for the prospective exposition. This text has already been published under the title *Modern Higher Algebra*. Its completion was followed shortly by the timely invitation of the Colloquium Committee of the American Mathematical Society to me to write these LECTURES embodying the desired exposition.

It has been most fortunately possible at this time to give a new treatment of the early parts of our subject simplifying not only the proofs in the theory of normal simple algebras but even the exposition of the structure theorems of Wedderburn. This does not evidence itself in the somewhat classical first two chapters. These contain the preliminary discussion of linear sets, direct products, direct sums, ideals in an algebra, and similar topics, with the additions and modifications made necessary by the fact that we are considering here algebras over an arbitrary field and that inseparable extension fields may exist. But the exposition given of the usual fundamental theorem, stating that every linear associative algebra is equivalent to a first algebra of square matrices and reciprocal to a second such algebra, is expanded here so as to have as consequence a result basic in the new treatment of the Wedderburn structure theory. While this result is not derived until its need appears in Chapter III the proof is so elementary that it might have been placed in the first chapter without change.

This basic theorem is that of R. Brauer on the structure of the direct product of a normal division algebra and its reciprocal algebra. It is combined with two theorems of J. H. M. Wedderburn to obtain as generalizations three tool theorems which are used throughout Chapters III and IV, and yield rather remarkable simplifications of the proofs of numerous fundamental results. In particular in Chapter III the foundation of the proof of the Wedderburn principal theorem on the structure of an algebra with a radical is simplified and the theorem itself then obtained.

While the first three chapters of this exposition do contain a considerable amount of new material their principal content is evidently an exposition in more modern form of the Wedderburn structure theorems which were first presented in book form in the two editions of the text on our subject by L. E. Dickson. I owe much to these expositions as well as to their author, who has been my teacher and the inspiration of all my research. The remaining eight chapters of the present text are composed principally of results derived since 1926 when the second (German) edition of Dickson's text was written. These results are due principally to R. Brauer, H. Hasse, E. Noether, and myself. Their exposition is begun in Chapter IV which contains the theory of the commutator subalgebra of a simple subalgebra of a normal simple algebra, the study of automorphisms of a simple algebra, splitting fields, and the index reduction factor theory.

The fifth chapter contains the foundation of the theory of crossed products and of their special case, cyclic algebras. The theory of exponents is derived there as well as the consequent factorization of normal division algebras into direct factors of prime-power degree.

Chapter VI consists of the recent study of the abelian group of cyclic systems which is applied in Chapter VII to yield the theory of the structure of direct products of cyclic algebras and the consequent properties of norms in cyclic fields. This chapter is closed with the recently developed theory of p-algebras.

In Chapter VIII an exposition is given of the theory of the representations of algebras. The treatment is somewhat novel in that while the recent expositions have used representation theorems to obtain a number of results on algebras, here the theorems on algebras are themselves used in the derivation of results on representations. The presentation has its inspiration in my work on the theory of Riemann matrices and is concluded by an introduction to the generalization (by H. Weyl and myself) of that theory.

In the ninth chapter the structure of rational division algebras is determined. The study begins with a detailed exposition of Hasse's theory of p-adic division algebras. The results are then extended so as to yield the theorems on rational division algebras without recourse to the theory of ideals in the integral sets of such algebras. This is believed to be the first time the extension has been made in a really simple fashion. The method is a greatly desirable one as the previous treatments used the results of a very voluminous theory which we are able to omit. It is necessary, of course, to assume without proof certain existence theorems from the theory of algebraic numbers. These presupposed theorems are indicated precisely, and I hope to be able to include their proofs in a future text on the theory of algebraic numbers and the arithmetic of algebras.

The theory of involutorial simple algebras arose in connection with the study of Riemann matrices but is now a separate branch of the theory of simple algebras with structure theorems on approximately the same level as those on arbitrary simple algebras. This theory is derived in Chapter X both for algebras over general fields and over the rational field. The results are also applied in the determination of the structure of the multiplication algebras of all generalized Riemann matrices, a result which is seen in Chapter XI to imply a complete solution of the principal problem on Riemann matrices.

This final reference is but one item in the last chapter which contains an exposition of a number of special results. In particular there are given new derivations of the structure of all normal division algebras of degrees three and four over any field. References to sources for the whole text are given in this chapter as well as indications of the literature on the subject and an extensive bibliography.

It is my hope that the form of this exposition will make it useful as a text on the theory of linear associative algebras as well as for its obvious purpose as a source book for young algebraists. Much of any success that there may have been in keeping the exposition completely correct and clear is due to the work of Dr. Sam Perlis who read the manuscript critically in each of the stages of its preparation. He not only assisted in keeping the exposition free of error but frequently indicated improvements resulting in greater clarity. I give him my great thanks.

I appreciate also the kind assistance of Professor Nathan Jacobson who suggested the proofs of two of the theorems as well as that of Mr. Morris Bloom who assisted in the preparation of the bibliography, and give thanks to Professor Saunders MacLane and Dr. Otto F. G. Schilling who were a critical audience for oral expositions of some of the proofs. Final thanks are due to Professor G. A. Bliss without whose encouragement the completion of these LECTURES would have been greatly delayed.

A. A. ALBERT

THE UNIVERSITY OF CHICAGO
March 5, 1939

TABLE OF CONTENTS

CHAPTER I

FUNDAMENTAL CONCEPTS

1. The notations. As was stated in the preface, our exposition is founded upon the author's *Modern Higher Algebra* and will use the notations, definitions, and theorems of that text. We shall also use script letters to represent sets of elements as well as the Gothic letters used in the former text.

The reference notation used in the *Modern Higher Algebra* will also be used here. Thus Theorem 7.5 will refer to Theorem 5 of Chapter VII, equation (4.29) to (29) of Chapter IV. A large number of references will be made to results of the *Modern Higher Algebra* by the use of the following notational device. We shall refer to Chapter AX, to Section A4.9, to Theorem A6.5, to equation (A5.28), or to page A196, and shall mean the respective references of our foundation text. References will also be made by title number to articles listed in the bibliography at the end of these LECTURES.

2. Linear sets over $\mathfrak{F}$. The theory of algebras is a theory of certain types of linear sets of finite order over a field $\mathfrak{F}$. The field $\mathfrak{F}$ will occupy the rôle of fundamental underlying coefficient field in our exposition, and we shall use the symbol $\mathfrak{F}$ with this as its meaning and without further restatement of the fact throughout these LECTURES. Later special types of fields $\mathfrak{F}$ will be considered, but, until we state otherwise, $\mathfrak{F}$ will be arbitrary.

The concept of a linear set $\mathfrak{A}$ of order n over $\mathfrak{F}$ is the usual one of Section A2.11. Thus in particular every linear set of order n over $\mathfrak{F}$ is equivalent to the set of all sequences $(\alpha_1, \cdots, \alpha_n)$ with α_i in $\mathfrak{F}$ and such that

$$(\alpha_1, \cdots, \alpha_n) + (\beta_1, \cdots, \beta_n) = (\alpha_1 + \beta_1, \cdots, \alpha_n + \beta_n),$$

$$\lambda(\alpha_1, \cdots, \alpha_n) = (\alpha_1, \cdots, \alpha_n)\lambda = (\lambda\alpha_1, \cdots, \lambda\alpha_n),$$

for all α_i, β_j and λ of $\mathfrak{F}$. We shall use the notations and properties of linear sets given in Section A2.11 and shall now obtain some additional results.

The linear set $\mathfrak{L}_0$ consisting of zero alone will be called the *zero set.* We shall write $\mathfrak{L}_0 = 0$ and say that $\mathfrak{L}_0$ has *order zero.* In what follows let $\mathfrak{A}$ be any linear set of order $n > 0$ over $\mathfrak{F}$. The *sum* $(\mathfrak{A}_1, \cdots, \mathfrak{A}_t)$ of linear subsets $\mathfrak{A}_i$ of $\mathfrak{A}$ is defined as the set of all $a_1 + \cdots + a_t$ for a_i in $\mathfrak{A}_i$. It is easily verified to be a linear subset over $\mathfrak{F}$ of $\mathfrak{A}$. Addition of linear sets is then an associative and commutative operation.

The intersection $[\mathfrak{B}, \mathfrak{C}]$ of two linear subsets of $\mathfrak{A}$ is the set of all quantities common to $\mathfrak{B}$ and $\mathfrak{C}$. It is clearly a linear subset over $\mathfrak{F}$ of $\mathfrak{A}$. Moreover, as we shall show, *the order of* $(\mathfrak{B}, \mathfrak{C})$ *is the sum of the orders of* $\mathfrak{B}$ *and* $\mathfrak{C}$ *minus the order of* $[\mathfrak{B}, \mathfrak{C}]$.

The expression of the quantities of $\mathfrak{A} = (\mathfrak{A}_1, \cdots, \mathfrak{A}_t)$ in the form $a =$

$a_1 + \cdots + a_t$ with a_i in $\mathfrak{A}_i$ is unique if and only if the order of $\mathfrak{A}$ is the sum of the orders of the $\mathfrak{A}_i$. This property is true if and only if $a = 0$ implies that every $a_i = 0$. In this case we call the $\mathfrak{A}_i$ *supplementary* in their sum $\mathfrak{A}$, call $\mathfrak{A}$ the *supplementary sum* of the $\mathfrak{A}_i$, and write

$$\mathfrak{A} = \mathfrak{A}_1 + \cdots + \mathfrak{A}_t . \tag{1}$$

We shall not use the symbol + for the sum of linear sets except when it is a supplementary sum. Observe that (1) holds if and only if $[\mathfrak{B}_i , \mathfrak{A}_{i+1}] = 0$ $(i = 1, \cdots , t)$, where $\mathfrak{B}_i = (\mathfrak{A}_1 , \cdots , \mathfrak{A}_i)$.

In the exercise of Section A2.11 a linear set $\mathfrak{A} = (u_1 , \cdots , u_n)$ over $\mathfrak{F}$ was considered with a proper linear subset $\mathfrak{B} = (v_1 , \cdots , v_m)$ and the statement was made that then $\mathfrak{A} = (v_1 , \cdots , v_n)$ over $\mathfrak{F}$ where $v_{m+1} , \cdots , v_n$ are in $\mathfrak{A}$. This result is important though trivially proved, and we state it as follows: *Let $\mathfrak{B}$ be a linear subset of $\mathfrak{A}$ over $\mathfrak{F}$. Then $\mathfrak{A} = \mathfrak{B} + \mathfrak{C}$ where $\mathfrak{C}$ is a linear subset of $\mathfrak{A}$ called a supplement of $\mathfrak{B}$ in $\mathfrak{A}$.* Note that $\mathfrak{C}$ is not unique and that in particular $\mathfrak{A} = \mathfrak{B} + \mathfrak{C}'$ where

$$\mathfrak{C}' = (v'_{m+1} , \cdots , v'_n), \qquad v'_i = v_i + b_i , \tag{2}$$

for any b_i in $\mathfrak{B}$.

To prove that the order of $(\mathfrak{B}, \mathfrak{C})$ is the sum of the orders of $\mathfrak{B}$ and $\mathfrak{C}$ minus the order of their intersection $\mathfrak{D}$, we have $\mathfrak{C} = \mathfrak{D} + \mathfrak{C}'$ as above. Then $(\mathfrak{B}, \mathfrak{C}) = (\mathfrak{B}, \mathfrak{C}')$. No quantity of $\mathfrak{C}'$ is in $\mathfrak{D}$ and hence no quantity of $\mathfrak{C}'$ is in $\mathfrak{B}$. It follows that $(\mathfrak{B}, \mathfrak{C}') = \mathfrak{B} + \mathfrak{C}'$, and the result we desire is an immediate consequence.

A set of quantities $b_1 , \cdots , b_r$ of $\mathfrak{A}$ is said to *span* the linear subset $\mathfrak{B}$ of $\mathfrak{A}$ consisting of all $\lambda_1 b_1 + \cdots + \lambda_r b_r$ for λ_i in $\mathfrak{F}$. The order of $\mathfrak{B}$ is then an integer $m \leqq r$, and in fact the b_i may be renumbered so that $\mathfrak{B} = (b_1 , \cdots , b_m)$ over $\mathfrak{F}$. Furthermore let $\mathfrak{A} = (u_1 , \cdots , u_n)$ over $\mathfrak{F}$ and thus

$$b_i = \sum_{j=1}^{n} \lambda_{ij} u_j \qquad (i = 1, \cdots , r), \tag{3}$$

with unique λ_{ij} in $\mathfrak{F}$. Then m is the number of linearly independent b_i. By Exercise 5, page A67, m is the rank of the matrix $L = (\lambda_{ij})$. Also, if $r = n$, then $\mathfrak{B} = \mathfrak{A}$ if and only if $m = n$, that is, the square matrix L is non-singular.

The above is our final preliminary study and we shall now begin our treatment of algebras themselves.

3. Algebras over $\mathfrak{F}$. The theory of algebras of order n over $\mathfrak{F}$ was treated in Chapter AX from its aspect as a theory of matrices. While certain of the resulting properties are of considerable importance and will be used later we shall not base our present exposition upon that treatment but shall tend rather to a more abstract discussion.*

Algebras of finite order over $\mathfrak{F}$ may be defined as rings which are linear sets

* However, if the present treatment be used as a course text, the material of sections A10.1–5 is to be considered as *essential* introductory material.

of finite order over $\mathfrak{F}$. An equivalent but more postulational definition, in view of the definition of linear set of Chapter AII, may be given as follows. We define ring and field by the postulates of Chapters AI, AII and then make the

DEFINITION. *Let $\mathfrak{F}$ be a field with unity quantity* 1, *$\mathfrak{A}$ be a ring, and let* (*the scalar products*) $\alpha a = a\alpha$ *be in $\mathfrak{A}$ for every α of $\mathfrak{F}$ and a of $\mathfrak{A}$. Then we call $\mathfrak{A}$ an algebra of order n over $\mathfrak{F}$ if for every α and β of $\mathfrak{F}$, a and b of $\mathfrak{A}$ we have*

I. $1a = a$, $\alpha(\beta a) = (\alpha\beta)a$, $(\alpha a)(\beta b) = (\alpha\beta)(ab)$;

II. $(\alpha + \beta)a = \alpha a + \beta a$; $\alpha(a + b) = \alpha a + \alpha b$;

III. *There exist quantities $u_1, \cdots, u_n$ in $\mathfrak{A}$ such that every quantity a of $\mathfrak{A}$ is uniquely expressible in the form*

$$a = \alpha_1 u_1 + \cdots + \alpha_n u_n \tag{4}$$

for α_i in $\mathfrak{F}$.

When no ambiguity is introduced by the following convention, we shall speak of algebras $\mathfrak{A}$ of order n over $\mathfrak{F}$ simply as *algebras*. Similarly we shall say that $\mathfrak{A}$ and $\mathfrak{A}'$ are *equivalent* when we mean that $\mathfrak{A}$ of order n over $\mathfrak{F}$ and $\mathfrak{A}'$ of order n over $\mathfrak{F}$ are equivalent over $\mathfrak{F}$ with respect to the definition of page A44.

Two algebras $\mathfrak{A}$ and $\mathfrak{A}'$ will be called, briefly, *reciprocal algebras* if they satisfy the condition given in the

DEFINITION. *Let $\mathfrak{A}$ and $\mathfrak{A}'$ be algebras of order n over $\mathfrak{F}$ and let there be a* (1-1) *correspondence $a \leftrightarrow a'$ between them such that*

$$(a + b)' = a' + b', \qquad (\lambda a)' = \lambda a', \qquad (ab)' = b'a'$$

for every a and b of $\mathfrak{A}$ and λ of $\mathfrak{F}$. Then we call $\mathfrak{A}$ and $\mathfrak{A}'$ reciprocal over $\mathfrak{F}$.

We shall designate the property that $\mathfrak{A}$ and $\mathfrak{A}'$ are reciprocal by writing $\mathfrak{A}^{-1}$ for $\mathfrak{A}'$. This notation will be used frequently in later chapters. Note that $\mathfrak{A}^{-1}$ is uniquely determined by $\mathfrak{A}$ in the sense of equivalence.

The fact that an algebra $\mathfrak{A}$ over $\mathfrak{F}$ is a linear set of finite order over $\mathfrak{F}$ will be used repeatedly. A subalgebra $\mathfrak{B}$ over $\mathfrak{F}$ of $\mathfrak{A}$ is any linear subset which is an algebra with respect to the operations defining $\mathfrak{A}$. Thus a linear subset $\mathfrak{B}$ of $\mathfrak{A}$ is a subalgebra of $\mathfrak{A}$ if and only if the product of any two quantities of $\mathfrak{B}$ is in $\mathfrak{B}$. Moreover $\mathfrak{B} = \mathfrak{A}$ if and only if $\mathfrak{B}$ and $\mathfrak{A}$ have the same order.

4. Products of linear subsets of an algebra. Let $\mathfrak{B} = (u_1, \cdots, u_m)$ and $\mathfrak{C} = (v_1, \cdots, v_t)$ over $\mathfrak{F}$ be linear subsets of an algebra $\mathfrak{A}$. We define the *product* $\mathfrak{B}\mathfrak{C}$ to be the linear subset over $\mathfrak{F}$ of $\mathfrak{A}$ spanned by the mt quantities $u_i v_j$. The order of $\mathfrak{B}\mathfrak{C}$ is at most mt, and the set $\mathfrak{B}\mathfrak{C}$ consists of all finite sums $b_1c_1 + \cdots + b_rc_r$ for b_i in $\mathfrak{B}$ and c_i in $\mathfrak{C}$. Thus the product is independent of the particular bases used in its definition.

The operation of multiplication of linear sets is associative but not necessarily commutative. We also have the distributive laws

$$(\mathfrak{B}, \mathfrak{C})\mathfrak{D} = (\mathfrak{B}\mathfrak{D}, \mathfrak{C}\mathfrak{D}), \qquad \mathfrak{D}(\mathfrak{B}, \mathfrak{C}) = (\mathfrak{D}\mathfrak{B}, \mathfrak{D}\mathfrak{C}), \tag{5}$$

that is, the usual distributive law of multiplication with addition. The reader should verify this as well as the laws

$$(6) \qquad [\mathfrak{B}, \mathfrak{C}]\mathfrak{D} \leqq [\mathfrak{B}\mathfrak{D}, \mathfrak{C}\mathfrak{D}], \qquad \mathfrak{D}[\mathfrak{B}, \mathfrak{C}] \leqq [\mathfrak{D}\mathfrak{B}, \mathfrak{D}\mathfrak{C}]$$

for multiplication and the operation of forming the intersection $[\mathfrak{B}, \mathfrak{C}]$ of $\mathfrak{B}$ and $\mathfrak{C}$.

If g is in $\mathfrak{A}$ and $\mathfrak{G} = (g)$ over $\mathfrak{F}$ is the linear set spanned by g then $\mathfrak{G}$ consists of all quantities αg for α in $\mathfrak{F}$. It follows that $\mathfrak{B}\mathfrak{G}$ consists of all bg for b in $\mathfrak{B}$ and we define

$$\mathfrak{B}g \equiv \mathfrak{B}\mathfrak{G}, \qquad g\mathfrak{B} \equiv \mathfrak{G}\mathfrak{B}.$$

Observe that the order of $\mathfrak{B}\mathfrak{C}$ above is mt if and only if $\mathfrak{B}\mathfrak{C} = \mathfrak{B}v_1 + \cdots + \mathfrak{B}v_t$ such that $\mathfrak{B}v_i$ has order m. A criterion for this latter property will be needed and is provided by

Theorem 1. *The order of $\mathfrak{B}g$ (of $g\mathfrak{B}$) is that of $\mathfrak{B}$ if and only if $bg \neq 0$ ($gb \neq 0$) for every $b \neq 0$ of $\mathfrak{B}$.*

For let $\mathfrak{B} = (u_1, \cdots, u_m)$ so that $\mathfrak{B}g$ is spanned by $u_1g, \cdots, u_mg$. The order of $\mathfrak{B}g$ is m if and only if the m quantities u_ig are linearly independent in $\mathfrak{F}$, that is, $\sum \alpha_i(u_ig) = (\sum \alpha_iu_i)g = bg = 0$ for α_i in $\mathfrak{F}$ if and only if the $\alpha_i = 0$. But $b = \sum \alpha_iu_i$ is the zero quantity of $\mathfrak{B}$ if and only if the α_i are all zero and we have proved our theorem.

COROLLARY. *Let $\mathfrak{B}$ and $\mathfrak{C} = (v_1, \cdots, v_t)$ be linear sets of orders m and t respectively over $\mathfrak{F}$. Then the order of $\mathfrak{B}\mathfrak{C}$ is mt if and only if $b_1v_1 + \cdots + b_tv_t \neq 0$ whenever the b_i are in $\mathfrak{B}$ and not all zero.*

The proof follows at once from the theorem above and the comment preceding it.

The product $\mathfrak{B}\mathfrak{B}$ of a linear subset $\mathfrak{B} \neq 0$ of $\mathfrak{A}$ by itself has already been defined. We write $\mathfrak{B}\mathfrak{B} = \mathfrak{B}^{\cdot 2}$ and call it the *square* of $\mathfrak{B}$. Define by induction the powers $\mathfrak{B}^{\cdot k} = \mathfrak{B}\mathfrak{B}^{\cdot k-1}$. This ordinary power will be used far less frequently than another (direct) power to be defined presently, and so we shall reserve the simpler notation $\mathfrak{B}^k$ for this other power. Notice that $\mathfrak{B}^{\cdot 2} \leqq \mathfrak{B}$ if and only if $\mathfrak{B}$ is a subalgebra of $\mathfrak{A}$.

If $\mathfrak{A}$ is an algebra we may form the power sequence

$$(7) \qquad \mathfrak{A} \geqq \mathfrak{A}^{\cdot 2} \geqq \cdots \geqq \mathfrak{A}^{\cdot k} \geqq \cdots.$$

This sequence consists of subalgebras of $\mathfrak{A}$ and possibly the zero set and, since $\mathfrak{A}$ has finite order, there is some place in the sequence where $\mathfrak{A}^{\cdot k} = \mathfrak{A}^{\cdot k+1}$. But then $\mathfrak{A}^{\cdot k} = \mathfrak{A}^{\cdot s}$ for every $s \geqq k$. Hence either $\mathfrak{A}^{\cdot 2} = \mathfrak{A}$ and we define $\alpha = 1$ or there is a (unique) integer $\alpha > 1$ such that $\mathfrak{A} > \mathfrak{A}^{\cdot 2} > \cdots > \mathfrak{A}^{\cdot \alpha} = \mathfrak{A}^{\cdot s}$ for every $s > \alpha$. In either case we shall call α the *index* of $\mathfrak{A}$.

It is clear that if $\mathfrak{A}$ has a unity quantity then $\mathfrak{A}^{\cdot 2} = \mathfrak{A}$. Most of our considerations will be those of algebras with unity quantities, and this should account for the relative unimportance, after the first few chapters, of the ordinary powers and the concept of index. When we study normal simple algebras, however, *we shall use the word index for another concept,* but there will be no confusion since these algebras will have unity elements.

5. Direct products. Let $\mathfrak{B}$ and $\mathfrak{C}$ be subalgebras of $\mathfrak{A}$ such that the order of $\mathfrak{B}\mathfrak{C}$ is the product of that of $\mathfrak{B}$ by that of $\mathfrak{C}$. Assume also that $bc = cb$ for every b of $\mathfrak{B}$ and c of $\mathfrak{C}$. Then we call $\mathfrak{B}\mathfrak{C}$ the *direct product* of $\mathfrak{B}$ and $\mathfrak{C}$ and introduce the notation

$$\mathfrak{B} \times \mathfrak{C} \tag{8}$$

for this algebra. This coincides, when $\mathfrak{A}$, $\mathfrak{B}$, $\mathfrak{C}$ are fields over $\mathfrak{F}$, with the concept of Section A7.4.

The definition above is not sufficiently general for our purposes. It is to be noticed however that, while we shall give a more general definition, we shall always attempt to form $\mathfrak{B} \times \mathfrak{C}$ so that the definition above is valid. Let us proceed to such a formulation.

Assume that $\mathfrak{B} = (u_1, \cdots, u_n)$ over $\mathfrak{F}$, $\mathfrak{C} = (v_1, \cdots, v_m)$ over $\mathfrak{F}$ are arbitrary algebras not necessarily having unity quantities and not necessarily distinct. Suppose that $u_iu_r = \sum_{\alpha=1}^{n} \gamma_\alpha^{(ir)} u_\alpha$, $v_jv_s = \sum_{\beta=1}^{m} \delta_\beta^{(js)} v_\beta$ are the respective multiplication tables of $\mathfrak{B}$ and $\mathfrak{C}$. We then define the *direct product* $\mathfrak{A} = (w_{11}, \cdots, w_{ij}, \cdots, w_{nm})$ over $\mathfrak{F}$ as the algebra with the multiplication table

$$w_{ij}w_{rs} = \sum_{\alpha=1,\cdots,n}^{\beta=1,\cdots,m} \gamma_\alpha^{(ir)} \delta_\beta^{(js)} w_{\alpha\beta}\,. \tag{9}$$

If we think of $w_{\alpha\beta}$ as the symbolic product $u_\alpha v_\beta$, we may notice then that $\mathfrak{B} \times \mathfrak{C}$ consists of all sums of a finite number of terms, each the symbolic product of a quantity of $\mathfrak{B}$ by a quantity of $\mathfrak{C}$. Then $\mathfrak{B} \times \mathfrak{C}$ is independent of the particular bases of $\mathfrak{B}$ and $\mathfrak{C}$ used in its definition, a fact which we shall prove now.

Notice first that we need only show $\mathfrak{A} = \mathfrak{B} \times \mathfrak{C}$ independent of the basis of $\mathfrak{B}$ and the desired result follows. If $(v_1, \cdots, v_m)$ is the basis of $\mathfrak{C}$ used above, every $a = \Sigma\, \alpha_{ij}w_{ij}$ of $\mathfrak{A}$ may be formally expressed as $\Sigma\, b_jv_j$ with the b_j in $\mathfrak{B}$. This is merely a formal expression since v_j and b_j are not contained in $\mathfrak{A}$. But if we express the b_j in terms of the basis $(u_1, \cdots, u_n)$ of $\mathfrak{B}$, use the distributive laws, and replace every u_iv_j by w_{ij}, we obtain the quantity a. Furthermore, the quantity a uniquely defines the quantities b_j of $\mathfrak{B}$, and conversely every expression Σb_jv_j defines a unique a of $\mathfrak{A}$. Under the natural definitions for addition and multiplication the set of all expressions Σb_jv_j forms an algebra equivalent to $\mathfrak{A}$. If we use a different basis for $\mathfrak{B}$ and get $\mathfrak{A}_0 = \mathfrak{B} \times \mathfrak{C}$, then $\mathfrak{A}_0$ is equivalent to the same algebra of formal expressions Σb_jv_j and hence is equivalent to $\mathfrak{A}$.

We shall be specially interested in the formation of the direct products of algebras with unity quantities. Hence let $\mathfrak{B}$ and $\mathfrak{C}$ be as above with respective unity quantities u_1, v_1. Then w_{11} is the unity quantity of $\mathfrak{A} = \mathfrak{B} \times \mathfrak{C}$. The subalgebra $\mathfrak{B}_0$ of $\mathfrak{A}$ with basis $w_{11}, \cdots, w_{n1}$ has the property $w_{i1}w_{r1} = \sum_{\alpha=1}^{n} \gamma_{\alpha}^{(ir)} w_{\alpha 1}$ and is equivalent to $\mathfrak{B}$. Similarly $\mathfrak{C}_0 = (w_{11}, \cdots, w_{1m})$ is equivalent to $\mathfrak{C}$, the unity quantities of $\mathfrak{A}$, $\mathfrak{B}_0$, $\mathfrak{C}_0$ coincide and $\mathfrak{A} = \mathfrak{B}_0 \times \mathfrak{C}_0$ in our earlier sense. Observe that w_{ij} is an abbreviation of the symbolic product $u_i v_j$, and that the inspiration of our definition of direct product as given in (9) is the symbolic definitive equation $w_{ij} = u_i v_j = v_j u_i$.

Matric connections between algebras and their direct products were given in Sections A10.4, A10.5, and we agreed there that if $\mathfrak{B}$ and $\mathfrak{C}$ are any two algebras with unity quantities then $\mathfrak{B} \times \mathfrak{C}$ is to be formed (replacing $\mathfrak{B}$ by $\mathfrak{B}_0$, $\mathfrak{C}$ by $\mathfrak{C}_0$) so that the unity quantities of $\mathfrak{B}$, $\mathfrak{C}$, $\mathfrak{B} \times \mathfrak{C}$ coincide, $\mathfrak{B}$ and $\mathfrak{C}$ are subalgebras of $\mathfrak{B} \times \mathfrak{C}$. There is no logical difficulty when we make this important convention. For, the properties of algebras to be considered in these LECTURES are unchanged whenever an algebra is replaced by any equivalent algebra. It is very important to keep this convention in mind as we shall repeatedly form direct products of algebras with this in mind and no restatement of the convention we have adopted.

In Chapter IV and later we shall use of the notion of *direct power* repeatedly. It is defined by induction in terms of what we shall call the *direct square* $\mathfrak{B}^2 = \mathfrak{B} \times \mathfrak{B}$. Thus the algebra formally indicated by $\mathfrak{B}^k = \mathfrak{B}^{k-1} \times \mathfrak{B}$ is the direct product of $\mathfrak{B}^{k-1}$ and an algebra equivalent to $\mathfrak{B}$.

6. The $\mathfrak{A}$-commutator of a subset. There is a particular type of subalgebra of an algebra $\mathfrak{A}$ which will be of frequent occurrence in our work and we shall describe it now. Let $\mathfrak{H}$ be any set of quantities of $\mathfrak{A}$ and $\mathfrak{A}^{\mathfrak{H}}$ be the set of all quantities of $\mathfrak{A}$ commutative with every quantity of $\mathfrak{H}$. If h is in $\mathfrak{H}$, α is in $\mathfrak{F}$, c_1 and c_2 are in $\mathfrak{A}^{\mathfrak{H}}$, then $c_1 h = hc_1$, $c_2 h = hc_2$, so that $c_1c_2h = c_1hc_2 = h(c_1c_2)$, $(c_1 + c_2)h = h(c_1 + c_2)$, $(\alpha c_1)h = h(\alpha c_1)$. It follows that $\mathfrak{A}^{\mathfrak{H}}$ is a subalgebra of $\mathfrak{A}$ which we shall call the *$\mathfrak{A}$-commutator* of $\mathfrak{H}$. We shall use the notation $\mathfrak{A}^{\mathfrak{H}}$ consistently for this concept.

The $\mathfrak{A}$-commutator of an algebra $\mathfrak{A}$ is the set $\mathfrak{Z} = \mathfrak{A}^{\mathfrak{A}}$ of all quantities z of $\mathfrak{A}$ such that $az = za$ for every a of $\mathfrak{A}$. We call this subalgebra of $\mathfrak{A}$ the *centrum* of $\mathfrak{A}$. When $\mathfrak{A}$ has a unity quantity (e) the subalgebra (e) over $\mathfrak{F}$ is in $\mathfrak{Z}$ and we make the

DEFINITION. *An algebra $\mathfrak{A}$ with a unity quantity e is called normal over $\mathfrak{F}$ if the centrum of $\mathfrak{A}$ is the algebra (e) over $\mathfrak{F}$.*

One exceedingly important example of a normal algebra has already been studied by the reader. It is the algebra $\mathfrak{M}$ consisting of all m-rowed square matrices with elements in $\mathfrak{F}$ and it was shown to be normal in Theorem A3.6.

7. Total matric algebras. The abstract algebra $\mathfrak{M}_m$ over $\mathfrak{F}$ equivalent to the set of all m-rowed square matrices with elements in $\mathfrak{F}$ appears frequently in the theory of algebras. We shall call any such algebra a *total matric algebra* of *degree m* over $\mathfrak{F}$. The order of $\mathfrak{M}_m$ is m^2, and we shall use this notation consistently with this meaning for the subscript m. We shall frequently write

$$\mathfrak{M}_m = (e_{ij}\,;\, i, j = 1, \cdots, m) \tag{10}$$

to indicate that e_{ij} is the quantity of $\mathfrak{M}_m$ corresponding to the matrix E_{ij} which has unity in the ith row and jth column and zeros elsewhere. Any m-rowed square matrix has the form $\mathfrak{A} = (\alpha_{ij}) = \sum_{i,j=1}^{m} \alpha_{ij}E_{ij}$, and it follows that the m^2 quantities e_{ij} form a basis of $\mathfrak{M}_m$ over $\mathfrak{F}$. The corresponding multiplication table

$$(11) \qquad e_{ij}e_{jk} = e_{ik}, \; e_{ij}e_{hk} = 0 \qquad (j \neq h; i, j, h, k = 1, \cdots, m),$$

is very convenient for use in many proofs. In Chapter AIII we called any basis as in (10), (11) an *ordinary matric basis* of $\mathfrak{M}_m$. We shall use this terminology. Note that

$$(12) \qquad e = e_{11} + \cdots + e_{mm}$$

corresponds to the identity matrix in the correspondence generated by $e_{ij} \leftrightarrow E_{ij}$. and e is the unity quantity of $\mathfrak{M}_m$.

One of the properties of quantities defined as in (10), (11) is their linear independence. We state this in

Theorem 2. *Let an algebra $\mathfrak{A}$ contain m^2 quantities e_{ij} not all zero and satisfying* (11). *Then the linear set $\mathfrak{M}_m$ spanned by the e_{ij} is a total matric subalgebra of order m^2 of $\mathfrak{A}$ with unity quantity e of* (12).

For, at least one $e_{pq} \neq 0$ and if $a = \sum_{i,j=1}^{m} \alpha_{ij}e_{ij} = 0$ then $e_{pi}ae_{jq} = \alpha_{ij}e_{pq} = 0$ if and only if $\alpha_{ij} = 0$. This proves the e_{ij} linearly independent in $\mathfrak{F}$ and hence our theorem.

The algebraic field $\mathfrak{K}$ of finite degree n over $\mathfrak{F}$ is clearly an example of an algebra of order n over $\mathfrak{F}$. We then let $\mathfrak{A}$ be a total matric algebra of degree m over $\mathfrak{K}$. If follows that $\mathfrak{A} = (e_{ij}; i, j = 1, \cdots, m)$ over $\mathfrak{K}$. But the linear set $\mathfrak{M}_m = (e_{ij}; i, j = 1, \cdots, m)$ over $\mathfrak{F}$ is clearly a total matric algebra and $\mathfrak{A} = \mathfrak{M}_m \times \mathfrak{K}$. In symbols

$$(13) \qquad \mathfrak{M}_m \text{ over } \mathfrak{K} = (\mathfrak{M}_m \text{ over } \mathfrak{F}) \times (\mathfrak{K} \text{ over } \mathfrak{F}).$$

We shall utilize this property frequently.

There is a property of total matric algebras which is of frequent use and we shall note it here. In symbols it is given as

$$\mathfrak{M}_n \times \mathfrak{M}_m = \mathfrak{M}_{mn},$$

that is, the direct product of two total matric algebras is a total matric algebra whose degree is the product of their respective degrees. This matric property was derived in Section A10.5, and we shall not attempt to give any more abstract proof of this result but refer the reader to our earlier exposition.

8. Automorphisms of an algebra. If an algebra $\mathfrak{A}$ is equivalent to itself by means of a correspondence

$$S: \qquad a \leftrightarrow a^S \qquad (a, a^S \text{ in } \mathfrak{A}),$$

we call S an *automorphism* of $\mathfrak{A}$. The concept is one of great importance. We observe that if $\mathfrak{A}$ and $\mathfrak{A}'$ are algebras equivalent by means of a correspondence

$$T: \qquad a' \leftrightarrow a = (a')^T \qquad (a' \text{ in } \mathfrak{A}',\ a \text{ in } \mathfrak{A}),$$

as well as by means of a second correspondence

$$U: \qquad a' \leftrightarrow a_0 = (a')^U \qquad (a' \text{ in } \mathfrak{A}',\ a_0 \text{ in } \mathfrak{A}),$$

then the correspondence

$$S: \qquad a = a'^T \leftrightarrow a_0 = (a')^U = a^S = [(a')^T]^S$$

is easily verified to be an automorphism of $\mathfrak{A}$. Hence every equivalence U is the product TS of a fixed equivalence T of $\mathfrak{A}$ and $\mathfrak{A}'$ by an automorphism S of $\mathfrak{A}$. Thus the problem of determining all equivalences of $\mathfrak{A}$ and $\mathfrak{A}'$ is essentially that of determining one equivalence and all automorphisms* of $\mathfrak{A}$.

If an algebra $\mathfrak{A}$ is reciprocal to itself by means of a correspondence S, we shall call S a *reciprocal automorphism* of $\mathfrak{A}$. Let S and T be two such reciprocal automorphisms. Then the correspondence

$$U: \qquad a \leftrightarrow a^U = (a^S)^T$$

is evidently a (1-1) correspondence of $\mathfrak{A}$ preserved under addition and such that $(\lambda a)^U = \lambda a^U$ for every λ of $\mathfrak{F}$. But

$$(ab)^{ST} = (b^S a^S)^T = a^{ST} b^{ST}$$

so that ST is an automorphism of $\mathfrak{F}$. The connection just given will be used in Chapter X.

If S is an equivalence (or reciprocal correspondence) of $\mathfrak{A}$ and $\mathfrak{A}'$, we may evidently write $\mathfrak{A}^S$ for $\mathfrak{A}'$ without confusion. Let us do this for the subalgebras of a given algebra. Thus let S be an automorphism (reciprocal automorphism) of an algebra $\mathfrak{A}$ so that S carries every subalgebra $\mathfrak{B}$ of $\mathfrak{A}$ into an equivalent (reciprocal) subalgebra which we designate by $\mathfrak{B}^S$. If $\mathfrak{B}^S \leqq \mathfrak{B}$ then $\mathfrak{B}^S = \mathfrak{B}$ and the correspondence $b \leftrightarrow b^S$ is an automorphism (reciprocal automorphism) of $\mathfrak{B}$. We shall say that this correspondence is *induced* by S of $\mathfrak{A}$. No ambiguity is introduced if we use the same notation S for this automorphism (reciprocal automorphism) of the subalgebra $\mathfrak{B}$ as we did for that of $\mathfrak{A}$, and we shall consistently do so.

If S is any automorphism (reciprocal automorphism) of an algebra $\mathfrak{A}$ the centrum $\mathfrak{Z}$ of $\mathfrak{A}$ has the property $\mathfrak{Z} = \mathfrak{Z}^S$. For if a is in $\mathfrak{A}$ and z is in $\mathfrak{Z}$ then $az = za$, $(az)^S = a^S z^S = (za)^S = z^S a^S$, and similarly for reciprocal automorphisms. But S is a (1-1) correspondence so that $az^S = za^S$ for every a of $\mathfrak{A}$, z^S is in $\mathfrak{Z}$, $\mathfrak{Z}^S \leqq \mathfrak{Z}$. Since they are equivalent (reciprocal) we have $\mathfrak{Z} = \mathfrak{Z}^S$.

It is clear that an automorphism of $\mathfrak{A}$ need not however leave the individual quantities of $\mathfrak{Z}$ invariant. There is an important class of automorphisms which do have this latter property. We let $\mathfrak{A}$ be an algebra with a unity quantity 1 and g be any *regular* quantity of $\mathfrak{A}$. Then g^{-1} is in $\mathfrak{A}$ such that $gg^{-1} = g^{-1}g = 1$.

* All concepts are of course relative to the base field $\mathfrak{F}$.

The correspondence

$$S_g: \qquad a \rightarrow a^{S_g} = gag^{-1} \qquad (a \text{ in } \mathfrak{A})$$

may trivially be shown to be an automorphism of $\mathfrak{A}$. We call S_g the *inner automorphism* of $\mathfrak{A}$ defined by g.

Note that if $gag^{-1} = hah^{-1}$ for every a then $a(g^{-1}h) = (g^{-1}h)a$, $g^{-1}h = k$ is a regular quantity of the centrum of $\mathfrak{A}$. The converse is trivial and we have shown that h and g define the same inner automorphism of $\mathfrak{A}$ if and only if h is the product gk of g by a regular quantity k of the centrum of $\mathfrak{A}$. We may remark that if $\mathfrak{A}_0$ and $\mathfrak{Z}_0$ are the multiplicative groups of regular quantities of $\mathfrak{A}$ and its centrum $\mathfrak{Z}$, respectively, then the quotient group $\mathfrak{A}_0/\mathfrak{Z}_0$ is equivalent to the group of all inner automorphisms of $\mathfrak{A}$ under the correspondence $g\mathfrak{Z}_0 \rightarrow S_{g^{-1}}$.

The inner automorphisms of $\mathfrak{A}$ clearly are seen to leave the quantities of its centrum $\mathfrak{Z}$ invariant. When $\mathfrak{Z}$ is a field we call such correspondences automorphisms over $\mathfrak{Z}$.

At this point we may remark that our exposition so far has consisted of no involved arguments but principally of the introduction of concepts and their rather immediate consequences. We now pass to somewhat deeper considerations wherein rather important results are obtained.

9. Linear transformations. The theory of linear transformations will enable us to connect the abstract linear algebra theory with theorems on matrices. There are a number of results obtainable most simply in this way and we shall begin their derivation in the present section.

Let $\mathfrak{L}$ be any linear set of order n over $\mathfrak{F}$. A correspondence

$$S: \qquad x \rightarrow x^S$$

carrying every x of $\mathfrak{L}$ into a unique x^S of $\mathfrak{L}$ will be called a transformation of $\mathfrak{L}$. We call S a linear transformation if

$$(14) \qquad (a + b)^S = a^S + b^S, \qquad (\alpha a)^S = \alpha a^S,$$

for every a and b of $\mathfrak{L}$, α of $\mathfrak{F}$. It may be readily verified that if α is in $\mathfrak{F}$ and S and T are any linear transformations of $\mathfrak{L}$ the correspondences

$$ST: \qquad a \rightarrow a^{ST} = (a^S)^T,$$

$$S + T: \qquad a \rightarrow a^{S+T} = a^S + a^T,$$

$$\alpha S: \qquad a \rightarrow a^{\alpha S} = \alpha a^S$$

are also linear transformations of $\mathfrak{L}$. We leave this verification to the reader, who may also verify at this point that the set of all linear transformations of $\mathfrak{L}$ is an algebra over $\mathfrak{F}$. However we shall arrive at this latter property in another fashion.

If $\mathfrak{L} = (u_1, \cdots, u_n)$ over $\mathfrak{F}$ then $u_i^S = \sum_{j=1}^n \xi_{ij}u_j$ with the ξ_{ij} in $\mathfrak{F}$. Hence if $a = \alpha_1 u_1 + \cdots + \alpha_n u_n$ is in $\mathfrak{L}$ the corresponding quantity

$$a^S = \sum_{j=1}^n \sum_{i=1}^n (\alpha_i \xi_{ij})u_j.$$

Designate the quantity a of $\mathfrak{L}$ by the vector (one-rowed matrix) $\underline{a} = (\alpha_1, \cdots, \alpha_n)$, with components the coordinates α_i of a. Then $\underline{a^S} = \underline{a}X$ where X is the matrix (ξ_{ij}) and the indicated product $\underline{a}X$ is the ordinary matrix product of the one by n matrix $\underline{a}$ and the n-rowed square matrix X. We shall call the matrix X uniquely determined by S and the given basis of $\mathfrak{L}$ *the matrix of the linear transformation* S with respect to this basis. Conversely it is evident that every n-rowed square matrix X defines a unique linear transformation S with a^S uniquely determined by $\underline{a^S} = \underline{a}X$.

If α is in $\mathfrak{F}$ and $\underline{a^T} = \underline{a}Y$ we see readily that

$$(15) \qquad \underline{a^{ST}} = \underline{a}XY, \quad \underline{a^{S+T}} = \underline{a^S} + \underline{a^T} = \underline{a}(X + Y), \quad \underline{a^{\alpha S}} = \alpha\underline{a^S} = \underline{a}(\alpha X).$$

It follows that the one-to-one correspondence $X \leftrightarrow S$ is an equivalence of the algebra of all n-rowed square matrices with elements in $\mathfrak{F}$ and the set of all linear transformations on a linear set $\mathfrak{L}$ of order n over $\mathfrak{F}$. Hence this latter algebra is a total matric algebra $\mathfrak{M}_n$ over $\mathfrak{F}$. The equivalence just defined is in fact the origin of the usual definitions of the elementary operations on square matrices.

We observe that the unity quantity 1 of $\mathfrak{M}_n$ is the identical transformation I, the scalars $\alpha \cdot 1$ of $\mathfrak{M}_n$ with α in $\mathfrak{F}$ have been defined as the transformations $a \rightarrow \alpha a$. Hence scalar multiplications in $\mathfrak{L}$ may be regarded as instances of linear transformations on $\mathfrak{L}$.

If $\underline{a^T} = \underline{a}B$ where B is non-singular we call T a *non-singular* linear transformation. Then the u_i^T form a basis of $\mathfrak{L}$ and the correspondence T is a (1-1) correspondence. Clearly each non-singular linear transformation carries every $a \neq 0$ of $\mathfrak{L}$ into $a^T \neq 0$.

Conversely every basis $(v_1, \cdots, v_n)$ over $\mathfrak{F}$ is obtained as the result $v_j = u_j^T$, where T is a non-singular linear transformation on $\mathfrak{L}$. Let $B = (\beta_{ji})$ be the matrix of T so that $v_j = \sum \beta_{ji} u_i$, and define $C = (\gamma_{ij}) = B^{-1}$. Then

$$u_i = \sum_{j=1}^{n} \gamma_{ij} v_j, \; a = \sum_{i=1}^{n} \alpha_i u_i = \sum_{j=1}^{n} \left(\sum_{i=1}^{n} \alpha_i \gamma_{ij} \right) v_j = \sum_{j=1}^{n} \beta_j v_j.$$

We define $\underline{\underline{a}} = (\beta_1, \cdots, \beta_n)$ and, as above, have $\underline{\underline{a^S}} = \underline{\underline{a}}X_0$ for every linear transformation S on $\mathfrak{L}$. But $\underline{\underline{a}} = \underline{a}C$, $\underline{\underline{a^S}} = \underline{a^S}C = \underline{a}XC = \underline{\underline{a}}C^{-1}XC$, $X_0 = C^{-1}XC = BXB^{-1}$. Thus *a change of basis accomplished by a linear transformation on $\mathfrak{L}$ with matrix B replaces the matrix X of any linear transformation S on $\mathfrak{L}$ by BXB^{-1}.*

It is clear that the equivalence $S \leftrightarrow X$ of the algebra of all linear transformations S on $\mathfrak{L}$ to the algebra of corresponding matrices X implies that $S \leftrightarrow X'$ is a reciprocal correspondence of the two algebras. This correspondence and property actually arise from an alternative definition of the concept of the matrix of a linear transformation. For let $\mathfrak{L} = (u_1, \cdots, u_n)$ over $\mathfrak{F}$, and write $U = (u_1, \cdots, u_n)$, a one by n matrix with elements in $\mathfrak{L}$. Then if $a =$

$\sum \alpha_i u_i$ with α_i in $\mathfrak{F}$ we have $a = \underline{a}U'$, where the prime indicates transpose, and our product is the ordinary row by column product. We define $U^S = (u_1^S, \cdots, u_n^S)$, and may write the n equations $u_i^S = \sum_{j=1}^n \xi_{ij}u_j$ as the single matrix equation $(U^S)' = XU'$. We did this essentially above, and then had $a^S = \underline{a}(U^S)' = (\underline{a}X)U'$, so that $\underline{a^S} = \underline{a}X$, and it was natural to call X the matrix of S. But our equations also have the form $u_i^S = \sum_{j=1}^n u_j\sigma_{ji}$ with $\sigma_{ji} = \xi_{ij}$, $(\sigma_{ji}) = X'$, that is, $U^S = UX'$. From this latter equation it would seem just as natural to regard X' as the matrix of S. However $U^{ST} = U(XY)' = UY'X'$, a result which we may interpret as $U^{ST} = (U^T)^S$. We shall consequently only use the definition which calls X the matrix of S.

We now begin a study of what are called the regular representations of an algebra. Let $\mathfrak{A}$ be an algebra with a unity quantity 1 and order n over $\mathfrak{F}$, so that $\mathfrak{A} = (u_1, \cdots, u_n)$ over $\mathfrak{F}$ and every x of $\mathfrak{A}$ is uniquely expressible in the form $x = \xi_1 u_1 + \cdots + \xi_n u_n$, with ξ_i in $\mathfrak{F}$. We form $xu_j = \sum_{i=1}^n u_i\sigma_{ij}$ with σ_{ij} in $\mathfrak{F}$, and define $X_x = (\sigma_{ij})$, an n-rowed square matrix. Then it was proved in Section A10.2 that the set $\mathfrak{B}$ of all the matrices X_x forms an algebra over $\mathfrak{F}$ equivalent to $\mathfrak{A}$ under the correspondence (right multiplication)

$$x \to X_x .$$

This was actually determined as a consequence of the equations $xU = UX_x$. We note in particular that X_1 is evidently the n-rowed identity matrix. The algebra $\mathfrak{B}$ was called the first regular representation of $\mathfrak{A}$ in Chapter AX, and the minimum and characteristic functions of the matrix X_x were called the minimum and (first) characteristic functions respectively of x. The important concept of scalar extension was also considered in Section A10.3, as well as earlier in our foundation text, and we shall assume that the reader is familiar with this concept. Let us study this representation further.

It may be verified readily that the correspondence

$$S_x: \qquad a \to a^{S_x} = xa \qquad (a \text{ in } \mathfrak{A}),$$

is a linear transformation on $\mathfrak{A}$ for every x of $\mathfrak{A}$. It is generated by the partial transformations

$$u_j \to u_j^{S_x} = xu_j .$$

But then $xU = (xu_1, \cdots, xu_n) = U^{S_x} = UX_x$, so that X_x is the transpose of what we have decided to call the matrix of the linear transformation S_x. We shall therefore not ask the reader to carry out the verification of Section A10.2 proving that $\mathfrak{A}$ is equivalent to the algebra consisting of all the X_x. But we shall show instead that $\mathfrak{A}$ is reciprocal to the algebra $\mathfrak{S}$ of all linear transformations S_x under the correspondence $x \to S_x$. For it seems clear now that the use of X_x in both Chapter AX and earlier expositions is somewhat unnatural.

Let y be in $\mathfrak{A}$ so that S_y is defined by

$$S_y: \qquad a \to a^{S_y} = ya.$$

We have

$S_yS_x:$ $$a \to (a^{S_y})^{S_x} = (ya)^{S_x} = x(ya) = xya,$$

and thus $S_{xy} = S_yS_x$. That $S_{\alpha x+\beta y} = \alpha S_x + \beta S_y$ follows from $(\alpha x + \beta y)a = \alpha(xa) + \beta(ya) = a^{\alpha S_x} + a^{\beta S_y} = a^{\alpha S_x+\beta S_y}$. This completes our proof. Note that, as in Section A10.2,

$$u_i^{S_{xy}} = x\,(yu_i) = x\sum_{j=1}^{n} \tau_{ji}u_j = \sum_{j,k=1}^{n} \tau_{ji}\sigma_{kj}u_k, \tag{16}$$

so that the matrix of S_{xy} is $X_y'X_x' = (X_xX_y)' = (X_{xy})'$, that is, as we have said, $x \to X_x$ is an equivalence.

A third representation is the so-called second regular representation of $\mathfrak{A}$, obtained by forming $u_ix = \sum \eta_{ij}u_j$, where the η_{ij} are in $\mathfrak{F}$. As in Exercise 6 of Section A10.2 we may prove that the corresponding algebra of n-rowed square matrices is reciprocal to $\mathfrak{A}$. This also follows as above from a consideration of what is essentially its transpose, that is, the seemingly more natural study of the algebra $\mathfrak{T}$ of all linear transformations (left multiplications)

$T_x:$ $$a \to a^{T_x} = ax.$$

Then $\mathfrak{T}$ is equivalent over $\mathfrak{F}$ to $\mathfrak{A}$ and in fact

$$a^{T_xT_y} = (ax)^{T_y} = (ax)y = a(xy) = a^{T_{xy}}.$$

We shall now leave the representations by matric algebras of Chapter AX and consider instead the algebras $\mathfrak{S}$ and $\mathfrak{T}$. They are both subalgebras of the total matric algebra $\mathfrak{M} = \mathfrak{M}_n$ of all linear transformations on $\mathfrak{A}$. If S_x is in $\mathfrak{S}$ and T_y is in $\mathfrak{T}$ then

$$a^{T_yS_x} = (a^{T_y})^{S_x} = (ay)^{S_x} = x(ay),$$

and

$$a^{S_xT_y} = (a^{S_x})^{T_y} = (xa)^{T_y} = (xa)y = x(ay), \tag{17}$$

since $\mathfrak{A}$ is associative. Hence $T_yS_x = S_xT_y$ for every S_x of $\mathfrak{S}$ and T_y of $\mathfrak{T}$. We have proved that, in the notation of Section 1.6,

$$\mathfrak{S} \leqq \mathfrak{M}^{\mathfrak{T}}, \qquad \mathfrak{T} \leqq \mathfrak{M}^{\mathfrak{S}}. \tag{18}$$

Conversely let S be any linear transformation commutative with all T_y of $\mathfrak{T}$. Then

$$a^{ST_y} = a^Sy = a^{T_yS} = (ay)^S, \tag{19}$$

and if we put $a = 1$, the unity quantity of $\mathfrak{A}$, we have

$$1^Sy = y^S \tag{20}$$

for every y of $\mathfrak{A}$. Then $1^Sa = a^S$, S is the transformation S_x where $x = 1^S$. Hence $\mathfrak{M}^{\mathfrak{T}} \leqq \mathfrak{S}$ and similarly $\mathfrak{M}^{\mathfrak{S}} \leqq \mathfrak{T}$. We have proved

Theorem 3. *Let $\mathfrak{A}$ be an algebra of order n over $\mathfrak{F}$ and with a unity quantity. Then $\mathfrak{A}$ is equivalent over $\mathfrak{F}$ to $\mathfrak{T}$, and reciprocal over $\mathfrak{F}$ to $\mathfrak{S}$, where $\mathfrak{S}$ and $\mathfrak{T}$ are subalgebras of a total matric algebra $\mathfrak{M}_n$ of order n^2 over $\mathfrak{F}$. Moreover $\mathfrak{S}$ is the $\mathfrak{M}_n$-commutator of $\mathfrak{T}$, $\mathfrak{T}$ is the $\mathfrak{M}_n$-commutator of $\mathfrak{S}$.*

Observe that when $\mathfrak{A}$ is a commutative algebra $\mathfrak{T} = \mathfrak{S}$ so that $\mathfrak{M}_n^{\mathfrak{S}} = \mathfrak{S}$. This is true, in particular, when $\mathfrak{A}$ is a field, and we shall observe this in another way later.

The property of Theorem 1.3 is so fundamental in our theory that we shall restate it here. Let $\mathfrak{A}$ be an algebra of order n over $\mathfrak{F}$ and with a unity quantity. Then $\mathfrak{A}$ and $\mathfrak{A}^{-1}$ are subalgebras of a total matric algebra $\mathfrak{M}_n$ whose unity quantity is the same as that of both $\mathfrak{A}$ and $\mathfrak{A}^{-1}$. Moreover $aa' = a'a$ for every a of $\mathfrak{A}$ and every a' of the algebra $\mathfrak{A}^{-1}$ reciprocal to $\mathfrak{A}$. In a later section we shall study a case in which these properties will imply that $\mathfrak{M}_n = \mathfrak{A} \times \mathfrak{A}^{-1}$.

10. Regular quantities of an algebra. In an arbitrary ring with a unity quantity there seems to be no particular construction of the inverse of a regular quantity. However the minimum function of a quantity of an *algebra* readily provides such a construction.

We let a be a non-zero quantity of an algebra $\mathfrak{A}$ with a unity quantity 1. Using our representation theory we consider the minimum function $\phi(\lambda) = \lambda^m + \alpha_1\lambda^{m-1} + \cdots + \alpha_m$, α_i in $\mathfrak{F}$, of a. We call our quantity *regular*, or *non-singular*, if there exists a quantity b such that either $ab = 1$ or $ba = 1$. We then prove

Theorem 4. *The quantity a is regular if and only if the constant term α_m of its minimum function is not zero. Thus $ab = 1$ implies that $b = a^{-1}$ is unique and a polynomial in a with coefficients in $\mathfrak{F}$ and hence $ba = 1$.*

For if $\alpha_m \neq 0$ the quantity

$$a^{-1} = (-\alpha_m^{-1})(a^{m-1} + \alpha_1 a^{m-2} + \cdots + \alpha_{m-1}) \tag{21}$$

has the property $aa^{-1} = a^{-1}a = 1$ and a is regular. Moreover $ab = 1$ implies that $a^{-1}(ab) = a^{-1} = (a^{-1}a)b = b$. The uniqueness of a^{-1} follows. Conversely if a is regular let us assume that $ab = 1$. Then $ca = 0$ implies that $cab = 0 = c\cdot 1 = c = 0$. If $\alpha_m = 0$ then $\phi(\lambda) = \lambda\psi(\lambda)$ where the degree of $\psi(\lambda)$ is less than m, and $c = \psi(a) \neq 0$, $ac = ca = 0$, a contradiction.

Of course we might have used above the property of arbitrary rings stating that a regular quantity is not a divisor of zero. This depended in Chapter AI upon our definition that a is regular if there exists a quantity b such that $ab = ba = 1$. Our definition has the presumably weaker hypothesis that either $ab = 1$ or $ba = 1$, but we have proved that b is a polynomial in a and hence the equivalence of our definition to that of Section A1.9.

11. Division algebras. Our investigations will lead us ultimately to the study of the structure of division algebras. These are algebras having properties closest to those of the number systems of elementary mathematics. Briefly an algebra $\mathfrak{D}$ over $\mathfrak{F}$ is a division algebra if its non-zero quantities form a group with respect to its operation of multiplication.

It is clear that *an algebra $\mathfrak{D}$ is a division algebra if and only if it has a unity quantity and every non-zero quantity of $\mathfrak{D}$ is regular.* The argument of Section 1.10 then states that *$\mathfrak{D}$ is a division algebra if and only if $\mathfrak{D}$ contains no divisors of zero.*

Theorem 1.4 implies that $\mathfrak{D}$ is a division algebra if and only if the minimum function $\phi(\lambda)$ of every non-zero quantity a of $\mathfrak{D}$ has non-zero constant term. But we may prove the better result of

Theorem 5. *An algebra $\mathfrak{D}$ over $\mathfrak{F}$ is a division algebra if and only if the minimum function of every non-zero quantity of $\mathfrak{D}$ is irreducible in $\mathfrak{F}$.*

For let the minimum function of $a \neq 0$ in $\mathfrak{D}$ have the form $\phi(\lambda) = \phi_1(\lambda) \cdot \phi_2(\lambda)$ where the $\phi_i(\lambda)$ are non-constant polynomials in λ. Then their degrees are less than that of $\phi(\lambda)$, and the definition of minimum function implies that $\phi_1(a) \neq 0$, $\phi_2(a) \neq 0$. However $\phi(a) = \phi_1(a) \cdot \phi_2(a) = 0$ which is impossible in $\mathfrak{D}$. The converse is evident.

The unity quantity 1 of any algebra $\mathfrak{A}$ with such a quantity may be replaced (by Theorem A1.9) by the unity quantity of $\mathfrak{F}$ and then $\mathfrak{F}$ is a subring of $\mathfrak{A}$. We agree to do so henceforth whenever we are dealing with a single algebra. Then every commutative division algebra of order m over $\mathfrak{F}$ is an algebraic field of degree m over $\mathfrak{F}$. Moreover if d is any quantity not in $\mathfrak{F}$ of a division algebra $\mathfrak{D}$ over $\mathfrak{F}$ the set $\mathfrak{F}[d]$ of all polynomials in d with coefficients in $\mathfrak{F}$ is a field of degree $m > 1$ over $\mathfrak{F}$. The integer m is the degree of the minimum function of d. The field $\mathfrak{F}[d]$ is a division subalgebra of $\mathfrak{D}$ and we shall show that indeed m divides the order of $\mathfrak{D}$ over $\mathfrak{F}$. This actually is a consequence of an important property given in

Theorem 6. *Let $\mathfrak{D}$ be a division subalgebra of an algebra $\mathfrak{A}$ whose unity quantity e coincides with that of $\mathfrak{D}$. Then if $\mathfrak{B}$ is any linear subset of order s of $\mathfrak{A}$ over $\mathfrak{F}$ such that $\mathfrak{D}\mathfrak{B} \leqq \mathfrak{B}$, and if m is the order of $\mathfrak{D}$ over $\mathfrak{F}$, we have $s = mt$,*

$$\mathfrak{B} = \mathfrak{D}u_1 + \cdots + \mathfrak{D}u_t . \tag{22}$$

In particular m divides the order $n = mq$ of $\mathfrak{A}$, and

$$\mathfrak{A} = \mathfrak{D}v_1 + \cdots + \mathfrak{D}v_q . \tag{23}$$

The final statement of our theorem is the case $\mathfrak{B} = \mathfrak{A}$ since $\mathfrak{D} \leqq \mathfrak{A}$, $\mathfrak{D}\mathfrak{A} \leqq \mathfrak{A}$. We observe that the set (22) is a supplementary sum so that the u_i are *left linearly independent* in $\mathfrak{D}$ with respect to the operation of multiplication in $\mathfrak{A}$. We are then stating that $\mathfrak{D}\mathfrak{B} \leqq \mathfrak{B}$ implies that $\mathfrak{B}$ is a *left linear set* of order t over $\mathfrak{D}$ with $u_1, \cdots, u_t$ as a basis. We shall call $\mathfrak{B}$ a *$\mathfrak{D}$-left linear set* and sometimes a *left linear set over $\mathfrak{D}$*. We shall also speak of $u_1, \cdots, u_t$ as a *$\mathfrak{D}$-basis* of $\mathfrak{B}$.

To prove our theorem consider the set $\mathfrak{D}u$ for any $u \neq 0$ in $\mathfrak{B}$. Evidently

$\mathfrak{D}u \leqq \mathfrak{B}$ and if $du = 0$ for d in $\mathfrak{D}$ and $d \neq 0$ we have $d^{-1}d = e$ for d^{-1} in $\mathfrak{D}$, $d^{-1}(du) = eu = u = 0$, a contradiction. By Theorem 1.1 $\mathfrak{D}u$ has order m. If $\mathfrak{B}_i = \mathfrak{D}u_1 + \cdots + \mathfrak{D}u_i < \mathfrak{B}$ there exists a non-zero u_{i+1} in $\mathfrak{B}$ and not in $\mathfrak{B}_i$. We form $\mathfrak{B}_{i+1} = (\mathfrak{B}_i, \mathfrak{D}u_{i+1})$ and have a set of order $m(i + 1)$ unless $[\mathfrak{B}_i, \mathfrak{D}u_{i+1}] \neq 0$ in which case $du_{i+1} = b_i \neq 0$ in $\mathfrak{B}_i$. But then $d \neq 0$, $u_{i+1} = d^{-1}b_i$ in $\mathfrak{B}_i$, a contradiction. Hence $\mathfrak{B}_{i+1} = \mathfrak{B}_i + \mathfrak{D}u_{i+1}$ of order $m(i + 1)$ over $\mathfrak{F}$. Since $\mathfrak{B}$ has finite order the process of forming $\mathfrak{B}_{i+1}$ must terminate and we have (22). We now obtain

Theorem 7. *Every subalgebra $\mathfrak{B}$ of a division algebra $\mathfrak{D}$ is a division algebra whose unity quantity coincides with that of $\mathfrak{D}$ and whose order divides that of $\mathfrak{D}$.*

For $\mathfrak{B}$ contains no divisors of zero and is a division algebra. The non-zero quantities of $\mathfrak{D}$ form a multiplicative group with those of $\mathfrak{B}$ as subgroup, and, by Theorem A1.6, their identity elements, that is, the unity quantities of $\mathfrak{B}$ and $\mathfrak{D}$, coincide. We apply Theorem 1.6 to complete our proof.

When the centrum $\mathfrak{K}$ of an algebra $\mathfrak{A}$ with a unity quantity is a field we may apply Theorem 1.6 to reduce the study of the structure of $\mathfrak{A}$ to the case where $\mathfrak{A}$ is normal. For, by that theorem $\mathfrak{A} = (v_1, \cdots, v_q)$ over $\mathfrak{K}$. Clearly $\mathfrak{A}$ is normal of order q over $\mathfrak{K}$. This is true, in particular, for division algebras, since the centrum of any division algebra is always a commutative division subalgebra and hence is a field. We remark that the unity quantity of $\mathfrak{A}$ is always in $\mathfrak{K}$ and that conversely, if the centrum $\mathfrak{K}$ of $\mathfrak{A}$ is a field, its unity quantity is the unity quantity of $\mathfrak{A}$. We state our result, in view of the convention made above, as

Theorem 8. *Let $\mathfrak{A}$ be an algebra of order n over $\mathfrak{F}$ with a unity quantity, and the centrum $\mathfrak{K}$ of $\mathfrak{A}$ be an algebraic field of degree m over $\mathfrak{F}$. Then $n = mq$ and $\mathfrak{A}$ is a normal algebra of order q over $\mathfrak{K}$.*

12. Scalar extension. If $\mathfrak{A}$ is an algebra over $\mathfrak{F}$ and $\mathfrak{K}$ is a scalar extension of finite degree over $\mathfrak{F}$ the field $\mathfrak{K}$ is a division algebra over $\mathfrak{F}$ and the scalar extension $\mathfrak{A}_{\mathfrak{K}}$ is the direct product $\mathfrak{A} \times \mathfrak{K}$ considered as an algebra over $\mathfrak{K}$. When $\mathfrak{A}$ has a unity quantity 1 and a in $\mathfrak{A}$ has $\lambda^m + \alpha_1\lambda^{m-1} + \cdots + \alpha_m = 0$ as its minimum equation the quantities $1, a, \cdots, a^{m-1}$ are quantities of $\mathfrak{A}$ linearly independent in $\mathfrak{F}$. But then they are quantities of $\mathfrak{A} \times \mathfrak{K}$ linearly independent in $\mathfrak{K}$. This is a repetition of the frequently repeated statement in the author's *Modern Higher Algebra* that the minimum function of a in $\mathfrak{A}$ is the same as its minimum function when considered as a quantity of any scalar extension $\mathfrak{A}_{\mathfrak{K}}$. We use this fact and have

Theorem 9. *Let $\mathfrak{D}$ be a division algebra of order $n > 1$ over $\mathfrak{F}$, d in $\mathfrak{D}$ be not in $\mathfrak{F}$. Then if ξ is a scalar root of the minimum function of d and $\mathfrak{K} = \mathfrak{F}(\xi)$ the algebra $\mathfrak{D}_{\mathfrak{K}}$ is not a division algebra.*

For, the minimum function of d is reducible in $\mathfrak{K}$ and $\mathfrak{D}_{\mathfrak{K}}$ cannot be a division algebra.

13. The minimum function of a division algebra. We have seen that every algebra $\mathfrak{A}$ of order n over $\mathfrak{F}$ with a unity quantity is equivalent to an algebra of n-rowed square matrices. Let us then assume that the quantities of $\mathfrak{A}$ are n-rowed square matrices and recall some of the properties derived in this way in Chapter AX.

If $\mathfrak{A} = (U_1, \cdots, U_n)$ over $\mathfrak{F}$ we may take $U_1 = I$ to be the n-rowed identity matrix. Let $\xi_1, \cdots, \xi_n$ be independent indeterminates over $\mathfrak{F}$, define $\mathfrak{K} = \mathfrak{F}(\xi_1, \cdots, \xi_n)$, $X = \xi_1 U_1 + \cdots + \xi_n U_n$ be the corresponding quantity of the scalar extension $\mathfrak{A}_{\mathfrak{K}}$ of $\mathfrak{A}$. We called X the *general quantity* of $\mathfrak{A}$, the polynomial

$$|\lambda I - X| = f(\lambda; \xi_1, \cdots, \xi_n)$$

the *characteristic function* of $\mathfrak{A}$. We also called the minimum function $g(\lambda; \xi_1, \cdots, \xi_n)$ of X the *minimum function* of $\mathfrak{A}$ and proved that it is a monic polynomial in λ with coefficients polynomials of $\mathfrak{F}[\xi_1, \cdots, \xi_n]$.

The degree r of $g(\lambda; \xi_1, \cdots, \xi_n)$ was called the *degree* of $\mathfrak{A}$ in Chapter AX but if $\mathfrak{A}$ is a field over $\mathfrak{F}$ the term degree has already been defined for many years and its meaning is that of the order of $\mathfrak{A}$ over $\mathfrak{F}$ which for inseparable fields may be seen to differ from the meaning above. We shall thus call r the *principal degree* of $\mathfrak{A}$. Whenever $\mathfrak{A}$ is *not* a field no confusion will arise if we delete the adjective "principal," and we shall do so.

The polynomial $f(\lambda; \xi_1, \cdots, \xi_n) = g_1 \cdots g_t$ where $g_i = g_i(\lambda; \xi_1, \cdots, \xi_n)$ are the invariant factors of $\lambda I - X$. They are monic polynomials in λ and, by Theorem A2.16, have coefficients in $\mathfrak{F}[\xi_1, \cdots, \xi_n]$. Moreover they may be chosen so that g_i divides g_{i-1} for $i = 2, \cdots, t$, and then g_1 is the minimum function of $\mathfrak{A}$. If we now replace the ξ_i by quantities α_i in $\mathfrak{F}$ we obtain a quantity $A = \alpha_1 U_1 + \cdots + \alpha_n U_n$ of A. Then $f(\lambda; \alpha_1, \cdots, \alpha_n)$ is clearly the characteristic function $|\lambda I - A|$ of A. We shall call $g(\lambda, A) = g(\lambda; \alpha_1, \cdots, \alpha_n)$ the *principal function* of A and the corresponding equation the *principal equation* of A. We clearly have

Theorem 10. *Every quantity A of an algebra is a root of its principal function $g(\lambda, A)$. The characteristic function $f(\lambda)$ of A is a product of $g(\lambda, A)$ by divisors of $g(\lambda, A)$, and $g(\lambda, A)$ is a product of the minimum function $\phi(\lambda)$ of A by divisors of $\phi(\lambda)$.*

We now pass to the case where $\mathfrak{A}$ is a division algebra and apply Theorems 1.5 and 1.10 to obtain immediately

Theorem 11. *Let A be a quantity of a division algebra. Then the principal and characteristic functions of A are powers of its (irreducible) minimum function.*

If $\xi_1, \cdots, \xi_m$ are independent indeterminates over $\mathfrak{F}$ and $\mathfrak{K} = \mathfrak{F}(\xi_1, \cdots, \xi_m)$, the quantities of the scalar extension $\mathfrak{A}_{\mathfrak{K}}$ of an algebra $\mathfrak{A} = (U_1, \cdots, U_n)$ over $\mathfrak{F}$ all have the form

$$A = \beta^{-1} \sum_{i_j=0}^{t_j} a_{i_1 \cdots i_m} \xi_1^{i_1} \cdots \xi_m^{i_m}$$

where $\beta \neq 0$ is in $\mathfrak{F}[\xi_1, \cdots, \xi_m]$ and the coefficients $a_{i_1 \cdots i_m}$ are in $\mathfrak{A}$. Then $A = 0$ if and only if these coefficients in $\mathfrak{A}$ are all zero. The degree of A is the maximum $i_1 + \cdots + i_m$ with coefficient not zero, and we said in Chapter AII that the degree of AB is the sum of the degrees of A and B unless $\mathfrak{A}$ has a divisor of zero. Hence if $\mathfrak{A}$ is a division algebra then $\mathfrak{A}_{\mathfrak{L}}$ contains no divisors of zero and is a division algebra. The converse is trivial and we have

Theorem 12. *Let $\xi_1, \cdots, \xi_m$ be independent indeterminates over $\mathfrak{F}$ and $\mathfrak{L} = \mathfrak{F}(\xi_1, \cdots, \xi_m)$. Then the scalar extension $\mathfrak{A}_{\mathfrak{L}}$ is a division algebra if and only if $\mathfrak{A}$ is a division algebra.*

We now apply Theorem 1.12 to $\mathfrak{A}_{\mathfrak{L}}$ for $m = n$ and have

Theorem 13. *The minimum function of a division algebra $\mathfrak{D}$ is irreducible in $\mathfrak{F}(\xi_1, \cdots, \xi_n)$ and the characteristic function of $\mathfrak{D}$ is a power of the minimum function.*

As a simple corollary of the results above and Theorem A7.30 we have

Theorem 14. *Let $\mathfrak{Q}$ be an infinite subset of a field $\mathfrak{F}$ and let the minimum function of a division algebra $\mathfrak{D} = (\bar{U}_1, \cdots, \bar{U}_n)$ over $\mathfrak{F}$ have degree r and be separable. Then every subfield $\mathfrak{F}(A)$ of $\mathfrak{D}$ has degree at most r and there exist quantities α_i in $\mathfrak{Q}$ such that the corresponding quantities $A = \alpha_1\bar{U}_1 + \cdots + \alpha_n\bar{U}_n$ generate separable subfields $\mathfrak{F}(A)$ of degree r over $\mathfrak{F}$ of $\mathfrak{D}$.*

For if $\mathfrak{F}(A)$ is a simple extension of $\mathfrak{F}$ defined by $A = \alpha_1\bar{U}_1 + \cdots + \alpha_n\bar{U}_n$ of $\mathfrak{D}$ we have $g(A; \alpha_1, \cdots, \alpha_n) = 0$, $\mathfrak{F}(A)$ has degree at most r. Conversely let $g(\lambda; \xi_1, \cdots, \xi_n)$ be separable so that the discriminant $D(\xi_1, \cdots, \xi_n)$ is a non-zero polymonial of $\mathfrak{F}[\xi_1, \cdots, \xi_n]$. By Theorem A7.30 there exist quantities $\alpha_1, \cdots, \alpha_n$ in $\mathfrak{F}$ such that $D(\alpha_1, \cdots, \alpha_n) \neq 0$, and thus $g(\lambda) = g(\lambda; \alpha_1, \cdots, \alpha_n)$ is separable. By Theorem 1.11 $g(\lambda) = [\phi(\lambda)]^q$ where $\phi(\lambda)$ is the minimum function of $A = \alpha_1\bar{U}_1 + \cdots + \alpha_n\bar{U}_n$ and is irreducible. Then the separability of $g(\lambda)$ implies that $q = 1$, $\phi(\lambda)$ has degree r and is separable, $\mathfrak{F}(A)$ is separable of degree r over $\mathfrak{F}$.

In Theorem A10.11 we considered quadrate algebras, that is, algebras $\mathfrak{A}$ of order n over $\mathfrak{F}$ such that the scalar extension $\mathfrak{A}_{\mathfrak{K}}$ is a total matric algebra for some field $\mathfrak{K}$ of finite degree over $\mathfrak{F}$. Then we proved that $n = \nu^2$, ν is the (principal) degree of $\mathfrak{A}$. Moreover, the minimum function of $\mathfrak{A}$ is separable. Theorem 1.14 gives

Theorem 15. *Let $\mathfrak{D}$ be a quadrate division algebra of order ν^2 over an infinite field $\mathfrak{F}$. Then there exist separable subfields of $\mathfrak{D}$ of degree ν over $\mathfrak{F}$.*

We shall later prove that every normal division algebra $\mathfrak{D}$ over $\mathfrak{F}$ is quadrate and shall give an alternative proof of the property in Theorem 1.15. Note that in Theorem 1.15 we could take the defining quantity $A = \alpha_1\bar{U}_1 + \cdots + \alpha_n\bar{U}_n$ of the separable field $\mathfrak{F}(A)$ as in Theorem 1.14 for $\mathfrak{L}$ any infinite subset of $\mathfrak{F}$. We shall use this property in Chapter X but shall not use the result above otherwise, preferring our alternative and more abstract proof given in Chapter IV.

14. The norm and trace functions. The coefficients of the characteristic function $|\lambda I - X| = \lambda^n - T(X)\lambda^{n-1} + \cdots + (-1)^n N(X)$ of the general quantity X of an algebra $\mathfrak{A}$ are invariants of $\mathfrak{A}$. We call $T(X)$ the *trace* of X or the *trace function* of $\mathfrak{A}$. It is the sum of the diagonal elements in the matrix X and is a linear polynomial in $\xi_1, \cdots, \xi_n$ so that $T(\alpha A + \beta B) = \alpha T(A) + \beta T(B)$ for every α and β of $\mathfrak{F}$, A and B of $\mathfrak{A}$. The quantity $N(X)$ is called the *norm* of X, or *norm function* of $\mathfrak{A}$, and $N(X) = |X|$ and is a homogeneous polynomial of degree n in $\xi_1, \cdots, \xi_n$. It is a multiplicative function, that is, $N(AB) = N(A)N(B)$ since this is true of the determinant of a product of two matrices. Moreover $N(\alpha A) = \alpha^n N(A) = \alpha^n |A|$ for every α of $\mathfrak{F}$ and A of $\mathfrak{A}$.

If $g(\lambda; \xi_1, \cdots, \xi_n)$ is the minimum function of $\mathfrak{A} = (U_1, \cdots, U_n)$ over $\mathfrak{F}$ then we have called $g(\lambda, A) = g(\lambda; \alpha_1, \cdots, \alpha_n)$ the principal polynomial of $A = \alpha_1 U_1 + \cdots + \alpha_n U_n$ and have seen in Chapter AX that $g(\lambda, A)$ is independent of the particular basis $U_1, \cdots, U_n$ of $\mathfrak{A}$ over $\mathfrak{F}$. Let $\mathfrak{A}$ have degree r so that $g(\lambda, A)$ is monic of degree r for every A of $\mathfrak{A}$, and let $g(\lambda; \xi_1, \cdots, \xi_n) = \lambda^r - T_P(X)\lambda^{r-1} + \cdots + (-1)^r N_P(X)$. The functions $T_P(X)$ and $N_P(X)$ have properties analogous to the trace and norm defined above, and we shall call them the *principal trace* and *principal norm* functions, respectively, of $\mathfrak{A}$, $T_P(A)$ and $N_P(A)$ the principal trace and principal norm of A for every A of $\mathfrak{A}$.

When $\mathfrak{A}$ is a division algebra $g(\lambda; \xi_1, \cdots, \xi_n)$ is irreducible, and the characteristic function of $\mathfrak{A}$ is $[g(\lambda; \xi_1, \cdots, \xi_n)]^k$, $n = kr$. Comparing coefficients we obtain $T(X) = kT_P(X)$, $N(X) = [N_P(X)]^k$. We now obtain

Theorem 16. *The principal norm function $N_P(X)$ of a division algebra $\mathfrak{D}$ of degree r is a homogeneous polynomial of degree r in $\xi_1, \cdots, \xi_n$, and has the properties, $N_P(AB) = N_P(A)N_P(B)$ for every A and B of $\mathfrak{D}$, $N_P(\alpha A) = \alpha^r N_P(A)$ for every α of $\mathfrak{F}$.*

Since $N(X) = [N_P(X)]^k$ is homogeneous of degree $n = kr$ it follows that $N_P(X)$ is homogeneous of degree r. The quantity $N_P(A)$ is obtained by replacing the ξ_i in $N_P(X)$ by quantities α_i in $\mathfrak{F}$, $N_P(\alpha A)$ is clearly obtained by replacing the ξ_i by $\alpha\alpha_i$, so that $N_P(\alpha A) = \alpha^r N_P(A)$ by the homogeneity of the principal norm function. The minimum function of the unity quantity $1 = I$ of $\mathfrak{D}$ is $\lambda - 1$ so that, by Theorem 1.11, $N_P(1) = 1$. We let $\xi_1, \cdots, \xi_n, \eta_1, \cdots, \eta_n$ be independent indeterminates over $\mathfrak{F}$ and have $[N_P(XY)]^k = N(XY) = [N_P(X)N_P(Y)]^k$, $N_P(XY) = \zeta N_P(X)N_P(Y)$ identically in the ξ_i, η_j, with $\zeta^k = 1$. Put $Y = 1$ and obtain $N_P(X) = \zeta N_P(X)$, $\zeta = 1$ as desired.

The linearity of $T(X) = kT_P(X)$ implies that $T_P(X)$ is also a linear function of X, provided that k is not a multiple of the characteristic of $\mathfrak{F}$. Further properties of the principal norm and trace of a division algebra will be obtained in Chapter VIII, where it will be shown in particular, for a certain type of algebra, that $T_P(X)$ is linear for $\mathfrak{F}$ of any characteristic, and in fact that $T_P(A)$ and $N_P(A)$ are actually the trace and determinant of a certain matrix.

15. A theorem of Wedderburn. There is an important tool theorem due to Wedderburn which we shall require in Chapter III. The result may be stated as

Theorem 17. *Let the unity quantity of a total matric subalgebra* $\mathfrak{M}$ *of an algebra* $\mathfrak{A}$ *coincide with that of* $\mathfrak{A}$. *Then*

$$\mathfrak{A} = \mathfrak{M} \times \mathfrak{C}, \qquad \mathfrak{C} = \mathfrak{A}^{\mathfrak{M}}. \tag{24}$$

For let $\mathfrak{M} = (e_{ij}\,; i, j = 1, \cdots, m)$ and a be any quantity of $\mathfrak{A}$. Define

$$a_{ij} = \sum_{k=1}^{m} e_{ki} a e_{jk}\,. \tag{25}$$

Then $a_{ij}e_{rs} = e_{ri}ae_{js} = e_{rs}a_{ij}$ and we have shown that the a_{ij} are all in the $\mathfrak{A}$-commutator $\mathfrak{C} = \mathfrak{A}^{\mathfrak{M}}$ of $\mathfrak{M}$. Also $1 = e_{11} + \cdots + e_{mm}$ so that

$$\sum_{i,j=1}^{m} a_{ij}e_{ij} = \sum_{i,j,k=1}^{m} e_{ki}ae_{jk}e_{ij} = \sum_{k,j=1}^{m} e_{kk}ae_{jj} = a, \tag{26}$$

$\mathfrak{A} \leqq \mathfrak{C}\mathfrak{M}$. But $\mathfrak{C}\mathfrak{M} = \mathfrak{M}\mathfrak{C} \leqq \mathfrak{A}$ so that $\mathfrak{A} = \mathfrak{M}\mathfrak{C}$. By the corollary to Theorem 1.1 it follows that $\mathfrak{A} = \mathfrak{M} \times \mathfrak{C}$ if and only if no $a = \sum_{i,j=1}^{m} a_{ij}e_{ij} = 0$, for a_{ij} in $\mathfrak{C}$, unless the a_{ij} are all zero. But $a = 0$ implies that $0 = \sum_{p=1}^{m} e_{pr}ae_{sp} = \sum_{p=1}^{m} a_{rs}e_{pp} = a_{rs}$ for every r and s, as desired.

We shall generalize this result in Chapter IV, and shall use the generalization as a fundamental tool for the study of normal simple algebras.

CHAPTER II

IDEALS AND NILPOTENT ALGEBRAS

1. Idempotent quantities of an algebra. The simplest type of algebra is the algebra of order one over $\mathfrak{F}$. Such an algebra is a linear set $\mathfrak{A} = (u)$ over $\mathfrak{F}$ which is an algebra if and only if $u^2 = \gamma u$ for γ in $\mathfrak{F}$. If $\gamma = 0$ the product $\alpha u \beta u = \alpha\beta u^2$ of any two quantities of $\mathfrak{A}$ is zero. Then $\mathfrak{A}$ is the case of order one of the algebras defined in the

DEFINITION. *An algebra $\mathfrak{A}$ over $\mathfrak{F}$ is called a zero algebra if $ab = 0$ for every a and b of $\mathfrak{A}$.*

Suppose next that $\gamma \neq 0$. Then $e = \gamma^{-1}u$ has the property $\mathfrak{A} = (u) = (e)$ over $\mathfrak{F}$, $e^2 = \gamma^{-2}(\gamma u) = \gamma^{-1}u = e$. The correspondence

$$\alpha e \leftrightarrow \alpha \qquad (\alpha \text{ in } \mathfrak{F}), \tag{1}$$

is an equivalence of $\mathfrak{A}$ and $\mathfrak{F}$, and we have proved

Theorem 1. *An algebra of order one over $\mathfrak{F}$ is either equivalent to $\mathfrak{F}$ or is a zero algebra over $\mathfrak{F}$.*

Quantities like e occur frequently in the theory of algebras. We call e an *idempotent* quantity, or, briefly, an *idempotent,* if $e \neq 0$ and $e^2 = e$. The unity quantity of an algebra is clearly an idempotent.

If e and u are any quantities of an algebra $\mathfrak{A}$ such that $eu = ue = 0$ we call e and u *orthogonal* quantities. The concept of orthogonality generalizes in an obvious way to sets of any number of quantities called *pairwise orthogonal* quantities. Clearly zero algebras consist of pairwise orthogonal quantities whose squares are zero. We shall frequently consider a set of pairwise orthogonal idempotents $e_1, \cdots, e_t$, that is, quantities such that

$$e_i^2 = e_i, \qquad e_ie_j = 0 \qquad (i \neq j;\, i, j = 1, \cdots, t). \tag{2}$$

For such quantities we have

Theorem 2. *Let $\mathfrak{B}$ be the linear subset of an algebra $\mathfrak{A}$ spanned by a set of pairwise orthogonal idempotent quantities $e_1, \cdots, e_t$ of $\mathfrak{A}$. Then $\mathfrak{B}$ is a subalgebra $\mathfrak{B} = (e_1, \cdots, e_t)$ of order t over $\mathfrak{F}$ of $\mathfrak{A}$ with $e = e_1 + \cdots + e_t$ as unity quantity. We call $\mathfrak{B}$ a* ***diagonal algebra*** *over $\mathfrak{F}$*

The verification that $\mathfrak{B}$ is a subalgebra of $\mathfrak{A}$ is straightforward and is left to the reader. If $b = \sum_{i=1}^{t} \beta_i e_i = 0$ for β_i in $\mathfrak{F}$, then $e_i b = \beta_i e_i = 0$ so that $\beta_i = 0$. Hence the e_i are linearly independent in $\mathfrak{F}$, $\mathfrak{B} = (e_1, \cdots, e_t)$ over $\mathfrak{F}$. Also $ee_i = e_ie = e_i^2 = e_i$ so that $eb = be = b$ for every b of $\mathfrak{B}$, e is the unity quantity of $\mathfrak{B}$. In particular we have shown that $e^2 = e$, a part of the theorem above stating that *the sum of pairwise orthogonal idempotents is idempotent.* Note that $\mathfrak{B}$ is equivalent to the algebra of all t-rowed diagonal square matrices with elements in $\mathfrak{F}$. This is, of course, our reason for calling $\mathfrak{B}$ a diagonal algebra.

The result of Theorem 2.2 has a partial converse which we state as

Theorem 3. *Let e and u be distinct idempotents such that $eu = ue = u$. Then $e = u + v$ where v is an idempotent orthogonal to u and $ev = ve = v$.*

For $v = e - u \neq 0$, $v^2 = e^2 - eu - ue + u^2 = e - u - u + u = v$. Also $ev = e^2 - eu = e - u = v$, $ve = v$, $uv = ue - u^2 = u - u = 0 = vu$.

The following result seems quite trivial but will be used later and hence stated here.

Theorem 4. *Let u and v be orthogonal idempotents and*

$$u = u_1 + \cdots + u_t, \qquad v = v_1 + \cdots + v_s \tag{3}$$

where the u_i are pairwise orthogonal idempotents and the v_i are pairwise orthogonal idempotents. Then the quantities of the set $u_1, \cdots, u_t, v_1, \cdots, v_s$ are pairwise orthogonal.

For $u_iu = u_i$, $vv_j = v_j$ so that $u_iv_j = u_iuvv_j = 0$. Similarly $v_ju_i = 0$.

2. Left ideals. A linear subset $\mathfrak{B}$ over $\mathfrak{F}$ of an algebra $\mathfrak{A}$ over $\mathfrak{F}$ is called a *left ideal* of $\mathfrak{A}$ if $\mathfrak{A}\mathfrak{B} \leqq \mathfrak{B}$. Since $\mathfrak{B}\mathfrak{B} \leqq \mathfrak{A}\mathfrak{B} \leqq \mathfrak{B}$ every non-zero left ideal of $\mathfrak{A}$ is a subalgebra of $\mathfrak{A}$. The distributive laws for sums (intersections) with products of linear sets then imply

LEMMA 1. *The sum and intersection of two left ideals of $\mathfrak{A}$ are left ideals of $\mathfrak{A}$.*

If a is in $\mathfrak{A}$ the set $\mathfrak{A}a$ is a left ideal of $\mathfrak{A}$. We then have

Theorem 5. *Let $\mathfrak{A}$ be an algebra with a unity quantity e. Then if $\mathfrak{A}$ is the supplementary sum*

$$\mathfrak{A} = \mathfrak{B}_1 + \cdots + \mathfrak{B}_t \tag{4}$$

of left ideals $\mathfrak{B}_i$, and thus $e = e_1 + \cdots + e_t$ with e_i in $\mathfrak{B}_i$, we have $\mathfrak{B}_i = \mathfrak{A}e_i$, and the e_i are pairwise orthogonal idempotents of $\mathfrak{A}$. Conversely the expression of e as a sum of t pairwise orthogonal idempotents e_i implies that $\mathfrak{A}$ is the supplementary sum (4) of its left ideals $\mathfrak{B}_i = \mathfrak{A}e_i$.

For $b_ie = b_i = b_ie_1 + \cdots + b_ie_t$. Since $\mathfrak{B}_j$ is a left ideal, b_ie_j is in $\mathfrak{B}_j$ for every b_i of $\mathfrak{B}_i$. Now (4) is a supplementary sum and this implies that $b_ie_j = 0$ for $i \neq j$, $b_ie_i = b_i$, $\mathfrak{B}_i \leqq \mathfrak{A}e_i$. In particular $e_i^2 = e_i$, $e_ie_j = 0$ for $i \neq j$ as desired. Also $\mathfrak{B}_i$ is a left ideal, e_i is in $\mathfrak{B}_i$, $\mathfrak{A}e_i \leqq \mathfrak{B}_i$ so that $\mathfrak{A}e_i = \mathfrak{B}_i$. Conversely if $e = e_1 + \cdots + e_t$ such that $e_i^2 = e_i$, $e_ie_j = 0$ for $i \neq j$, we define $\mathfrak{B}_i = \mathfrak{A}e_i$. Then $\mathfrak{A} = \mathfrak{A}e = \mathfrak{A}(e_1 + \cdots + e_t) = (\mathfrak{B}_1, \cdots, \mathfrak{B}_t)$. To prove this sum supplementary we let $a = a_1 + \cdots + a_t$ with a_i in $\mathfrak{B}_i$. Then $ae_i = a_i$ so that the a_i are uniquely determined by a. This proves our theorem.

3. Ideals of $\mathfrak{A}$. We call a linear subset $\mathfrak{B}$ of $\mathfrak{A}$ a *right ideal* of $\mathfrak{A}$ if $\mathfrak{B}\mathfrak{A} \leqq \mathfrak{B}$. Clearly the results of the preceding section on left ideals have analogues for right ideals.

A linear subset $\mathfrak{B}$ of $\mathfrak{A}$ is called a *two-sided ideal* or, simply, an *ideal* of $\mathfrak{A}$, if $\mathfrak{B}$ is both a left and a right ideal of $\mathfrak{A}$. The set 0 is such an ideal called the *zero ideal.* Moreover $\mathfrak{A}$ itself and all the subalgebras $\mathfrak{A}^{\cdot 2}$, $\mathfrak{A}^{\cdot 3}$, $\cdots$ are ideals of $\mathfrak{A}$. If $\mathfrak{B}$ is any linear subset of $\mathfrak{A}$, the set $\mathfrak{A}\mathfrak{B}\mathfrak{A}$ is clearly an ideal of $\mathfrak{A}$ and in particular $\mathfrak{A}b\mathfrak{A}$ is an ideal of $\mathfrak{A}$ for every b of $\mathfrak{A}$.

LEMMA 2. *If $\mathfrak{B}$ is a left ideal of $\mathfrak{A}$ and $\mathfrak{C}$ is any linear subset of $\mathfrak{A}$ the product $\mathfrak{B}\mathfrak{C}$ is a left ideal of $\mathfrak{A}$.*

For $\mathfrak{A}(\mathfrak{B}\mathfrak{C}) = (\mathfrak{A}\mathfrak{B})\mathfrak{C} \leqq \mathfrak{B}\mathfrak{C}$.

We have the analogue of Lemma 2.2 for right ideals and then have

LEMMA 3. *Let $\mathfrak{B}$ be a left ideal of $\mathfrak{A}$ and $\mathfrak{C}$ be a right ideal. Then $\mathfrak{B}\mathfrak{C}$ is an ideal of $\mathfrak{A}$.*

LEMMA 4. *The product $\mathfrak{B}\mathfrak{C}$ and intersection $[\mathfrak{B}, \mathfrak{C}]$ of two ideals $\mathfrak{B}$ and $\mathfrak{C}$ of $\mathfrak{A}$ are ideals of $\mathfrak{A}$, and $\mathfrak{B}\mathfrak{C} \leqq [\mathfrak{B}, \mathfrak{C}]$.*

For $\mathfrak{B}\mathfrak{C}$ is an ideal by Lemma 2.3 and $[\mathfrak{B}, \mathfrak{C}]$ by Lemma 2.1 and its right-ideal analogue. Since $\mathfrak{B}$ and $\mathfrak{C}$ are ideals we have $\mathfrak{B}\mathfrak{C} \leqq \mathfrak{B}$, $\mathfrak{B}\mathfrak{C} \leqq \mathfrak{C}$ so that $\mathfrak{B}\mathfrak{C} \leqq [\mathfrak{B}, \mathfrak{C}]$.

4. Nilpotent algebras. A quantity $y \neq 0$ of an algebra $\mathfrak{A}$ is called *nilpotent* if $y^\rho = 0$ for some integer ρ. The least such integer is called the *index* of y.

We call an algebra $\mathfrak{N}$ *nilpotent if* $\mathfrak{N}^{\cdot\rho} = 0$ for some exponent ρ. Clearly the least such ρ is what we have already called the index of $\mathfrak{N}$. Then $\mathfrak{N}$ is nilpotent of index α if and only if $y_1 \cdots y_\alpha = 0$ for all y_i of $\mathfrak{N}$, and there exist quantities x_j in $\mathfrak{N}$ such that $x_1 \cdots x_{\alpha-1} \neq 0$.

We now have the

Theorem 6. *All subalgebras of a nilpotent algebra $\mathfrak{N}$ are nilpotent. Moreover $\mathfrak{N}$ has a non-zero proper ideal if and only if $\mathfrak{N}$ is not a zero algebra of order one.*

For if $\mathfrak{N}_0 < \mathfrak{N}$ and $\mathfrak{N}^{\cdot\alpha} = 0$, then $\mathfrak{N}_0^\alpha = 0$. An algebra of order one has no non-zero proper subalgebra. Conversely let $\mathfrak{N}$ be nilpotent of order $n > 1$. If $\mathfrak{N}$ is a zero algebra then $\mathfrak{N} = (u_1, \cdots, u_n)$ over $\mathfrak{F}$, $u_iu_j = 0$ for $i, j = 1, \cdots, n$, the algebra $(u_1, \cdots, u_{n-1})$ is a non-zero proper ideal of $\mathfrak{N}$. Otherwise $\mathfrak{N}^{\cdot 2} \neq 0$ and is a non-zero proper ideal of $\mathfrak{N}$.

If a left (right or two-sided) ideal $\mathfrak{B}$ of an algebra $\mathfrak{A}$ is either zero or a nilpotent algebra we call $\mathfrak{B}$ a nilpotent left (right or two-sided) ideal. Thus *all ideals of a nilpotent algebra are nilpotent ideals.*

5. The radical of an algebra. The existence of a maximal nilpotent ideal of an algebra will be obtained as a consequence of

LEMMA 5. *The sum of two nilpotent left ideals $\mathfrak{B}$ and $\mathfrak{C}$ of $\mathfrak{A}$ is a nilpotent left ideal of $\mathfrak{A}$.*

For by Lemma 2.1 $(\mathfrak{B}, \mathfrak{C})$ is a left ideal of $\mathfrak{A}$. The quantities of $(\mathfrak{B}, \mathfrak{C})^{\cdot\alpha}$ are sums of products $a = a_1 \cdots a_\alpha$ with factors in either $\mathfrak{B}$ or $\mathfrak{C}$. Let β of the factors be in $\mathfrak{B}$ and combine factors to the left of these with them to obtain $a = b_1 \cdots b_\beta a_0$ with b_i in $\mathfrak{B}$ and a_0 in $\mathfrak{C}$. Similarly $a = c_1 \cdots c_\gamma a_{00}$ with $\alpha = \beta + \gamma$, c_i in $\mathfrak{C}$. We let $\alpha = \alpha_1 + \alpha_2 - 1$ where α_1 is the index of $\mathfrak{B}$, α_2 the index of $\mathfrak{C}$. If $\beta \geqq \alpha_1$ then $a = 0$. Otherwise $\beta < \alpha_1, \gamma = \alpha_1 + \alpha_2 - 1 - \beta \geqq \alpha_2$, $a = 0$, $(\mathfrak{B}, \mathfrak{C})$ is nilpotent of index at most $\alpha_1 + \alpha_2 - 1$.

We next prove

LEMMA 6. *If $\mathfrak{L}$ is a nilpotent left ideal of $\mathfrak{A}$ the sum $(\mathfrak{L}, \mathfrak{L}\mathfrak{A})$ is a nilpotent ideal of $\mathfrak{A}$.*

For if α is the index of $\mathfrak{L}$ we have $(\mathfrak{L}\mathfrak{A})^{\cdot\alpha} = \mathfrak{L}\cdot(\mathfrak{A}\mathfrak{L})^{\cdot\alpha-1}\mathfrak{A} \leqq \mathfrak{L}\mathfrak{L}^{\alpha-1}\mathfrak{A} = 0$ and $\mathfrak{L}\mathfrak{A}$ is a nilpotent left ideal of $\mathfrak{A}$. By Lemma 2.5 $(\mathfrak{L}, \mathfrak{L}\mathfrak{A})$ is a nilpotent left ideal of $\mathfrak{A}$. But $(\mathfrak{L}, \mathfrak{L}\mathfrak{A})\mathfrak{A} = (\mathfrak{L}\mathfrak{A}, \mathfrak{L}\mathfrak{A}) \leqq (\mathfrak{L}, \mathfrak{L}\mathfrak{A})$ so that $(\mathfrak{L}, \mathfrak{L}\mathfrak{A})$ is an ideal of $\mathfrak{A}$.

As an immediate consequence we have

Theorem 7. *Every nilpotent left, right, or two-sided ideal of an algebra $\mathfrak{A}$ is contained in a unique maximal nilpotent ideal of $\mathfrak{A}$ called its* ***radical****.*

For let $\mathfrak{N}$ be a nilpotent ideal of $\mathfrak{A}$ of largest possible order. If $\mathfrak{N}_0$ is also a nilpotent ideal, the algebra $(\mathfrak{N}, \mathfrak{N}_0)$ is a nilpotent left ideal of $\mathfrak{A}$ by Lemma 2.5. The right-side analogue of Lemma 2.5 implies that $(\mathfrak{N}, \mathfrak{N}_0)$ is a nilpotent ideal of $\mathfrak{A}$ whose order is greater than or equal to that of $\mathfrak{N}$. But $\mathfrak{N}$ has maximal order, $\mathfrak{N} \geqq (\mathfrak{N}, \mathfrak{N}_0)$, $\mathfrak{N} \geqq \mathfrak{N}_0$. If $\mathfrak{L}$ is a nilpotent left ideal of $\mathfrak{A}$ we have $(\mathfrak{L}, \mathfrak{L}\mathfrak{A}) \leqq \mathfrak{N}$ by Lemma 2.6, $\mathfrak{L} \leqq \mathfrak{N}$. Similarly $\mathfrak{N}$ contains all nilpotent right ideals of $\mathfrak{A}$.

6. The existence of an idempotent. In a nilpotent algebra $\mathfrak{N}$ of index α the αth power of every quantity of $\mathfrak{N}$ is zero. Thus a nilpotent algebra is composed of zero and nilpotent quantities. Conversely, if every non-zero quantity of an algebra $\mathfrak{N}$ is nilpotent, $\mathfrak{N}$ is nilpotent. This is a consequence of the deeper

Theorem 8. *Every non-nilpotent algebra $\mathfrak{A}$ contains an idempotent quantity.*

For algebras of order one the result is a consequence of Theorem 2.1. We make an induction on the order n of $\mathfrak{A}$ and assume the theorem true for algebras of order less than n. If $\mathfrak{A}a = \mathfrak{A}$ for some quantity a of $\mathfrak{A}$ then, by Theorem 1.1, we have $xa = 0$ only if $x = 0$. Also $a = ea$ for $e \neq 0$ in $\mathfrak{A}$, $ea = e^2a$, $(e^2 - e)a = 0$, $e^2 = e$, e is idempotent. Hence let every $\mathfrak{A}a < \mathfrak{A}$. The left ideals $\mathfrak{A}a$ have order less than $\mathfrak{A}$ and the hypothesis of our induction states that either one $\mathfrak{A}a$, and hence $\mathfrak{A}$, contains an idempotent, or every $\mathfrak{A}a$ is nilpotent. But then $\mathfrak{A}^2$ is the sum of nilpotent left ideals $\mathfrak{A}u_i$ and Lemma 2.5 implies that $\mathfrak{A}^2$ is nilpotent. Hence so is $\mathfrak{A}$. This completes the induction.

A left ideal $\mathfrak{B} \neq 0$ of $\mathfrak{A}$ is called *left simple* if there exists no left ideal $\mathfrak{C}$ of $\mathfrak{A}$ such that $0 < \mathfrak{C} < \mathfrak{B}$. By the method used above we may prove a result sometimes used in the theory of representations of algebras

Theorem 9. *A left simple left ideal $\mathfrak{L}$ of an algebra $\mathfrak{A}$ is either a zero algebra or $\mathfrak{L} = \mathfrak{A}e$ for an idempotent e of $\mathfrak{L}$.*

For if $\mathfrak{L}^{\cdot 2} \neq 0$ there exists at least one y in $\mathfrak{L}$ such that $\mathfrak{L}y \neq 0$. The set $\mathfrak{L}y$ is a left ideal of $\mathfrak{A}$ and since $\mathfrak{L}$ is left simple, $\mathfrak{L}y = \mathfrak{L}$. By the proof above $\mathfrak{L}$ contains an idempotent $e \neq 0$. But $\mathfrak{A}e \leqq \mathfrak{L}$, $\mathfrak{A}e$ contains $e^2 = e \neq 0$, $\mathfrak{A}e \neq 0$, $\mathfrak{A}e = \mathfrak{L}$.

7. Properly nilpotent quantities. A quantity $y \neq 0$ of an algebra $\mathfrak{A}$ is called *properly nilpotent* if both ay and ya are zero or nilpotent for every a of $\mathfrak{A}$. Now $(ay)^{\alpha+1} = a(ya)^{\alpha}y$ and $(ya)^{\alpha+1} = y(ay)^{\alpha}a$, so that ya is a nilpotent or zero if and only if ay is nilpotent or zero. We have proved

LEMMA 7. *A quantity $y \neq 0$ of $\mathfrak{A}$ is properly nilpotent if and only if for every a of $\mathfrak{A}$ one of the products ay, ya is zero or nilpotent.*

We use this result and prove

Theorem 10. *The set $\mathfrak{N}_0$ consisting of zero and all properly nilpotent quantities of an algebra $\mathfrak{A}$ is its radical $\mathfrak{N}$.*

For if x is in $\mathfrak{N}$ and a is in $\mathfrak{A}$ we have ax in $\mathfrak{N}$ is zero or nilpotent, x is in $\mathfrak{N}_0$, $\mathfrak{N} \leqq \mathfrak{N}_0$. Conversely if y is in $\mathfrak{N}_0$ the left ideal $\mathfrak{A}y$ consists only of zero and nilpotent quantities and is nilpotent by Theorem 2.8. Then by Theorem 2.7, $\mathfrak{A}y \leqq \mathfrak{N}$. However $\mathfrak{N}$ is an algebra and $\mathfrak{A}(\lambda_1 y_1 + \lambda_2 y_2) \leqq \mathfrak{N}$ for every λ_1 and λ_2 of $\mathfrak{F}$, y_1 and y_2 of $\mathfrak{N}_0$. Hence $\lambda_1 y_1 + \lambda_2 y_2$ is properly nilpotent, $\mathfrak{N}_0$ is a linear set, $\mathfrak{A}\mathfrak{N}_0 \leqq \mathfrak{N} \leqq \mathfrak{N}_0$. Thus $\mathfrak{N}_0^{\cdot 2} \leqq \mathfrak{N}$, $\mathfrak{N}_0$ is a nilpotent left ideal of $\mathfrak{A}$, $\mathfrak{N}_0 \leqq \mathfrak{N}$, $\mathfrak{N}_0 = \mathfrak{N}$.

8. The Peirce decomposition. We shall leave the study of nilpotent algebras and shall now assume that $\mathfrak{A}$ has an idempotent e. Define $\mathfrak{L}_e$ to be the set of all quantities x of $\mathfrak{A}$ such that $xe = 0$. Since if $ye = 0$ then $a(ye) = 0$ and $(\alpha x + \beta y)e = 0$ for a in $\mathfrak{A}$, α and β in $\mathfrak{F}$, the set $\mathfrak{L}_e$ is a left ideal of $\mathfrak{A}$. We write

$$a = ae + (a - ae) \tag{5}$$

for any a of $\mathfrak{A}$. Since $e^2 = e$ we have $(a - ae)e = 0$, $a - ae$ is in $\mathfrak{L}_e$. Also ae is in $\mathfrak{A}e$. If b were in both $\mathfrak{L}_e$ and $\mathfrak{A}e$ then $be = 0$, $be = b$ so that $b = 0$. Hence $\mathfrak{L}_e$ and $\mathfrak{A}e$ are supplementary in $\mathfrak{A}$ and we have

LEMMA 8. *If e is any idempotent of $\mathfrak{A}$ we may express $\mathfrak{A}$ as the supplementary sum*

$$\mathfrak{A} = \mathfrak{A}e + \mathfrak{L}_e \tag{6}$$

of left ideals $\mathfrak{A}e$, $\mathfrak{L}_e$ of $\mathfrak{A}$.

Equation (5) is called the *left-sided Peirce decomposition of a*, (6) the *left-sided Peirce decomposition of $\mathfrak{A}$ relative to e.*

Define $\mathfrak{R}_e$ as the right ideal of all x in $\mathfrak{A}$ such that $ex = 0$, $\mathfrak{C}_e$ as the intersection of $\mathfrak{R}_e$ and $\mathfrak{L}_e$. Notice that $\mathfrak{C}_e$ is the set of all quantities z such that $ez = ze = 0$.

We call $\mathfrak{C}_e$ the set of all quantities *orthogonal* to e, thus to the algebra $e\mathfrak{A}e$. We now obtain the *two-sided Peirce decomposition.**

Theorem 11. *Let e be an idempotent of an algebra $\mathfrak{A}$. Then $\mathfrak{A} = e\mathfrak{A}e + e\mathfrak{L}_e + \mathfrak{R}_e e + \mathfrak{C}_e$.*

Note that $e\mathfrak{A}e$, $e\mathfrak{L}_e$, $\mathfrak{R}_e e$, and $\mathfrak{C}_e$ are all algebras. Also $e\mathfrak{L}_e$ is the set of all quantities b of $\mathfrak{A}$ such that $eb = b$, $be = 0$, $\mathfrak{R}_e$ is the set of all quantities c of $\mathfrak{A}$ such that $ce = c$, $ec = 0$, while $e\mathfrak{A}e$ is the set of all d of $\mathfrak{A}$ such that $ed = de = d = ede$. It follows that if $\mathfrak{B}$ is any ideal of $\mathfrak{A}$, $e\mathfrak{B}e$ is the intersection of $e\mathfrak{A}e$ and $\mathfrak{B}$. For, $[e\mathfrak{A}e, \mathfrak{B}]$ consists of quantities $d = ede$ with d in $\mathfrak{B}$ and hence in $e\mathfrak{B}e$, $[e\mathfrak{A}e, \mathfrak{B}] \leqq e\mathfrak{B}e$. Since $\mathfrak{B}$ is an ideal, $e\mathfrak{B}e \leqq \mathfrak{B}$, $e\mathfrak{B}e \leqq e\mathfrak{A}e$, hence $e\mathfrak{B}e = [e\mathfrak{A}e, \mathfrak{B}]$.

For proof we have the two-sided Peirce decomposition of a quantity given by

$$a = eae + e(a - ae) + (a - ea)e + (a - ea - ae + eae). \tag{7}$$

The quantity eae is in $e\mathfrak{A}e$, $e(a - ae)$ in $e\mathfrak{L}_e$, $(a - ea)e$ in $\mathfrak{R}_e e$, $a - ea - ae + eae$ in $\mathfrak{C}_e$. If $0 = a_1 + a_2 + a_3 + a_4$ with a_1 in $e\mathfrak{A}e$, a_2 in $e\mathfrak{L}_e$, a_3 in $\mathfrak{R}_e e$, a_4 in $\mathfrak{C}_e$, we have $e0e = 0 = ea_1e = a_1$, $e0 = a_2 = 0$, $0e = a_3 = 0$ so that $a_4 = 0$. Hence the sum in our theorem is supplementary.

We now connect the radical of a non-nilpotent algebra with the Peirce decomposition. We prove

Theorem 12. *Let $\mathfrak{A}$ have radical $\mathfrak{N}$ and e be an idempotent of $\mathfrak{A}$. Then the radical of $e\mathfrak{A}e$ is the intersection $e\mathfrak{N}e$ of $\mathfrak{N}$ and $e\mathfrak{A}e$, the radical of $\mathfrak{C}_e$ is the intersection of $\mathfrak{N}$ and $\mathfrak{C}_e$.*

For, $\mathfrak{N}$ is an ideal of $\mathfrak{A}$ and $[e\mathfrak{A}e, \mathfrak{N}] = e\mathfrak{N}e$. Let y be in the radical $\mathfrak{N}_0$ of $e\mathfrak{A}e$ so that $y = eye$. Then if α is the index of $\mathfrak{N}_0$ we have

$$(ay)^{\alpha+1} = ay(eaey)^{\alpha} = 0,$$

since $eaey$ is in $\mathfrak{N}_0$ for every a of $\mathfrak{A}$. Hence ay is zero or nilpotent, y is properly nilpotent in $\mathfrak{A}$, $\mathfrak{N}_0 \leqq \mathfrak{N}$ by Theorem 2.10, $\mathfrak{N}_0 \leqq [e\mathfrak{A}e, \mathfrak{N}]$. Conversely, the quantities of $[e\mathfrak{A}e, \mathfrak{N}]$ are properly nilpotent in $\mathfrak{A}$, hence properly nilpotent in $e\mathfrak{A}e$. Thus they are in $\mathfrak{N}_0$ by Theorem 2.10 and $\mathfrak{N}_0 = [e\mathfrak{A}e, \mathfrak{N}]$.

We next let $\mathfrak{N}_e$ be the radical of $\mathfrak{C}_e$. Now $[\mathfrak{N}, \mathfrak{C}_e]$ is an ideal of $\mathfrak{C}_e$ since $\mathfrak{N}$ is an ideal of $\mathfrak{A}$. Also $[\mathfrak{N}, \mathfrak{C}_e]$ is nilpotent and contained in $\mathfrak{N}_e$ by Theorem 2.7. If $z \neq 0$ is in $\mathfrak{N}_e$ then $ze = ez = 0$, $zaz = z(a - ae - ea + eae)z = za_ez$ where a_e is the component of a in (7) which is in $\mathfrak{C}_e$. But then $(az)^{\alpha+1} = az(az)^{\alpha} = az(a_ez)^{\alpha} = 0$ if α is the index of $\mathfrak{N}_e$. It follows that z is properly nilpotent in $\mathfrak{A}$, z is in $\mathfrak{N}$, $\mathfrak{N}_e \leqq [\mathfrak{N}, \mathfrak{C}_e]$ as desired. This proves our theorem.

9. Principal idempotents. An idempotent e of an algebra $\mathfrak{A}$ is called a *principal idempotent* of $\mathfrak{A}$ if $\mathfrak{A}$ has no idempotent u orthogonal to e. In the Peirce decomposition of $\mathfrak{A}$ relative to e the set $\mathfrak{C}_e$ consists of all quantities orthogonal to e. Combining this result with Theorem 2.8 we have

* See Section A4.7 for the illuminating example of this decomposition in the case of the algebra of n-rowed square matrices.

LEMMA 9. *An idempotent e is principal in $\mathfrak{A}$ if and only if $\mathfrak{C}_e$ is zero or a nilpotent algebra.*

If e is idempotent and u is an idempotent of $e\mathfrak{A}e$, then $eu = ue = u$ so that e is a principal idempotent of $e\mathfrak{A}e$. Also if $e \neq u$ then by Theorem 2.3, $e - u$ is an idempotent orthogonal to u. Hence e is the only principal idempotent of $e\mathfrak{A}e$. This also implies that the only principal idempotent of an algebra with a unity quantity is the unity quantity.

Theorem 13. *If u is a non-principal idempotent of $\mathfrak{A}$ there exists a principal idempotent $e = u + v$ such that $eu = ue = u$ and v is an idempotent orthogonal to u.*

For by Lemma 2.9 if u is non-principal there exists an idempotent v in $\mathfrak{C}_u$. Then v is orthogonal to u and $e = u + v$ is idempotent and such that $ev = ve = v$, $eu = ue = u$. Since $ev = v \neq 0$ the quantity v is in $\mathfrak{C}_u$ but not in $\mathfrak{C}_e$. However every x in $\mathfrak{C}_e$ is in $\mathfrak{C}_u$ since $xe = ex = 0$ implies that $xu = xeu = 0$, $ux = uex = 0$. Thus $\mathfrak{C}_e < \mathfrak{C}_u$. We may now choose v so that $\mathfrak{C}_e$ has the least possible order. If e were not principal there would be an idempotent w in $\mathfrak{C}_e$ such that $g = e + w$ is idempotent, e is orthogonal to w, $\mathfrak{C}_g < \mathfrak{C}_e$. But then $g = u + (v + w)$ and we have a contradiction providing that we show that $v + w$ is an idempotent of $\mathfrak{C}_u$. This is true since w and v are in $\mathfrak{C}_u$, so is their sum, $vw = vew = 0 = wv$, $(v + w)^2 = v^2 + w^2 = v + w$.

Theorem 2.13 may be combined with Theorem 2.8 and we state the result as

Theorem 14. *Every non-nilpotent algebra contains a principal idempotent.*

In the Peirce decomposition $\mathfrak{A} = e\mathfrak{A}e + e\mathfrak{L}_e + \mathfrak{R}_e e + \mathfrak{C}_e$ of Theorem 2.11 the set $\mathfrak{L}_e$ is a left ideal, $\mathfrak{R}_e$ is a right ideal, $\mathfrak{C}_e$ is their intersection. We define $\mathfrak{T}_e = e\mathfrak{L}_e + \mathfrak{R}_e e + \mathfrak{C}_e$, and have

$$\mathfrak{A} = e\mathfrak{A}e + \mathfrak{T}_e, \qquad \mathfrak{T}_e \leqq (\mathfrak{L}_e, \mathfrak{R}_e). \tag{8}$$

But then we may prove

Theorem 15. *If e is a principal idempotent of $\mathfrak{A}$ the set $\mathfrak{T}_e$ is contained in the radical of $\mathfrak{A}$.*

For by Lemma 2.9, $\mathfrak{C}_e$ is zero or a nilpotent algebra. We have $\mathfrak{R}_e\mathfrak{L}_e \leqq \mathfrak{C}_e$, and if α is the index of $\mathfrak{C}_e$ we have $(\mathfrak{R}_e\mathfrak{L}_e)^{\cdot\alpha} = 0$. But then $(\mathfrak{L}_e\mathfrak{R}_e)^{\cdot\alpha+1} = \mathfrak{L}_e(\mathfrak{R}_e\mathfrak{L}_e)^{\cdot\alpha}\mathfrak{R}_e = 0$. By Lemma 2.3, $\mathfrak{L}_e\mathfrak{R}_e$ is an ideal of $\mathfrak{A}$, by Theorem 2.7, $\mathfrak{L}_e\mathfrak{R}_e \leqq \mathfrak{N}$. Now $\mathfrak{L}_e e = 0$ and $\mathfrak{L}_e\mathfrak{A} = (\mathfrak{L}_e\mathfrak{R}_e e, \mathfrak{L}_e\mathfrak{C}_e) \leqq \mathfrak{L}_e\mathfrak{R}_e$ since $\mathfrak{R}_e e \leqq \mathfrak{R}_e$. Similarly $\mathfrak{A}\mathfrak{R}_e \leqq \mathfrak{L}_e\mathfrak{R}_e \leqq \mathfrak{N}$, and this implies that $\mathfrak{L}_e^{\cdot 2} \leqq \mathfrak{N}$, $\mathfrak{R}_e^{\cdot 2} \leqq \mathfrak{N}$, $\mathfrak{L}_e \leqq \mathfrak{N}$, $\mathfrak{R}_e \leqq \mathfrak{N}$, $\mathfrak{T}_e \leqq \mathfrak{N}$.

10. Primitive idempotents. An idempotent e of an algebra $\mathfrak{A}$ is called primitive in $\mathfrak{A}$ if e is not the sum $u + v$ of orthogonal idempotents u and v. We may easily derive the equivalent definitions contained in

LEMMA 10. *An idempotent e is primitive in $\mathfrak{A}$ if and only if $\mathfrak{A}$ contains no idempotent $u \neq e$ such that $eu = ue = u$, that is, e is the only idempotent of $e\mathfrak{A}e$.*

For by Theorem 2.3 if e is primitive no $u \neq e$ can exist such that $eu = ue = u$. Conversely, if e were not primitive so that $e = u + v$ then $u \neq e$, $eu = ue = u$, and our proof is complete. The final remark follows since $eu = ue = u$ implies that $u = eue$ is in $e\mathfrak{A}e$.

We now derive

Theorem 16. *Every non-primitive idempotent e of an algebra $\mathfrak{A}$ is the sum of a finite number of pairwise orthogonal primitive idempotents e_i of $\mathfrak{A}$ such that $e_i e = ee_i = e_i$.*

For, by our definition e is a sum of at least two orthogonal idempotents. If e were expressed as a sum of r pairwise orthogonal idempotents, the linear set spanned by them is a (diagonal) subalgebra of $\mathfrak{A}$ of order r by Theorem 2.2. It follows that the maximum such r is an integer t not greater than the order of $\mathfrak{A}$. Then let $e = e_1 + \cdots + e_t$ for this t and pairwise orthogonal idempotents e_i. If any e_i is not primitive we have $e_i = u + v$ where u and v are orthogonal idempotents. Then by Theorems 2.2 and 2.4 the idempotents $e_1, \cdots, e_{i-1}, u, v, e_{i+1}, \cdots, e_t$ are $t + 1$ pairwise orthogonal idempotents whose sum is e. This contradicts our definition of t. Hence the e_i are all primitive.

By Theorems 2.13, 2.16 we have

Theorem 17. *Let u be a non-principal idempotent of $\mathfrak{A}$. Then there exists a principal idempotent e such that $e = e_1 + \cdots + e_t$ for pairwise orthogonal primitive idempotents e_i such that $u = e_1 + \cdots + e_s$, $s < t$.*

11. Difference algebras. An algebra $\mathfrak{A}$ is an additive abelian group and every (additive abelian) subgroup $\mathfrak{B}$ of $\mathfrak{A}$ is a normal divisor of $\mathfrak{A}$. We let $\mathfrak{B}$ be a linear subset of $\mathfrak{A}$ so that the quotient group $\mathfrak{A}/\mathfrak{B}$ exists. Since we are now discussing additive groups we shall prefer to call this group the *difference* group and in referring to it shall use the notation

$$\mathfrak{A} - \mathfrak{B}.$$

The quantities of $\mathfrak{A} - \mathfrak{B}$ are the cosets of $\mathfrak{B}$ with respect to addition. Each such coset $[a]$ defined by a in $\mathfrak{A}$ is the set of all elements $a + b$ for b ranging over the quantities of $\mathfrak{B}$. The quantity a is called a *representative* of $[a]$, and $[a] = [a_1]$ if and only if $a - a_1$ is in $\mathfrak{B}$.

We define

$$[a_1] + [a_2] = [a_1 + a_2], \qquad \lambda[a] = [\lambda a]$$

for every a, a_1, a_2 of $\mathfrak{A}$ and λ of $\mathfrak{F}$, and see now that $\mathfrak{A} - \mathfrak{B}$ is a linear set over $\mathfrak{F}$. In fact let $\mathfrak{C}$ be the supplement of $\mathfrak{B}$ in $\mathfrak{A}$ as in Section 1.2, $\mathfrak{C} = (u_1, \cdots, u_t)$ over $\mathfrak{F}$. Then clearly $\mathfrak{B} = [0]$,

$$\mathfrak{A} - \mathfrak{B} = ([u_1], \cdots, [u_t]) \text{ over } \mathfrak{F}, \tag{9}$$

and $\mathfrak{A} - \mathfrak{B}$ over $\mathfrak{F}$ is a linear set equivalent to $\mathfrak{C}$ over $\mathfrak{F}$.

The linear set $\mathfrak{A} - \mathfrak{B}$ is not necessarily a ring. However we may prove

Theorem 18. *Let $\mathfrak{B}$ be an ideal of order m over $\mathfrak{F}$ in an algebra $\mathfrak{A}$ of order n over $\mathfrak{F}$ and define*

$$[a_1][a_2] = [a_1a_2] \qquad (a_1, a_2 \text{ in } \mathfrak{A}), \tag{10}$$

in the difference group $\mathfrak{A} - \mathfrak{B}$. Then $\mathfrak{A} - \mathfrak{B}$ is an algebra of order $t = n - m$ over $\mathfrak{F}$ called the **difference algebra** of $\mathfrak{A}$ **modulo** $\mathfrak{B}$.

For if $[a_1] = [a_{10}]$ and $[a_2] = [a_{20}]$ then $a_{10} = a_1 + b_1$, $a_{20} = a_2 + b_2$ for b_i in $\mathfrak{B}$. Thus $a_{10}a_{20} = a_1a_2 + b_3$ where $b_3 = a_1b_2 + b_1a_2 + b_1b_2$ is in the ideal $\mathfrak{B}$. But then $[a_{10}a_{20}] = [a_1a_2]$ and (10) is independent of the representatives a_1, a_2 and defines a unique product in $\mathfrak{A} - \mathfrak{B}$. It is easily seen that the postulates of a ring are now satisfied and that $\mathfrak{A} - \mathfrak{B}$ is an algebra over $\mathfrak{F}$.

The algebra $\mathfrak{A} - \mathfrak{B}$ has the basis (9) where $\mathfrak{C} = (u_1, \cdots, u_t)$ over $\mathfrak{F}$. If $\mathfrak{C}$ is an algebra we have $u_iu_j = \sum \gamma_{ijk}u_k$, $[u_i][u_j] = \sum \gamma_{ijk}[u_k]$ and the correspondence $\sum \xi_iu_i \leftrightarrow \sum \xi_i[u_i]$ is a (1-1) correspondence and implies

Theorem 19. *Let $\mathfrak{A} = \mathfrak{B} + \mathfrak{C}$ where $\mathfrak{B}$ is an ideal and $\mathfrak{C}$ a subalgebra of $\mathfrak{A}$. Then $\mathfrak{A} - \mathfrak{B}$ is an algebra equivalent to $\mathfrak{C}$.*

As in the theory of groups we have the following important result whose proof is quite trivial. We leave the verification to the reader.

Theorem 20. *Let $\mathfrak{B}$ be an ideal of an algebra $\mathfrak{A}$, $\mathfrak{A}_0 = \mathfrak{A} - \mathfrak{B}$. Then there is a (1-1) correspondence between the subalgebras $\mathfrak{C}_0$ of $\mathfrak{A}_0$ and the subalgebras $\mathfrak{C} \geqq \mathfrak{B}$ of $\mathfrak{A}$ such that $\mathfrak{C} \leftrightarrow \mathfrak{C}_0 = \mathfrak{C} - \mathfrak{B}$. Moreover $\mathfrak{C}_0$ is an ideal of $\mathfrak{A}_0$ if and only if the corresponding $\mathfrak{C}$ is an ideal of $\mathfrak{A}$, and in this case we have the equivalence over $\mathfrak{F}$,*

$$\mathfrak{A} - \mathfrak{C} \cong \mathfrak{A}_0 - \mathfrak{C}_0 .$$

12. Direct sums. An algebra $\mathfrak{A}$ is defined to be the *direct sum* of algebras $\mathfrak{B}_i$, and we write

$$\mathfrak{A} = \mathfrak{B}_1 \oplus \cdots \oplus \mathfrak{B}_t , \tag{11}$$

if the $\mathfrak{B}_i$ are subalgebras of $\mathfrak{A}$ such that $\mathfrak{A} = \mathfrak{B}_1 + \cdots + \mathfrak{B}_t$ and $\mathfrak{B}_i\mathfrak{B}_j = 0$ for $i \neq j$. Then $\mathfrak{A}\mathfrak{B}_i \leqq \mathfrak{B}_i$, $\mathfrak{B}_i\mathfrak{A} \leqq \mathfrak{B}_i$ and the $\mathfrak{B}_i$ are ideals of $\mathfrak{A}$. Conversely if $\mathfrak{A} = \mathfrak{B}_1 + \cdots + \mathfrak{B}_t$ for ideals $\mathfrak{B}_i$ of $\mathfrak{A}$, the intersection of distinct $\mathfrak{B}_i$, $\mathfrak{B}_j$ is zero. By Lemma 2.4 $\mathfrak{B}_i\mathfrak{B}_j = 0$ $(i \neq j)$, and $\mathfrak{A}$ is the direct sum of the $\mathfrak{B}_i$.

If an algebra $\mathfrak{A}$ is not expressible as the direct sum $\mathfrak{B}_1 \oplus \cdots \oplus \mathfrak{B}_t$ of $t > 1$ subalgebras $\mathfrak{B}_i$ we call $\mathfrak{A}$ an *irreducible* algebra. Otherwise we call $\mathfrak{A} = \mathfrak{B}_1 \oplus \cdots \oplus \mathfrak{B}_t$ *reducible,* and call the $\mathfrak{B}_i$ *components* of $\mathfrak{A}$. Note that if $\mathfrak{A}$ has a unity quantity then, by Theorem 2.5, so does every component $\mathfrak{B}_i$ of $\mathfrak{A}$. Every right, left, or two-sided ideal of a component $\mathfrak{B}_i$ of $\mathfrak{A}$ is trivially a corresponding ideal of $\mathfrak{A}$. In fact we may prove

Theorem 21. *Let an algebra $\mathfrak{A}$ with a unity quantity e be expressed as a direct sum $\mathfrak{A} = \mathfrak{B}_1 \oplus \cdots \oplus \mathfrak{B}_t$. Then a linear subset $\mathfrak{C}$ of $\mathfrak{A}$ is a right, left, or two-sided ideal of $\mathfrak{A}$ if and only if $\mathfrak{C} = \mathfrak{C}_1 \oplus \cdots \oplus \mathfrak{C}_t$, where $\mathfrak{C}_i$ is correspondingly a right, left, or two-sided ideal of $\mathfrak{B}_i$ and $\mathfrak{C}_i = [\mathfrak{C}, \mathfrak{B}_i]$.*

For by Theorem 2.5 $e = e_1 + \cdots + e_t$ for idempotents e_i in $\mathfrak{B}_i = \mathfrak{A}e_i = e_i\mathfrak{A}$, and e_i is the unity quantity of $\mathfrak{B}_i$. Define $\mathfrak{C}_i = \mathfrak{C}e_i$. If $\mathfrak{C}$ is a right ideal of $\mathfrak{A}$ then $\mathfrak{C}e_i = \mathfrak{C}_i \leqq \mathfrak{C}$. Every quantity c of $\mathfrak{A}$ is uniquely expressible in the form $c = c_1 + \cdots + c_t$ with c_i in $\mathfrak{A}e_i$, and if c is in $\mathfrak{C}$ then the $c_i = ce_i$ are in its respective subalgebras $\mathfrak{C}_i$. It follows that $\mathfrak{C} = \mathfrak{C}_1 + \cdots + \mathfrak{C}_t$. But $\mathfrak{C}_i \leqq \mathfrak{B}_i$, $\mathfrak{C}_i\mathfrak{C}_j \leqq \mathfrak{B}_i\mathfrak{B}_j = 0$ for $i \neq j$, $\mathfrak{C} = \mathfrak{C}_1 \oplus \cdots \oplus \mathfrak{C}_t$. We see that $\mathfrak{C}_i \leqq [\mathfrak{B}_i, \mathfrak{C}]$, and since $\mathfrak{B}_i = \mathfrak{B}_i e_i$ we have $[\mathfrak{B}_i, \mathfrak{C}] = [\mathfrak{B}_i, \mathfrak{C}]e_i \leqq \mathfrak{C}_i$, $\mathfrak{C}_i = [\mathfrak{B}_i, \mathfrak{C}]$. Finally $\mathfrak{C}_i = [\mathfrak{B}_i, \mathfrak{C}]$ is a right ideal of $\mathfrak{A}$ by Lemma 2.1 and hence is a right ideal of $\mathfrak{B}_i$. Conversely a direct sum $\mathfrak{C}$ of right ideals $\mathfrak{C}_i$ of $\mathfrak{B}_i$ has the property $\mathfrak{C}_i\mathfrak{B}_i \leqq \mathfrak{C}_i$, $\mathfrak{C}_i\mathfrak{B}_j = 0$ for $i \neq j$, $\mathfrak{C}\mathfrak{A} = (\mathfrak{C}_1\mathfrak{B}_1, \cdots, \mathfrak{C}_t\mathfrak{B}_t) \leqq \mathfrak{C}$. The results for left and two-sided ideals are derived analogously.

As an immediate consequence we have the

Corollary. *Let $\mathfrak{A}$ be as in Theorem 2.21 and $\mathfrak{N}$ be its radical. Then $\mathfrak{N} = \mathfrak{N}_1 \oplus \cdots \oplus \mathfrak{N}_t$ where $\mathfrak{N}_i = [\mathfrak{N}, \mathfrak{B}_i]$ is the radical of $\mathfrak{B}_i$.*

For $\mathfrak{N}_i \leqq \mathfrak{N}$ is nilpotent and by the above is a nilpotent ideal of $\mathfrak{B}_i$. Theorem 2.7 implies that $\mathfrak{N}_i$ is contained in the radical $\mathfrak{N}_{i0}$ of $\mathfrak{B}_i$. Evidently $\mathfrak{N}_{i0}$ is a nilpotent ideal of $\mathfrak{A}$, $\mathfrak{N}_{i0} \leqq \mathfrak{N}$, $\mathfrak{N}_{i0} \leqq [\mathfrak{N}, \mathfrak{B}_i]$, $\mathfrak{N}_{i0} = \mathfrak{N}_i$.

13. Reduction to irreducible components. In considering reducible algebras we may wish to know whether the irreducible components are unique. We obtain a solution of this question for our most important case in

Theorem 22. *Every reducible algebra with a unity quantity is expressible as a direct sum of irreducible components uniquely apart from the order of its components.*

For, the finiteness of the order of $\mathfrak{A}$ implies the existence of at least one decomposition $\mathfrak{A} = \mathfrak{B}_1 \oplus \cdots \oplus \mathfrak{B}_t$ with irreducible $\mathfrak{B}_i$. Suppose also that $\mathfrak{A} = \mathfrak{C}_1 \oplus \cdots \oplus \mathfrak{C}_s$ with the $\mathfrak{C}_i$ irreducible. Then each $\mathfrak{C}_i$ is an ideal of $\mathfrak{A}$. By Theorem 2.21, $\mathfrak{C}_i = \mathfrak{C}_{i1} \oplus \cdots \oplus \mathfrak{C}_{it}$, $\mathfrak{C}_{ij} = [\mathfrak{C}_i, \mathfrak{B}_j]$, and since $\mathfrak{C}_i$ is irreducible we have all the $\mathfrak{C}_{ij} = 0$ for $j \neq j_i$, $\mathfrak{C}_i = \mathfrak{C}_{ij_i} \leqq \mathfrak{B}_{j_i}$. By symmetry every $\mathfrak{B}_j \leqq \mathfrak{C}_{k_j}$. A direct sum is supplementary and $\mathfrak{C}_i \leqq \mathfrak{B}_{j_i} \leqq \mathfrak{C}_{k_{j_i}}$ is possible only if $\mathfrak{C}_i = \mathfrak{C}_{k_{j_i}} = \mathfrak{B}_{j_i}$. Thus every $\mathfrak{C}_i = \mathfrak{B}_i$ for the $\mathfrak{C}_i$ numbered properly. The direct sum of the $\mathfrak{C}_i$ is equal to that of the $\mathfrak{B}_i$ and we must have $s = t$.

Observe that Theorems 2.21 and 2.22 are both false if $\mathfrak{A}$ does not have a unity quantity. An example proving this is given by $\mathfrak{A} = \mathfrak{B}_1 \oplus \mathfrak{B}_2 = \mathfrak{C}_1 \oplus \mathfrak{C}_2$, where $\mathfrak{B}_1 = \mathfrak{C}_1 = (u_1)$, $u_1^2 = 0$, $\mathfrak{B}_2 = (u_2, u_3)$, $u_2^2 = u_3$, $u_2^3 = 0$, $\mathfrak{C}_2 = (u_1 + u_2, u_3) \neq \mathfrak{B}_2$.

If $\mathfrak{A} = \mathfrak{B} \oplus \mathfrak{C}$ we have seen that $\mathfrak{B}$ is an ideal of $\mathfrak{A}$. A partial converse is proved in

Theorem 23. *Let an ideal* $\mathfrak{B}$ *of an algebra* $\mathfrak{A}$ *have a unity quantity* e. *Then* $\mathfrak{A} = \mathfrak{B} \oplus \mathfrak{C}_e$. *Moreover, in any expression* $\mathfrak{A} = \mathfrak{B} \oplus \mathfrak{C}$ *we have* $\mathfrak{C} = \mathfrak{C}_e$.

For $\mathfrak{B}$ is an ideal and all the sets $e\mathfrak{A}e$, $e\mathfrak{L}_e$, $\mathfrak{R}_e e$ of Theorem 2.11 are in $\mathfrak{B}$. Thus $\mathfrak{A} = \mathfrak{B}_1 + \mathfrak{C}_e$ with $\mathfrak{B}_1 \leqq \mathfrak{B}$. Every quantity of $\mathfrak{B}$ has the form $b = b_1 + c$ with b_1 in $\mathfrak{B}_1$ and c in $\mathfrak{C}_e$. Then $be = b = b_1 e + ce = b_1$ since b_1 is in $\mathfrak{B}$, $ce = 0$. Hence $\mathfrak{B} = \mathfrak{B}_1$, $\mathfrak{A} = \mathfrak{B} + \mathfrak{C}_e$. Also $e\mathfrak{C}_e = \mathfrak{C}_e e = 0$, $\mathfrak{B} = \mathfrak{B}e = e\mathfrak{B}$, $\mathfrak{C}_e\mathfrak{B} = \mathfrak{B}\mathfrak{C}_e = 0$, $\mathfrak{A} = \mathfrak{B} \oplus \mathfrak{C}_e$. If $\mathfrak{A} = \mathfrak{B} \oplus \mathfrak{C}$ every quantity of $\mathfrak{C}$ is orthogonal to e, $\mathfrak{C} \leqq \mathfrak{C}_e$, $\mathfrak{B} \oplus \mathfrak{C} = \mathfrak{B} \oplus \mathfrak{C}_e$ and $\mathfrak{C}$ and $\mathfrak{C}_e$ have the same order. Hence $\mathfrak{C} = \mathfrak{C}_e$ as desired.

In the case of algebras $\mathfrak{A}$ with a unity quantity this result furnishes an alternative definition of the term "component of $\mathfrak{A}$" as any ideal $\mathfrak{B}$ of $\mathfrak{A}$ such that $\mathfrak{B}$ has a unity quantity. Thus we see that the unique irreducible components of Theorem 2.22 are actually the totality of irreducible components of the algebra.

14. The centrum of a direct sum. The elementary relation between the centrum of a direct sum and the centra of its components is given by

Theorem 24. *Let* $\mathfrak{A} = \mathfrak{B}_1 \oplus \cdots \oplus \mathfrak{B}_t$ *have centrum* $\mathfrak{C}$ *and* $\mathfrak{C}_i$ *be the centrum of* $\mathfrak{B}_i$. *Then* $\mathfrak{C}_i = [\mathfrak{B}_i, \mathfrak{C}]$, $\mathfrak{C} = \mathfrak{C}_1 \oplus \cdots \oplus \mathfrak{C}_t$.

For if c_i is in $\mathfrak{C}_i$ and $a = b_1 + \cdots + b_t$ for b_i in $\mathfrak{B}_i$ then $c_i a = c_i b_i = b_i c_i = ac_i$ and c_i is in $\mathfrak{C}$. Hence $\mathfrak{C}_i \leqq \mathfrak{C}$, $\mathfrak{C}_i \leqq [\mathfrak{B}_i, \mathfrak{C}]$. Every quantity c of $\mathfrak{C}$ has the property that $cb_i = b_i c$ for b_i in $\mathfrak{B}$, and if also c is in $[\mathfrak{B}_i, \mathfrak{C}]$, it is evident that c is in $\mathfrak{C}_i$, $\mathfrak{C}_i = [\mathfrak{B}_i, \mathfrak{C}]$. Now every c of $\mathfrak{C}$ has the form $c = c_1 + \cdots + c_t$ with c_i in $\mathfrak{B}_i$. Also $cb_i = b_i c = c_i b_i = b_i c_i$, so that c_i is in $\mathfrak{C}_i$, c_i is in $\mathfrak{C}$, $\mathfrak{C} = \mathfrak{C}_1 + \cdots + \mathfrak{C}_t$, $\mathfrak{C} = \mathfrak{C}_1 \oplus \cdots \oplus \mathfrak{C}_t$.

The result given above has a partial converse which we state as

Theorem 25. *Let* $\mathfrak{A}$ *have a unity quantity* e *and the centrum of* $\mathfrak{A}$ *be* $\mathfrak{C} = \mathfrak{C}_1 \oplus \cdots \oplus \mathfrak{C}_t$. *Then if* $\mathfrak{B}_i = \mathfrak{A}\mathfrak{C}_i$ *we have* $\mathfrak{A} = \mathfrak{B}_1 \oplus \cdots \oplus \mathfrak{B}_t$, $\mathfrak{C}_i = [\mathfrak{B}_i, \mathfrak{C}]$, $\mathfrak{C}_i$ *is the centrum of* $\mathfrak{B}_i$.

For e is in $\mathfrak{C}$ and is the unity quantity of $\mathfrak{C}$. By Theorem 2.5 we have $e = e_1 + \cdots + e_t$, $\mathfrak{C}_i = \mathfrak{C}e_i$ has e_i as unity quantity. The centrum $\mathfrak{C}$ contains all the e_i since $e_i \leqq \mathfrak{C}_i \leqq \mathfrak{C}$. Apply Theorem 2.5 again and have $\mathfrak{A} = \mathfrak{B}_1 + \cdots + \mathfrak{B}_t$ with $\mathfrak{B}_i = \mathfrak{A}e_i = e_i\mathfrak{A}$. The $\mathfrak{B}_i$ are ideals of $\mathfrak{A}$ and thus $\mathfrak{A} = \mathfrak{B}_1 \oplus \cdots \oplus \mathfrak{B}_t$. Now $\mathfrak{B}_i = \mathfrak{A}e_i \geqq \mathfrak{A}\mathfrak{C}_i e_i = \mathfrak{A}\mathfrak{C}_i$, $\mathfrak{A}\mathfrak{C}_i \geqq \mathfrak{A}e_i \geqq \mathfrak{B}_i$, $\mathfrak{B}_i = \mathfrak{A}\mathfrak{C}_i$. The centrum of $\mathfrak{B}_i$ is clearly a subset of $\mathfrak{C}$ and hence in $\mathfrak{C}_i = \mathfrak{C}e_i$, $\mathfrak{C}_i$ is contained in the centrum of $\mathfrak{B}_i$, $\mathfrak{C}_i$ is the centrum of $\mathfrak{B}_i$. By Theorem 2.24 we have $\mathfrak{C}_i = [\mathfrak{B}_i, \mathfrak{C}]$.

There are certain types of algebras $\mathfrak{A}$ with radical the zero ideal but such that the radical of a scalar extension $\mathfrak{A}_\mathfrak{K}$ is not zero. This will be seen to be due to the fact that the centrum of $\mathfrak{A}$ has the same property, and actually only in the case of modular fields $\mathfrak{F}$. We are thus lead to a study of the structure of algebras which are inseparable fields over $\mathfrak{F}$ and to a study of scalar extensions of arbitrary algebraic fields of finite degree over $\mathfrak{F}$.

15. Scalar extensions of separable fields. A separable field $\mathfrak{Z}$ of degree n over $\mathfrak{F}$ is a simple extension $\mathfrak{Z} = \mathfrak{F}(x)$ defined, as in Section A7.11, by a root x of an irreducible separable equation $\phi(\lambda) = 0$. Let $\mathfrak{W}$ be a scalar root field of $\phi(\lambda)$ so that

$$\phi(\lambda) = (\lambda - \xi_1) \cdots (\lambda - \xi_n) \tag{12}$$

with $\mathfrak{W} = \mathfrak{F}(\xi_1, \cdots, \xi_n)$. We shall derive the result of Exercise 1 on page A239 on the structure of $\mathfrak{Z}_{\mathfrak{W}}$ by a consideration of the diagonal algebra $\mathfrak{A} = (e_1, \cdots, e_n)$ over $\mathfrak{W}$. Here our notation indicates that the e_i are pairwise orthogonal idempotents of $\mathfrak{A}$.

The algebra $\mathfrak{A}$ over $\mathfrak{W}$ contains the field $\mathfrak{F}[x_0]$ of all polynomials in $x_0 = \xi_1 e_1 + \cdots + \xi_n e_n$ with coefficients in $\mathfrak{F}$. The *pairwise orthogonality* of the e_i implies that *$\mathfrak{F}(x_0)$ is equivalent over $\mathfrak{F}$ to $\mathfrak{Z}$*, and hence *there is no loss of generality if we take $x_0 = x$*. Moreover it implies that

$$\psi(x) = \psi(\xi_1)e_1 + \cdots + \psi(\xi_n)e_n , \tag{13}$$

for every polynomial $\psi(\lambda)$ with coefficients scalars with respect to $\mathfrak{A}$. Note that $e_1 x = e_1 \xi_1$, a result used in the proof of Theorem 5.6.

Define $\psi_i(\lambda) \equiv (\lambda - \xi_i)^{-1}\phi(\lambda)$ so that $\psi_i(\xi_i) \neq 0$, $\psi_i(\xi_j) = 0$ for $i \neq j$. Then

$$e_i = \psi_i(x)[\psi_i(\xi_i)]^{-1}. \tag{14}$$

Hence every e_i is in the scalar extension $\mathfrak{Z}_{\mathfrak{W}}$. But $\mathfrak{Z}$ has order n over $\mathfrak{F}$, $\mathfrak{Z}_{\mathfrak{W}}$ has order n over $\mathfrak{W}$, $\mathfrak{Z}_{\mathfrak{W}} \leqq \mathfrak{A}$ of order n over $\mathfrak{W}$. Hence $\mathfrak{Z}_{\mathfrak{W}} = \mathfrak{A}$ over $\mathfrak{W}$.

Theorem 26. *If $\mathfrak{Z}$ is a separable field of finite degree over $\mathfrak{F}$ and $\mathfrak{W}$ is its scalar root field, the algebra $\mathfrak{Z}_{\mathfrak{W}}$ is a diagonal algebra over $\mathfrak{W}$.*

In Chapter V we shall use an extension of the property above for the case where $\mathfrak{Z}$ is normal over $\mathfrak{F}$. We shall thus prove

Theorem 27. *Let $\mathfrak{Z} = \mathfrak{F}(x)$ be normal over $\mathfrak{F}$ with automorphism group $\mathfrak{G} = (S_1, S_2, \cdots, S_n)$, $S_1 = I$, the identity automorphism. Assume also that $\mathfrak{W}$ is a scalar extension $\mathfrak{F}(\xi)$ of $\mathfrak{F}$ equivalent over $\mathfrak{F}$ to $\mathfrak{Z}$ under a correspondence generated by $x \leftrightarrow \xi$ so that $x^{S_i} = \theta_i(x) \leftrightarrow \theta_i(\xi) \equiv \xi^{T_i}$. Then the algebra*

$$\mathfrak{Z}_{\mathfrak{W}} = (e, e^{S_2}, \cdots, e^{S_n}) = (e, e^{T_2}, \cdots, e^{T_n}) \tag{15}$$

for pairwise orthogonal idempotents $e^{S_i} = e^{T_i^{-1}}$.

Notice that we have extended the automorphisms S_i of $\mathfrak{Z}$ to be automorphisms of $\mathfrak{Z} \times \mathfrak{W}$ by the definition $(\sum z_k w_k)^{S_i} = \sum {z_k}^{S_i} w_k$, and similarly for each T_i.

It is clear that the polynomial $\psi_i(\lambda)$ used in (14) has the property $\psi_i(\lambda) = [\psi_1(\lambda)]^{T_i}$. Hence $e_i = e_1^{T_i}$. But also

$$x^{S_i^{-1}} = \xi^{T_i^{-1}} e_1 + \cdots + \xi e_i + \cdots + \xi^{T_n T_i^{-1}} e_n , \tag{16}$$

so that

$$(17) \qquad e_1^{S_i^{-1}} = \psi_1(x^{S_i^{-1}})[\psi_1(\xi)]^{-1} = e_i = e_1^{T_i}.$$

As S_i ranges over all the automorphisms of $\mathfrak{G}$ so does S_i^{-1} and hence $e_1 = e$ has the properties of our theorem.*

We shall now pass to a consideration of some properties of inseparable fields.

16. Inseparable fields. The study of inseparable fields of Chapter AVII was a brief one, the subject being introduced there to provide an adequate basis for the Galois theory. The modern study of algebras over a general field requires a more extensive study of such fields and we shall proceed to obtain results on this subject adequate for our treatment of algebras. Let us begin by summarizing some of the results of Chapter AVII.

An irreducible polynomial $f(\lambda)$ with coefficients in $\mathfrak{F}$ is called *separable* if it has no multiple roots, otherwise *inseparable.* Then $f(\lambda)$ is inseparable if and only if $\mathfrak{F}$ has finite characteristic p and $f(\lambda) = g(\lambda^p)$. Henceforth in this section let p designate the characteristic of $\mathfrak{F}$, p a rational prime integer.

We call a quantity x of a field $\mathfrak{K}$ of finite degree over $\mathfrak{F}$ *separable* if its minimum function $f(\lambda)$ is separable. Otherwise call x *inseparable.* Then $f(\lambda) = h(\lambda^{p^e})$ where $h(\lambda)$ is separable, $x^{p^e} = y$ is a separable quantity of $\mathfrak{K}$.

We consider fields $\mathfrak{K}$ of finite degree n over $\mathfrak{F}$. Call $\mathfrak{K}$ *separable over* $\mathfrak{F}$ if every quantity of $\mathfrak{K}$ is separable. Theorem A7.26 then states that $\mathfrak{K}$ is separable over $\mathfrak{F}$ if and only if $\mathfrak{K} = \mathfrak{F}(x)$, where x is a separable quantity of degree n over $\mathfrak{F}$. Moreover every field $\mathfrak{F}(x_1, \cdots, x_t)$, where the x_i are separable quantities of finite degree over $\mathfrak{F}$, is separable over $\mathfrak{F}$. We call $\mathfrak{K}$ *inseparable* if it is not separable.

An inseparable field $\mathfrak{K}$ is called a *pure inseparable extension* of $\mathfrak{F}$ if there exists an integer $p^e \geqq 1$ such that x^{p^e} is in $\mathfrak{F}$ for every x of $\mathfrak{K}$. Then, as we shall show, every x of $\mathfrak{K}$ is a root of an irreducible polynomial $f(\lambda) = \lambda^{p^f} - g, f \leqq e$, g in $\mathfrak{F}$. Such polynomials are *pure inseparable polynomials* and we have

LEMMA 11. *Let g be in a field $\mathfrak{F}$ of characteristic p. Then a polynomial $f(\lambda) = \lambda^{p^e} - g$ is irreducible in $\mathfrak{F}$ if and only if $g \neq h^p$ for any h of $\mathfrak{F}$.*

For, if $f(\lambda)$ is irreducible and $g = h^p$ for h in $\mathfrak{F}$ then $f(\lambda) = (\lambda^{p^{e-1}} - h)^p$, a contradiction. Conversely, let $g \neq h^p$ for any h of $\mathfrak{F}$. Let $\mathfrak{K}$ over $\mathfrak{F}$ be a field which contains a root ξ of $f(\lambda) = 0$, $n = p^e$. Then $f(\lambda) = (\lambda - \xi)^n$. If $\phi(\lambda)$ is the minimum function of ξ over $\mathfrak{F}$ it is irreducible in $\mathfrak{F}$, divides $f(\lambda)$, and we must have $\phi(\lambda) = (\lambda - \xi)^m$ for some integer $m \leqq n$. Then ξ^m differs at most in sign from the constant term of $\phi(\lambda)$ and is in $\mathfrak{F}$. Write $m = p^f\mu$, where μ is prime to p, and have $\mu s = 1 + rp^e$ for integers r and s, $\xi^{ms} =$

* Note that the result of Theorem 2.27 may be used to prove that if $\mathfrak{F}$ is an infinite field then $\mathfrak{Z}$ has a basis $(u, u^{S_2}, \cdots, u^{S_n})$. For, we write $u = \lambda_1 + \lambda_2 x + \cdots + \lambda_n x^{n-1}$ and form $u^{S_i} = \lambda_{1i} + \lambda_{2i}x + \cdots + \lambda_{ni}x^{n-1}$. The determinant of the coefficients of these equations is a polynomial $\Delta(\lambda_1, \cdots, \lambda_n) \not\equiv 0$ since it is not zero for $u = e$. But then we apply Theorem A7.30 to choose $\lambda_{10}, \cdots, \lambda_{n0}$ in $\mathfrak{F}$ such that $\Delta(\lambda_{10}, \cdots, \lambda_{n0}) \neq 0$, and $u_0 = \lambda_{10} + \lambda_{20}x + \cdots + \lambda_{n0}x^{n-1}$ is the desired quantity. The special cyclic case of this result was carried out in the proof of the lemma on page A201.

$\xi^{p^f}\cdot(\xi^{rp^f})^{p^e}$. Since ξ^{p^e} is in $\mathfrak{F}$ the quantity ξ^{ms} is in $\mathfrak{F}$ if and only if $h_0 = \xi^{p^f}$ is in $\mathfrak{F}$. If f were less than e we could have $h_0^{p^{e-f}} = g = h^p$ for h in $\mathfrak{F}$, a contradiction. Hence $f = e$, $m = p^f\mu \leqq p^e$, $m = p^e = n$, $f(\lambda) = \phi(\lambda)$ is irreducible in $\mathfrak{F}$.

We now formulate an almost immediate corollary.

Lemma 12. *Let $\mathfrak{K}$ be separable over $\mathfrak{F}$, $\mathfrak{K}_0 = \mathfrak{F}(\xi)$ be pure inseparable of degree p^e over $\mathfrak{F}$. Then $\mathfrak{K}(\xi)$ has degree p^e over $\mathfrak{K}$ and is the direct product $\mathfrak{K} \times \mathfrak{K}_0$.*

For by Lemma 2.11 if $f(\lambda) = \lambda^{p^e} - g$ is the minimum function of ξ the degree of $\mathfrak{K}(\xi)$ over $\mathfrak{K}$ if less than p^e if and only if $g = h^p$ for h in $\mathfrak{K}$. But then h cannot be in $\mathfrak{F}$, h is an inseparable quantity of a field $\mathfrak{K}$ containing no inseparable quantities, a contradiction.

Let $\mathfrak{K}$ be a field of degree n over $\mathfrak{F}$ of characteristic p so that $\mathfrak{K} = (u_1, \cdots, u_n)$ over $\mathfrak{F}$. If $k = \alpha_1 u_1 + \cdots + \alpha_n u_n$ is in $\mathfrak{K}$ then $k^p = \alpha_1^p u_1^p + \cdots + \alpha_n^p u_n^p$ is in $\mathfrak{K}^{(p)} = \mathfrak{F}(u_1^p, \cdots, u_n^p)$. Thus the field for which we have introduced the notation $\mathfrak{K}^{(p)}$ is the extension of $\mathfrak{F}$ obtained by adjoining to it the pth powers of all of the quantities of $\mathfrak{K}$. Define $\mathfrak{K}^{(p^{i+1})} = (\mathfrak{K}^{(p^i)})^{(p)}$ and obtain a sequence of fields $\mathfrak{K} \geqq \mathfrak{K}^{(p)} \geqq \cdots$. Only a finite number of these inequalities can be proper since $\mathfrak{K}$ has finite degree over $\mathfrak{F}$. If $\mathfrak{K}^{(p^i)} = \mathfrak{K}^{(p^{i+1})}$ then our definition implies that $\mathfrak{K}^{(p^i)} = \mathfrak{K}^{(p^j)}$ for all $j > i$. Hence

$$\mathfrak{K} > \mathfrak{K}^{(p)} > \cdots > \mathfrak{K}^{(p^e)} = \mathfrak{K}^{(p^i)}$$

for every $i \geqq e$. We call the integer p^e defined in this way the *exponent* of $\mathfrak{K}$ over $\mathfrak{F}$. In the literature on such fields e itself has been called the exponent of $\mathfrak{K}$, but we shall find our present definition more convenient. We shall also call the degree of $\mathfrak{K}^{(p^e)}$ over $\mathfrak{F}$ the *reduced degree* of $\mathfrak{K}$ over $\mathfrak{F}$. We now prove

Theorem 28. *If $\mathfrak{K}$ is separable over $\mathfrak{F}$ then $\mathfrak{K} = \mathfrak{K}^{(p)}$.*

For let $\mathfrak{K} > \mathfrak{K}^{(p)}$ so that some x of $\mathfrak{K}$ is not in $\mathfrak{K}^{(p)}$. Then $x^p = y$ is in $\mathfrak{K}^{(p)}$, the polynomial $\lambda^p - y$ is irreducible in $\mathfrak{K}^{(p)}$, and the minimum function of x over $\mathfrak{F}$ has $\lambda^p - y$ as a factor. But this is impossible since x is in $\mathfrak{K}$ and hence is separable.

The converse of the theorem above states that *if $\mathfrak{K} = \mathfrak{K}^{(p)}$ then $\mathfrak{K}$ is separable.* These results are equivalent to the statement that *$\mathfrak{K}$ is inseparable over $\mathfrak{F}$ if and only if $\mathfrak{K} > \mathfrak{K}^{(p)}$* This is the case $e > 0$ of the following

Theorem 29. *Let $\mathfrak{K} = \mathfrak{F}(x_1, \cdots, x_t)$ have finite degree over $\mathfrak{F}$ so that $x_i^{p^{e_i}} = y_i$ is separable over $\mathfrak{F}$ for integers $e_i \geqq 0$ chosen to be the least possible. Then if e is the maximum of the e_i the exponent of $\mathfrak{K}$ over $\mathfrak{F}$ is p^e, $\mathfrak{K}_0 = \mathfrak{F}(y_1, \cdots, y_t)$ is the field of all quantities of $\mathfrak{K}$ separable over $\mathfrak{F}$, and hence it is the maximal separable subfield of $\mathfrak{K}$. Moreover $\mathfrak{K}_0 = \mathfrak{K}^{(p^e)}$, $\mathfrak{K}$ is a pure inseparable extension of degree at least p^e and exponent p^e over $\mathfrak{K}_0$.*

Note that this result characterizes the exponent of $\mathfrak{K}$ over $\mathfrak{F}$ as the least integer p^e such that a^{p^e} is separable over $\mathfrak{F}$ for every a of $\mathfrak{K}$.

The hypothesis that the y_i are separable implies that $\mathfrak{K}_0$ is separable. By

Lemma 2.11 our assumption about e_i implies that $\mathfrak{F}(x_i)$ is a pure inseparable extension of degree p^{e_i} over $\mathfrak{F}(y_i)$ so that, by Lemma 2.12, $\mathfrak{K}_0(x_i)$ is a pure inseparable extension of degree p^{e_i} of $\mathfrak{K}_0$. Hence $\mathfrak{K}$ is pure inseparable of degree at least p^e over $\mathfrak{K}_0$. By Theorem 2.28 we have $\mathfrak{K}_0 = \mathfrak{K}_0^{(p^e)}$. Evidently $\mathfrak{K}_0 \geqq \mathfrak{K}^{(p^e)} \geqq \mathfrak{K}_0^{(p^e)} = \mathfrak{K}_0$ so that $\mathfrak{K}_0 = \mathfrak{K}^{(p^e)}$. Now some x_i, say x, has $e_i = e$, $x^{p^e} = y$ in $\mathfrak{K}_0$, $\mathfrak{K}_0(x)$ has degree p^e over $\mathfrak{K}_0$, $\mathfrak{K}_0$ does not contain $x^{p^{e-1}}$. Hence $\mathfrak{K}_0 < \mathfrak{K}^{(p^{e-1})}$, that is, $\mathfrak{K}^{(p^{e-1})} > \mathfrak{K}^{(p^e)}$, and since $\mathfrak{K}^{(p^e)} = \mathfrak{K}_0 = \mathfrak{K}_0^{(p)}$ we have proved that p^e is the exponent of $\mathfrak{K}$. If w is any quantity of $\mathfrak{K}$ separable over $\mathfrak{F}$ and $\mathfrak{W} = \mathfrak{F}(w)$ we have $\mathfrak{W} = \mathfrak{W}^{(p^e)} \leqq \mathfrak{K}^{(p^e)} = \mathfrak{K}_0$ by Theorem 2.28 as desired.

Observe that we have now shown that if every x of $\mathfrak{K}$ not in $\mathfrak{F}$ is inseparable over $\mathfrak{F}$ then $\mathfrak{K}$ is pure inseparable over $\mathfrak{F}$, $\mathfrak{F}(x)$ defines a pure inseparable subfield over $\mathfrak{F}$ of $\mathfrak{K}$. We may also obtain an immediate corollary of our result stated as

Theorem 30. *Let $\mathfrak{K}$ be separable over $\mathfrak{K}_1$ and $\mathfrak{K}_1$ be separable over $\mathfrak{F}$. Then $\mathfrak{K}$ is separable over $\mathfrak{F}$.*

For, $\mathfrak{K}$ is a simple extension of $\mathfrak{K}_1$, $\mathfrak{K}_1$ is a simple extension of $\mathfrak{F}$ so that $\mathfrak{K}$ is the extension of $\mathfrak{F}$ by the adjunction of two quantities (only one of which we know is separable over $\mathfrak{F}$). By Theorem 2.29 with $t = 2$ we have $\mathfrak{K} \geqq \mathfrak{K}_0 \geqq \mathfrak{K}_1$, so that $\mathfrak{K}$ must be separable over $\mathfrak{K}_0$. But then $e = 0$, $\mathfrak{K} = \mathfrak{K}_0$ is separable over $\mathfrak{F}$.

The result above may also be proved by noting that $\mathfrak{K}_1 = \mathfrak{F}(x_1)$ with x_1 separable over $\mathfrak{F}$, $\mathfrak{K} = \mathfrak{K}_1(x_2) = \mathfrak{K}^{(p^e)}$ when $\mathfrak{K}$ is considered as a field over $\mathfrak{K}_1$. Then $\mathfrak{K} = \mathfrak{K}_1(x_2^{p^e})$ where we may choose e so that $x_2^{p^e} = y_2$ is separable over $\mathfrak{F}$ and thus $\mathfrak{K} = \mathfrak{F}(x_1, y_2)$ is separable over $\mathfrak{F}$ by the lemma on page A169. This lemma was used of course, without explicit reference to it, in the proof above.

For our next result we shall generalize Lemma 2.12 and obtain

Theorem 31. *Let $\mathfrak{Z}$ be separable over $\mathfrak{F}$ of characteristic p, $\mathfrak{K} = \mathfrak{F}(x_1, \cdots, x_t)$ be a pure inseparable extension of $\mathfrak{F}$. Then the composite $\mathfrak{Z}(x_1, \cdots, x_t)$ is the direct product $\mathfrak{Z} \times \mathfrak{K}$ over $\mathfrak{F}$. Moreover, $\mathfrak{Z}$ is the maximal separable subfield of $\mathfrak{Z} \times \mathfrak{K}$, and the exponent of $\mathfrak{K}$ over $\mathfrak{F}$ is the exponent of $\mathfrak{Z} \times \mathfrak{K}$ over $\mathfrak{Z}$.*

For Lemma 2.12 is clearly the case $t = 1$ of our theorem. We make an induction on t and thus assume that if $\mathfrak{K}_{t-1} = \mathfrak{F}(x_1, \cdots, x_{t-1})$ then $\mathfrak{Z}_{t-1} = \mathfrak{Z}(x_1, \cdots, x_{t-1}) = \mathfrak{K}_{t-1} \times \mathfrak{Z}$. Now $\mathfrak{K}_t = \mathfrak{K}_{t-1}(x_t) = \mathfrak{K}$, $\mathfrak{Z}_{t-1} = \mathfrak{K}_{t-1} \times \mathfrak{Z}$ is separable over $\mathfrak{K}_{t-1}$ since it is obtained by adjoining a separable quantity over $\mathfrak{F}$, and hence over $\mathfrak{K}_{t-1}$, to $\mathfrak{K}_{t-1}$. By the case $t = 1$ we have $\mathfrak{Z}_t = \mathfrak{Z}(x_1, \cdots, x_t) = \mathfrak{Z}_{t-1}(x_t) = \mathfrak{K}_t \times \mathfrak{Z}_{t-1}$ over $\mathfrak{K}_{t-1}$. If n is the degree of $\mathfrak{K}_t$ over $\mathfrak{F}$, m is the degree of $\mathfrak{K}_{t-1}$ over $\mathfrak{F}$, and q is the degree of $\mathfrak{K}_t$ over $\mathfrak{K}_{t-1}$, we have $n = mq$. Let r be the degree of $\mathfrak{Z}$ over $\mathfrak{F}$ so that the degree of $\mathfrak{Z}_{t-1} = \mathfrak{K}_{t-1} \times \mathfrak{Z}$ over $\mathfrak{K}_{t-1}$ is r. Hence the degree of $\mathfrak{Z}_t = \mathfrak{K}_t \times \mathfrak{Z}_{t-1}$ over $\mathfrak{K}_{t-1}$ is qr, its degree over $\mathfrak{F}$ is $qrm = nr$. But for fields $\mathfrak{K}_t$, $\mathfrak{Z}$ the composite $\mathfrak{Z}_t$ is then the direct product $\mathfrak{K}_t \times \mathfrak{Z}$ as desired.

To derive the final statement of our theorem observe that the minimum function of x_i over $\mathfrak{Z}$ divides its minimum function $\lambda^{p^{e_i}} - g_i$ over $\mathfrak{F}$, g_i in $\mathfrak{F}$, and hence has the form $\lambda^{p^{f_i}} - y_i$ with y_i in $\mathfrak{Z}$. But $y_i = x_i^{p^{f_i}}$ is in $\mathfrak{K}$ as well as in $\mathfrak{Z}$, and the intersection of these fields has been shown to be $\mathfrak{F}$. Thus y_i is in $\mathfrak{F}$, $y_i = g_i$, $f_i = e_i$. The maximum of the integers p^{f_i} was shown, in Theorem 2.29, to be the exponent of $\mathfrak{W} = \mathfrak{Z} \times \mathfrak{K}$ over $\mathfrak{Z}$, and has now been shown to be the maximum of the p^{e_i}, that is, the exponent p^e of $\mathfrak{K}$ over $\mathfrak{F}$. If $\mathfrak{W}_0$ is the maximal separable subfield of $\mathfrak{W}$ then $\mathfrak{W}_0 = \mathfrak{W}^{(p^e)} \leqq \mathfrak{Z}^{(p^e)}\mathfrak{K}^{(p^e)} = \mathfrak{Z}$, while clearly $\mathfrak{Z} \leqq \mathfrak{W}_0$, $\mathfrak{W}_0 = \mathfrak{Z}$. This completes our proof.

In Chapter VI we shall be interested in the possible order of an automorphism of a certain type of algebra. We treat the case of a field here and prove

Theorem 32. *Let $\mathfrak{Y}_0$ of degree m over $\mathfrak{F}$ be the maximal separable subfield of a field $\mathfrak{Y}$ of degree n over $\mathfrak{F}$, and S be an automorphism over $\mathfrak{F}$ of $\mathfrak{Y}$. Then $\mathfrak{Y}_0 = \mathfrak{Y}_0^S$, the order of S in $\mathfrak{Y}$ is its order μ in $\mathfrak{Y}_0$, and μ divides m and consequently n.*

For, if y_0 is any separable quantity of $\mathfrak{Y}$ the quantity y_0^S is also separable. It follows that $\mathfrak{Y}_0^S \leqq \mathfrak{Y}_0$, $\mathfrak{Y}_0^S = \mathfrak{Y}_0$. The field $\mathfrak{Y}_0$ is a simple extension $\mathfrak{F}(\eta)$ of $\mathfrak{F}$ and the order of S in $\mathfrak{Y}_0$ is clearly the largest integer μ such that $\eta, \eta^S, \cdots, \eta^{S^{\mu-1}}$ are all distinct. Then the usual argument on cyclic groups implies that $\eta^{S^\mu} = \eta$. Also $f(\lambda) = (\lambda - \eta) \cdots (\lambda - \eta^{S^{\mu-1}})$ has coefficients in the field $\mathfrak{Y}_1$ of all quantities x in $\mathfrak{Y}_0$ such that $x = x^S$. If $\phi(\lambda)$ is the (irreducible) minimum function of η over $\mathfrak{Y}_1$ we have $0 = \phi(\eta) = \phi(\eta^{S^k})$ so that every root η^{S^k} of $f(\lambda)$ is a root of $\phi(\lambda)$. But $\phi(\lambda)$ divides $f(\lambda)$, $\phi(\lambda) = f(\lambda)$, μ is the degree of $\mathfrak{Y}_0$ over $\mathfrak{Y}_1$ and divides m. Now m divides n and thus μ divides n. If x is in $\mathfrak{Y}$, $T = S^\mu$, and p^e is the exponent of $\mathfrak{Y}$ we have x^{p^e} in $\mathfrak{Y}_0$, $(x^{p^e})^T = (x^T)^{p^e} = x^{p^e}$, $x^T = x$. Thus T is the identity automorphism of $\mathfrak{Y}$ and μ is the order of S in $\mathfrak{Y}$.

As an immediate consequence of this result and the definition of a cyclic field we have the

Corollary. *Let $\mathfrak{Y}$ be a field of degree n over $\mathfrak{F}$ and n be the order of an automorphism S over $\mathfrak{F}$ of $\mathfrak{Y}$. Then $\mathfrak{Y}$ is cyclic over $\mathfrak{F}$ with generating automorphism S.*

We shall now pass on to the theory of scalar extensions of an algebra relative to the consequent scalar extensions of its centrum.

17. Scalar extensions of the centrum. The results we are about to obtain are of fundamental importance in our treatment of the Wedderburn Principal Theorem on Algebras. We first derive

Theorem 33. *Let the centrum of an algebra $\mathfrak{A}$ with a unity quantity e contain a separable subfield $\mathfrak{X} = \mathfrak{F}(x)$ of finite degree m over (e) over $\mathfrak{F}$, and let $\mathfrak{K} = \mathfrak{F}(\xi_1, \cdots, \xi_m)$ be the corresponding scalar root field. Then*

$$\mathfrak{A}_{\mathfrak{K}} = (\mathfrak{A}_1)_{\mathfrak{K}} \oplus \cdots \oplus (\mathfrak{A}_m)_{\mathfrak{K}},$$

where $\mathfrak{A}_i$ is an algebra over $\mathfrak{K}_i = \mathfrak{F}(\xi_i)$ with e_i of (2.14) as unity quantity, and $\mathfrak{A}_i$ is equivalent over $\mathfrak{F}$ to $\mathfrak{A}$ under the correspondence $ae_i \leftrightarrow a = ae$ for every a of $\mathfrak{A}$. Hence in particular $xe_i = \xi_i e_i \leftrightarrow x$.

Notice that we are not stating that $\mathfrak{A}_{\mathfrak{K}}$ is a direct sum of algebras equivalent over $\mathfrak{F}$ to $\mathfrak{A}$, but rather to the scalar extension by $\mathfrak{K}$ of such a direct sum. For proof write $\mathfrak{A} = (u_1, \cdots, u_n)$ over the field $\mathfrak{X}$ by Theorem 1.6. The e_i of (2.14) are in $\mathfrak{X}_{\mathfrak{K}}$ and hence in the centrum of $\mathfrak{A}_{\mathfrak{K}}$, so we may apply Theorem 2.5 to write $\mathfrak{A}_{\mathfrak{K}} = \mathfrak{A}_{\mathfrak{K}}e_1 \oplus \cdots \oplus \mathfrak{A}_{\mathfrak{K}}e_m$, $\mathfrak{A}_{\mathfrak{K}}e_i = (u_1e_i, \cdots, u_me_i)$ over $\mathfrak{K}$. Evidently $xe_i = x_i = \xi_ie_i$. But then

$$(u_je_i)(u_ke_i) = (u_ju_k)e_i = \sum_{r=1}^{n} [g_r^{(jk)}(x_i)](u_re_i),$$

where $u_ju_k = \sum_{r=1}^{n} g_r^{(jk)} u_r$ is the multiplication table of $\mathfrak{A}$. It is clear that $\mathfrak{A}_{\mathfrak{K}}e_i$ is the scalar extension $(\mathfrak{A}_i)_{\mathfrak{K}}$ of the algebra $\mathfrak{A}_i = (u_1e_i, \cdots, u_ne_i)$ over $\mathfrak{K}_i = \mathfrak{F}(\xi_i)$. Moreover $\mathfrak{A}_i$ may be considered as an algebra over $\mathfrak{F}$ with basis $\xi_i^ku_je_i$ $(j = 1, \cdots, n; k = 0, \cdots, m-1)$, and is equivalent over $\mathfrak{F}$ to $\mathfrak{A}$ over $\mathfrak{F}$ under the correspondence

$$a = \sum \alpha_{jk}x^ku_j \leftrightarrow \sum \alpha_{jk}\xi_i^ku_je_i,$$

for all α_{jk} in $\mathfrak{F}$.

The result above will be applied to certain algebras $\mathfrak{A}$ with a unity quantity and centrum the separable field $\mathfrak{X}$, and will be seen to be essential in our proof of Theorem 3.21.

If the centrum of an algebra $\mathfrak{A}$ contains a nilpotent quantity z, the radical of $\mathfrak{A}$ is not zero and contains z. For $z^t = 0$, $(za)^t = z^ta^t = 0$, z is properly nilpotent in $\mathfrak{A}$. We use this result for considering algebras with centrum an inseparable field and prove

Theorem 34. *Let the centrum $\mathfrak{C}$ of an algebra $\mathfrak{A}$ with a unity quantity e contain an inseparable subfield $\mathfrak{Y}$ whose unity quantity is e. Then there exists a scalar extension $\mathfrak{K}$ of $\mathfrak{F}$ such that $\mathfrak{Y}_{\mathfrak{K}}$ contains a properly nilpotent quantity of $\mathfrak{A}_{\mathfrak{K}}$.*

For, the field $\mathfrak{F}$ then has characteristic p and $\mathfrak{Y}$ contains a quantity y such that $y^p = x$, the field $\mathfrak{F}(x)$ is separable over $\mathfrak{F}$, and y is not in $\mathfrak{F}(x)$. We form the scalar root field $\mathfrak{F}(\xi_1, \cdots, \xi_m)$ as in Theorem 2.33, and let $\eta^p = \xi_1$, $\mathfrak{K} = \mathfrak{F}(\eta, \xi_2, \cdots, \xi_m)$. Then $\mathfrak{A}_{\mathfrak{K}} = \mathfrak{A}_{1\mathfrak{K}} \oplus \cdots \oplus \mathfrak{A}_{m\mathfrak{K}}$, where $\mathfrak{A}_1$ contains $y_1 = ye_1$ such that $y_1^p = \xi_1e_1$. Now y_1 is not in $\mathfrak{F}(\xi_1e_1)$ and $(y_1 - \eta e_1)^p = \xi_1e_1 - \eta^pe_1 = 0$. Hence $z = y_1 - \eta e_1$ is nilpotent and, since z is in the centrum of $\mathfrak{A}_{1\mathfrak{K}}$, it is in the centrum of $\mathfrak{A}_{\mathfrak{K}}$, z is properly nilpotent.

Note that our result is a consequence of Theorem A10.22 applied to our quantity y as the matrix A of that theorem.

CHAPTER III

THE STRUCTURE THEOREMS OF WEDDERBURN

1. Semi-simple algebras. The principal structure theorems on algebras over a non-modular field are due to J. H. M. Wedderburn and were first given in [425]. We shall give an exposition of the generalization of Wedderburn's theorems to algebras over any field. Our treatment will be a modification and, the author believes, a simplification of those of [425], [136], [142], and [119]. We begin with the

DEFINITION. *An algebra $\mathfrak{A}$ is called semi-simple if its radical $\mathfrak{N}$ is the zero ideal.*

If $\mathfrak{K}$ is a scalar extension field of $\mathfrak{F}$ the radical of $\mathfrak{A}_{\mathfrak{K}}$ contains $\mathfrak{N}_{\mathfrak{K}}$, where $\mathfrak{N}$ is the radical of $\mathfrak{A}$. For it is trivial to show that $\mathfrak{N}_{\mathfrak{K}}$ is a nilpotent ideal of $\mathfrak{A}_{\mathfrak{K}}$. Hence if the algebra $\mathfrak{A}_{\mathfrak{K}}$ is semi-simple for some $\mathfrak{K}$ then $\mathfrak{A}$ is itself semi-simple.

A semi-simple algebra $\mathfrak{A}$ is clearly not nilpotent. By Theorem 2.14, $\mathfrak{A}$ has a principal idempotent quantity. We apply Theorem 2.15 and obtain

Theorem 1. *Every semi-simple algebra has a unity quantity.*

As a consequence of Theorem 2.20 we may prove

Theorem 2. *Let $\mathfrak{N}$ be a nilpotent ideal of $\mathfrak{A} > \mathfrak{N}$. Then $\mathfrak{A} - \mathfrak{N}$ is semi-simple if and only if $\mathfrak{N}$ is the radical of $\mathfrak{A}$.*

For let $\bar{\mathfrak{N}}$ be the radical of $\mathfrak{A} - \mathfrak{N}$, $\mathfrak{N}_1$ be the corresponding subalgebra of $\mathfrak{A}$ such that $\bar{\mathfrak{N}} = \mathfrak{N}_1 - \mathfrak{N}$. By Theorem 2.20, $\mathfrak{N}_1$ is an ideal of $\mathfrak{A}$. Now $\bar{\mathfrak{N}}^{\cdot \alpha} = 0$ so that $\mathfrak{N}_1^{\cdot \alpha} \leqq \mathfrak{N}$, $(\mathfrak{N}_1^{\cdot \alpha})^{\cdot \beta} = 0$, $\mathfrak{N}_1$ is a nilpotent ideal of $\mathfrak{A}$. If $\mathfrak{N}$ is the radical of $\mathfrak{A}$ then $\mathfrak{N}_1 \leqq \mathfrak{N}$, $\bar{\mathfrak{N}} = 0$, $\mathfrak{A} - \mathfrak{N}$ is semi-simple. Conversely if $\mathfrak{A} - \mathfrak{N}$ is semi-simple and $\mathfrak{N}_1$ is the radical of $\mathfrak{A}$, we have $\mathfrak{N}_1 - \mathfrak{N}$ a nilpotent ideal of $\mathfrak{A} - \mathfrak{N}$, $\mathfrak{N}_1 - \mathfrak{N} = 0$, $\mathfrak{N}_1 = \mathfrak{N}$.

Analogous results may be obtained for algebras defined in the

DEFINITION. *An algebra $\mathfrak{A}$ is called simple if the only proper ideal of $\mathfrak{A}$ is the zero ideal and if $\mathfrak{A}$ is not a zero algebra of order one.*

In Theorem 2.6 we showed that the only nilpotent algebra without a non-zero proper ideal is the zero algebra of order one. This latter trivial case of an algebra has none of the interesting properties of all other algebras without non-zero proper ideals and it is desirable to delete it from the list of simple algebras. Note that we then have

Theorem 3. *A simple algebra is semi-simple.*

We thus have the following immediate consequence of Theorem 3.1.

COROLLARY. *A simple algebra has a unity quantity.*

A proper ideal $\mathfrak{B}$ of $\mathfrak{A}$ is called *maximal* if no proper ideal $\mathfrak{C}$ of $\mathfrak{A}$ contains $\mathfrak{B}$. The analogue of Theorem 3.2 is then given by

Theorem 4. *Let $\mathfrak{B}$ be an ideal of $\mathfrak{A}$. Then $\mathfrak{B}$ is maximal if and only if $\mathfrak{A} - \mathfrak{B}$ is either simple or a zero algebra of order one.*

The proof of Theorem 3.4 is trivial and we leave it to the reader.

It is interesting to consider the structure of $e\mathfrak{A}e$ in the light of the above definitions, where e is an idempotent of $\mathfrak{A}$. We first apply Theorem 2.12 with $\mathfrak{N} = 0$ and have

Theorem 5. *If e is an idempotent of a semi-simple algebra $\mathfrak{A}$, the algebra $e\mathfrak{A}e$ is semi-simple.*

We next prove

Theorem 6. *If e is an idempotent of a simple algebra $\mathfrak{A}$ the algebra $e\mathfrak{A}e$ is simple.*

For let $\mathfrak{B} \neq 0$ be an ideal of $e\mathfrak{A}e$. Then $(e\mathfrak{A}e)\mathfrak{B}(e\mathfrak{A}e) = e(\mathfrak{A}\mathfrak{B}\mathfrak{A})e \leqq \mathfrak{B}$. Now $\mathfrak{B} \neq 0$, $\mathfrak{A}\mathfrak{B}\mathfrak{A} \geqq \mathfrak{B}$, $\mathfrak{A}\mathfrak{B}\mathfrak{A}$ is a non-zero ideal of $\mathfrak{A}$. Since $\mathfrak{A}$ is simple we have $\mathfrak{A}\mathfrak{B}\mathfrak{A} = \mathfrak{A}$, $e\mathfrak{A}e \leqq \mathfrak{B} \leqq e\mathfrak{A}e$, $\mathfrak{B} = e\mathfrak{A}e$. Hence $e\mathfrak{A}e$ is simple.

If $\mathfrak{D}$ is a division algebra with unity quantity e and u is an idempotent of $\mathfrak{D}$ we have $uu^{-1} = e$, $u^2 = u$, $u^{-1}(u^2) = eu = u = u^{-1}u = e$. Hence the only idempotent of $\mathfrak{D}$ is its unity quantity. We use this result in the derivation of an important tool for our structure theory.

Theorem 7. *Let e be an idempotent of a simple algebra $\mathfrak{A}$. Then $e\mathfrak{A}e$ is a division algebra if and only if e is primitive in $\mathfrak{A}$.*

For, if $e\mathfrak{A}e$ is a division algebra, e is the only idempotent of $e\mathfrak{A}e$, e is primitive in $\mathfrak{A}$ by Lemma 2.10. Conversely, let e be primitive in $\mathfrak{A}$ so that $\mathfrak{D} = e\mathfrak{A}e$ is a simple algebra with e as unity quantity by Theorem 3.6. If $a \neq 0$ is in $\mathfrak{D}$ then $\mathfrak{D}a$ is a non-zero left ideal of $\mathfrak{D}$. However $\mathfrak{D}$ has zero radical by Theorem 3.3 so that $\mathfrak{D}a$ is not nilpotent. Thus $\mathfrak{D}a$ contains an idempotent and this must be $e = ba$ for b in $\mathfrak{D}$. The non-zero quantities of $\mathfrak{D}$ now form a multiplicative group and $\mathfrak{D}$ is a division algebra.

2. Reduction to simple components. Semi-simple algebras are really merely direct sums of simple algebras. We shall prove this fact as a consequence of

LEMMA 1. *A semi-simple algebra $\mathfrak{A}$ is irreducible if and only if it is simple.*

For let $\mathfrak{A}$ be irreducible but not simple so that $\mathfrak{A}$ has a non-zero proper ideal $\mathfrak{B}$. Since $\mathfrak{A}$ has a unity quantity, $\mathfrak{A}\mathfrak{B} = \mathfrak{B}\mathfrak{A} = \mathfrak{B}$. Also if $\mathfrak{N}$ is the radical of $\mathfrak{B}$, the set $\mathfrak{A}\mathfrak{N}\mathfrak{A} \leqq \mathfrak{B}$, $\mathfrak{A}\mathfrak{N}\mathfrak{A}$ is an ideal of $\mathfrak{A}$. Now $\mathfrak{B}\mathfrak{N}\mathfrak{B} \leqq \mathfrak{N}$ and is nilpotent, $\mathfrak{B}\mathfrak{N}\mathfrak{B}$ is an ideal of $\mathfrak{A}$, $\mathfrak{B}\mathfrak{N}\mathfrak{B} = 0$. Hence $(\mathfrak{A}\mathfrak{N}\mathfrak{A})^3 \leqq (\mathfrak{A}\mathfrak{N}\mathfrak{A})\cdot\mathfrak{N}\cdot(\mathfrak{A}\mathfrak{N}\mathfrak{A}) \leqq \mathfrak{B}\mathfrak{N}\mathfrak{B} = 0$, $\mathfrak{A}\mathfrak{N}\mathfrak{A}$ is a nilpotent ideal of $\mathfrak{A}$, $\mathfrak{A}\mathfrak{N}\mathfrak{A} = 0$. Since $\mathfrak{A}$ has a unity quantity, $\mathfrak{A}\mathfrak{N}\mathfrak{A} \geqq \mathfrak{N}$, $\mathfrak{N} = 0$, $\mathfrak{B}$ is semi-simple. It has a unity quantity e by Theorem 3.1, $\mathfrak{A} = \mathfrak{B} \oplus \mathfrak{C}_e$ by Theorem 2.23, $\mathfrak{A} > \mathfrak{B}$, $\mathfrak{C}_e \neq 0$, a contradiction. The converse was noted at the beginning of Section 2.12.

We now prove the principal structure theorem on semi-simple algebras.

Theorem 8. *An algebra $\mathfrak{A}$ is semi-simple if and only if $\mathfrak{A}$ is either simple or is expressible as a direct sum of simple components. These components are unique apart from their order.*

For let $\mathfrak{A}$ be semi-simple. By Theorem 2.22, $\mathfrak{A}$ is either irreducible or $\mathfrak{A} = \mathfrak{A}_1 \oplus \cdots \oplus \mathfrak{A}_t$ for the $\mathfrak{A}_i$ irreducible ideals of $\mathfrak{A}$ with respective unity quantities e_i such that $e_1 + \cdots + e_t$ is the unity quantity of $\mathfrak{A}$. By Lemma 3.1 if $\mathfrak{A}$ is irreducible it is simple. Otherwise $t > 1$ and $\mathfrak{A}$ is not simple. Now $\mathfrak{A}_i = e_i\mathfrak{A}e_i$ and $\mathfrak{A}_i$ is semi-simple by Theorem 3.5. Apply Lemma 3.1 to see that each $\mathfrak{A}_i$ is simple. The uniqueness of the $\mathfrak{A}_i$ follows from Theorem 2.22.

Conversely, let $\mathfrak{A}$ be a direct sum of simple algebras so that the sum of the unity quantities of its components is the unity quantity of $\mathfrak{A}$. Each component has zero radical and so does $\mathfrak{A}$ by the corollary to Theorem 2.21.

Note that any diagonal algebra $\mathfrak{A} = (e_1, \cdots, e_n)$ over $\mathfrak{F}$, for pairwise orthogonal idempotents e_i, is semi-simple. Hence *if $\mathfrak{A}$ is an algebra over $\mathfrak{F}$ and there exists a scalar extension $\mathfrak{K}$ of $\mathfrak{F}$ such that $\mathfrak{A}_\mathfrak{K}$ is a diagonal algebra then $\mathfrak{A}$ is a commutative semi-simple algebra.*

3. Structure of simple algebras. The principal result on simple algebras is given by

Theorem 9. *Every simple algebra $\mathfrak{A}$ is expressible as $\mathfrak{A} = \mathfrak{M} \times \mathfrak{D}$, where $\mathfrak{M}$ is a total matric algebra and $\mathfrak{D}$ is a division algebra. Conversely, every such direct product is simple. Moreover, if the unity quantity of $\mathfrak{A}$ is expressed as a sum $e_1 + \cdots + e_m$ of pairwise orthogonal primitive idempotents e_i of $\mathfrak{A}$ then $\mathfrak{M}$ may be taken to be $(e_{ij};\, i, j = 1, \cdots, m)$ with $e_{ii} = e_i$ for $i = 1, \cdots, m$.*

For by Theorem 2.16 $\mathfrak{A}$ has a unity quantity $1 = e_1 + \cdots + e_m$, where the e_i are pairwise orthogonal primitive idempotents of $\mathfrak{A}$. The algebras $e_i\mathfrak{A}e_i$ are division algebras by Theorem 3.7. Since $\mathfrak{A}e_i\mathfrak{A}$ is an ideal of $\mathfrak{A}$ containing e_i we have $\mathfrak{A}e_i\mathfrak{A} \neq 0$, $\mathfrak{A}e_i\mathfrak{A} = \mathfrak{A}$ for $i = 1, \cdots, m$.

Define $\mathfrak{A}_{ij} = e_i\mathfrak{A}e_j$. Now $e_je_j = e_j$, $e_je_h = 0$ for $j \neq h$ and we have

$$(1)\quad \mathfrak{A}_{ij}\mathfrak{A}_{jk} = e_i(\mathfrak{A}e_je_j\mathfrak{A})e_k = \mathfrak{A}_{ik};\ \mathfrak{A}_{ij}\mathfrak{A}_{hk} = 0 \quad (j \neq h;\, i, j, h, k = 1, \cdots, m).$$

It follows that the product $\mathfrak{A}_{ij}\mathfrak{A}_{ij}$ is zero or $\mathfrak{A}_{ij}$, the $\mathfrak{A}_{ij}$ are subalgebras of $\mathfrak{A}$ with the properties

$$(2)\quad \begin{aligned} &e_ia_{ij} = a_{ij}e_j = a_{ij}, \qquad e_ha_{ij} = a_{ij}e_k = 0 \\ &\qquad\qquad (h \neq i;\, k \neq j;\, i, j, h, k = 1, \cdots, m), \end{aligned}$$

for every a_{ij} of $\mathfrak{A}_{ij}$. If $j > 1$ we have $\mathfrak{A}_{1j}\mathfrak{A}_{j1} = \mathfrak{A}_{11}$, $\mathfrak{A}_{11} \geqq \mathfrak{A}_{1j}e_{j1} \neq 0$ for some e_{j1} of $\mathfrak{A}_{j1}$. Then $a_{1j}e_{j1} = a_j \neq 0$ in $\mathfrak{A}_{11}$ for some a_{1j} of $\mathfrak{A}_{1j}$. Since $\mathfrak{A}_{11}$ is a division algebra there exists a quantity b_j in $\mathfrak{A}_{11}$ such that $b_ja_j = e_1$. By

(1) the quantity $b_j a_{1j} = e_{1j}$ is in $\mathfrak{A}_{1j}$, and since $e_1 e_1 = e_1$ we have proved the existence of quantities $e_{1j} = b_j a_{1j}$ in $\mathfrak{A}_{1j}$, e_{j1} in $\mathfrak{A}_{j1}$, such that

$$e_{1j} e_{j1} = e_{11} = e_1 \qquad (j = 1, \cdots, m). \tag{3}$$

Define

$$e_{ij} = e_{i1} e_{1j} \qquad (i, j = 1, \cdots, m), \tag{4}$$

and consequently obtain

$$e_{ij} e_{jk} = e_{i1} e_{1j} e_{j1} e_{1k} = e_{i1} e_1 e_{1k} = e_{ik} \quad (i, j, k = 1, \cdots, m). \tag{5}$$

Also e_{ij} is in $e_i \mathfrak{A} e_j$, and by (2) we have

$$e_{ij} e_{hk} = 0 \qquad (j \neq h; i, j, h, k = 1, \cdots, m). \tag{6}$$

Since $e_{11} = e_1 \neq 0$ we may apply Theorem 1.2 to see that $\mathfrak{M} = (e_{ij}; i, j = 1, \cdots, m)$ is a total matric subalgebra of order m^2 over $\mathfrak{F}$ of $\mathfrak{A}$. Then by (5) $(e_{ii})^2 = e_{ii} \neq 0$ is an idempotent of $e_i \mathfrak{A} e_i$, and since e_i is primitive, it is the only idempotent of $e_i \mathfrak{A} e_i$, $e_{ii} = e_i$,

$$e_{11} + \cdots + e_{mm} = e_1 + \cdots + e_m = 1$$

is the unity quantity of both $\mathfrak{A}$ and $\mathfrak{M}$. By Theorem 1.17 $\mathfrak{A} = \mathfrak{M} \times \mathfrak{D}$ where $\mathfrak{D} = \mathfrak{A}^{\mathfrak{M}}$. Clearly $\mathfrak{A}$ is the supplementary sum of the algebras $e_{ij}\mathfrak{D}$ and since $\mathfrak{M} \times \mathfrak{D}$ is a direct product, the order of $e_i \mathfrak{A} e_i = e_{ii} \mathfrak{D}$ is the same as that of $\mathfrak{D}$. The correspondence $e_i d \leftrightarrow d$ for d in $\mathfrak{D}$ is thus a (1-1) correspondence which trivially defines an equivalence of $e_{ii}\mathfrak{D}$ and $\mathfrak{D}$ since e_i is an idempotent such that $e_i d = d e_i$ for every d of $\mathfrak{D}$. But, as we have already stated, $e_i \mathfrak{A} e_i$ is a division algebra, so that $\mathfrak{D}$ is a division algebra.

Conversely let $\mathfrak{A} = \mathfrak{M} \times \mathfrak{D}$ and $\mathfrak{B}$ be a non-zero ideal of $\mathfrak{A}$. Then $\mathfrak{B}$ contains a non-zero quantity $b = \sum_{i,j=1}^{m} d_{ij} e_{ij}$ with quantities d_{ij} in $\mathfrak{D}$ and not all zero. Hence if $d_{pq} \neq 0$ it has an inverse in $\mathfrak{A}$ and, since $\mathfrak{B}$ is an ideal,

$$a = a d_{pq}^{-1} \sum_{i=1}^{m} e_{ip} b e_{qi}$$

is in $\mathfrak{B}$ for every a of $\mathfrak{A}$. We have proved that $\mathfrak{B} \geqq \mathfrak{A}$, $\mathfrak{B} = \mathfrak{A}$, $\mathfrak{A}$ is simple.

We shall use the obvious

COROLLARY. *A commutative semi-simple algebra is a direct sum of fields.*

A division algebra $\mathfrak{D}$ is to be regarded as a direct product $\mathfrak{M} \times \mathfrak{D}$ where $\mathfrak{M}$ has order one. Similarly a total matric algebra $\mathfrak{M}$ is a direct product $\mathfrak{M} \times \mathfrak{D}$ where $\mathfrak{D}$ has order one. Thus division algebras and total matric algebras are instances of simple algebras, all simple algebras of order one are equivalent.

We have already defined the concept of a normal algebra. Thus $\mathfrak{A}$ is called a *normal simple* algebra if it is simple and has $\mathfrak{F}$ as its centrum. In particular we have seen that if $\mathfrak{A}$ is a total matric algebra over $\mathfrak{F}$ it is a normal simple algebra. We shall now derive some elementary properties of normal algebras.

4. Direct products of normal algebras. In the present section we shall consider direct products $\mathfrak{A} = \mathfrak{B} \times \mathfrak{C}$ where $\mathfrak{A}$, $\mathfrak{B}$, $\mathfrak{C}$ will always be assumed to have the same unity quantity. With this convention we prove

Theorem 10. *Let $\mathfrak{B}$ be normal over $\mathfrak{F}$, $\mathfrak{A} = \mathfrak{B} \times \mathfrak{C}$. Then $\mathfrak{C}$ is the $\mathfrak{A}$-commutator $\mathfrak{A}^{\mathfrak{B}}$ of $\mathfrak{B}$, and the centrum of $\mathfrak{A}$ coincides with that of $\mathfrak{C}$.*

For, let $\mathfrak{C} = (u_1, \cdots, u_t)$ over $\mathfrak{F}$. Every quantity a of $\mathfrak{A}$ is uniquely expressible in the form $a = b_1u_1 + \cdots + b_tu_t$ for b_i in $\mathfrak{B}$. If a is in $\mathfrak{A}^{\mathfrak{B}}$ then $ab = ba$ for every b of $\mathfrak{B}$, $u_ib = bu_i$ since $\mathfrak{A} = \mathfrak{B} \times \mathfrak{C}$, $ab - ba = \sum_{i=1}^{t} (b_ib - bb_i)u_i = 0$ if and only if $b_ib = bb_i$ for every b of $\mathfrak{B}$. Then the b_i are in the centrum of the normal algebra $\mathfrak{B}$ and hence in $\mathfrak{F}$, a is in $\mathfrak{C}$, $\mathfrak{A}^{\mathfrak{B}} \leqq \mathfrak{C}$. Clearly $\mathfrak{C} \leqq \mathfrak{A}^{\mathfrak{B}}$, $\mathfrak{C} = \mathfrak{A}^{\mathfrak{B}}$. Every quantity k of the centrum of $\mathfrak{A}$ is certainly in $\mathfrak{A}^{\mathfrak{B}}$ and hence in $\mathfrak{C}$. But $kc = ck$ for every c of $\mathfrak{C}$, k is in the centrum of $\mathfrak{C}$, $\mathfrak{A}^{\mathfrak{A}} \leqq \mathfrak{C}^{\mathfrak{C}}$. The reversed inequality is obvious and our theorem true.

As an immediate corollary we have

Theorem 11. *The direct product $\mathfrak{B} \times \mathfrak{C}$ of two normal algebras $\mathfrak{B}$ and $\mathfrak{C}$ is a normal algebra $\mathfrak{A}$ such that $\mathfrak{C}$ is the $\mathfrak{A}$-commutator of $\mathfrak{B}$, $\mathfrak{B}$ is the $\mathfrak{A}$-commutator of $\mathfrak{C}$.*

The converse of Theorem 3.11 is also valid and is stated as

Theorem 12. *Let $\mathfrak{A} = \mathfrak{B} \times \mathfrak{C}$ be normal. Then $\mathfrak{B}$ and $\mathfrak{C}$ are normal.*

For, if b is in the centrum of $\mathfrak{B}$ and $\mathfrak{A} = \mathfrak{B} \times \mathfrak{C}$ then b is evidently in the centrum of $\mathfrak{A}$. Since $\mathfrak{A}$ is normal b is in $\mathfrak{F}$, $\mathfrak{B}$ is normal. Similarly $\mathfrak{C}$ is normal.

We have now shown that if $\mathfrak{A}$ is normal and $\mathfrak{A} = \mathfrak{B} \times \mathfrak{C}$ then not only is $\mathfrak{B}$ normal but $\mathfrak{C}$ is the uniquely determined $\mathfrak{A}$-commutator of $\mathfrak{B}$. Hence, while the expression of a normal algebra as the direct product of two normal algebras $\mathfrak{B}$ and $\mathfrak{C}$ is by no means unique, if either factor is given the other is determined.

The result of Theorem 3.10 is of particular interest to us for the case where $\mathfrak{A}$ is a simple algebra and $\mathfrak{B}$ is a total matric algebra. We apply it to obtain

Theorem 13. *Let $\mathfrak{A}$ be a simple algebra so that $\mathfrak{A} = \mathfrak{M} \times \mathfrak{D}$ where $\mathfrak{M}$ is a total matric algebra and $\mathfrak{D}$ is a division algebra. Then the centrum of $\mathfrak{A}$ is the field $\mathfrak{K}$ which is the centrum of $\mathfrak{D}$. Thus $\mathfrak{A}$ is normal simple over $\mathfrak{K}$ and $\mathfrak{A}$ over $\mathfrak{K}$ is the direct product of $\mathfrak{M}$ and the normal division algebra $\mathfrak{D}$ over $\mathfrak{K}$.*

Notice here that $\mathfrak{A}$ over $\mathfrak{K}$ has *not* been constructed as being the scalar extension of $\mathfrak{A}$ over $\mathfrak{F}$, but rather we have used Theorem 1.6 to express $\mathfrak{A}$ as an algebra over $\mathfrak{K}$ of generally smaller order than $\mathfrak{A}$ over $\mathfrak{F}$, where $\mathfrak{K}$ is actually a commutative division subalgebra over $\mathfrak{F}$ of $\mathfrak{A}$. This process is not that of extension of the coefficient field but that of reduction of the order of the linear set $\mathfrak{A}$ when $\mathfrak{K} > \mathfrak{F}$.

5. A fundamental property of normal simple algebras. We shall obtain a result in this section which is basic for our theory of normal simple algebras and which will be used to obtain great simplifications of the proofs of the results in the literature. We first prove

LEMMA 2. *Let $\mathfrak{D}$ be a normal division subalgebra of $\mathfrak{A}$, $\mathfrak{C}$ be a subalgebra of $\mathfrak{A}$ such that $cd = dc$ for every c of $\mathfrak{C}$, d of $\mathfrak{D}$. Then if the unity quantity of $\mathfrak{D}$ is that of $\mathfrak{C}$ we have $\mathfrak{D}\mathfrak{C} = \mathfrak{D} \times \mathfrak{C}$.*

For if $\mathfrak{C} = (u_1, \cdots, u_t)$ over $\mathfrak{F}$ then $\mathfrak{D}\mathfrak{C}$ is the sum $(\mathfrak{D}u_1, \cdots, \mathfrak{D}u_t)$. Renumbering the u_i if necessary we may use the method of proof of Theorem 1.6 applied to the algebra $\mathfrak{D}\mathfrak{C}$ and obtain $\mathfrak{D}\mathfrak{C} = \mathfrak{D}u_1 + \cdots + \mathfrak{D}u_r$ for $r \leqq t$. If $r = t$ then we have the result $\mathfrak{D}\mathfrak{C} = \mathfrak{D} \times \mathfrak{C}$ as desired. Otherwise u_{r+1} in $\mathfrak{D}\mathfrak{C}$ has the form $u_{r+1} = d_1u_1 + \cdots + d_ru_r$ for the d_i in $\mathfrak{D}$. Our assumption implies that $du_{r+1} - u_{r+1}d = 0$ for every d of $\mathfrak{D}$, $\sum_{i=1}^{r} (dd_i - d_id)u_i = 0$. Since $\mathfrak{D}\mathfrak{C} = \mathfrak{D}u_1 + \cdots + \mathfrak{D}u_r$ we have $dd_i - d_id = 0$, the d_i are in the centrum of $\mathfrak{D}$ and hence in $\mathfrak{F}$, u_{r+1} is in $(u_1, \cdots, u_r)$ over $\mathfrak{F}$. This implies that $u_1, \cdots, u_t$ are linearly dependent in $\mathfrak{F}$, a contradiction.

We now use our lemma to prove the fundamental

LEMMA 3. *Let $\mathfrak{D}$ be a normal division algebra over $\mathfrak{F}$, $\mathfrak{D}^{-1}$ be reciprocal to $\mathfrak{D}$. Then the direct product*

$$\mathfrak{M} = \mathfrak{D} \times \mathfrak{D}^{-1}$$

is a total matric algebra.

For let $\mathfrak{D}$ have order n over $\mathfrak{F}$. By Theorem 1.3 $\mathfrak{T}\mathfrak{S} = \mathfrak{S}\mathfrak{T} \leqq \mathfrak{M}_n$ where $\mathfrak{T}$ is equivalent to $\mathfrak{D}$, $\mathfrak{S}$ is reciprocal to $\mathfrak{D}$. By Lemma 3.2, $\mathfrak{T}\mathfrak{S} = \mathfrak{T} \times \mathfrak{S} = \mathfrak{D} \times \mathfrak{D}^{-1}$ in the sense of equivalence. The order of $\mathfrak{M}_n$ is n^2 and that is the order of $\mathfrak{T} \times \mathfrak{S} \leqq \mathfrak{M}_n$. Hence $\mathfrak{M}_n = \mathfrak{D} \times \mathfrak{D}^{-1}$.

This lemma will be of basic importance in our theory of algebras. It is interesting to observe that it is entirely independent of any material in these LECTURES after Theorem 1.6. However it has more meaning and rather immediate application in its present position. We shall not refer to it later but to

Theorem 14. *The direct product of a normal simple algebra and its reciprocal algebra is a total matric algebra.*

This final expression of our fundamental result depends upon the structure Theorem 3.9 and its amplification, Theorem 3.13. Observe first that a total matric algebra $\mathfrak{M}$ is self-reciprocal under the correspondence carrying each quantity $a = \sum \alpha_{ij}e_{ij}$ of $\mathfrak{M}$ into what is, effectively, its transpose $\sum \alpha_{ij}e_{ji}$. Now $\mathfrak{A} = \mathfrak{M} \times \mathfrak{D}$ and, as we readily see, $\mathfrak{A}^{-1} = \mathfrak{M}^{-1} \times \mathfrak{D}^{-1} = \mathfrak{M} \times \mathfrak{D}^{-1}$, $\mathfrak{A} \times \mathfrak{A}^{-1} = \mathfrak{M} \times \mathfrak{M} \times (\mathfrak{D} \times \mathfrak{D}^{-1})$ is a direct product of total matric algebras and, by Theorem A10.8, is a total matric algebra. Observe that the order of $\mathfrak{A} \times \mathfrak{A}^{-1}$ is of course the square of the order of $\mathfrak{A}$.

Since we shall deal with $\mathfrak{A} \times \mathfrak{A}^{-1}$ frequently, it is important for the reader to notice that if $\mathfrak{A}$ is normal so is $\mathfrak{A}^{-1}$, and likewise the properties of being semi-simple, simple, a division algebra, or a total matric algebra hold for $\mathfrak{A}^{-1}$ when they are true for $\mathfrak{A}$.

6. Normal simple algebras. There are certain elementary but important properties of normal simple algebras which are easily derived by the use of

Theorem 3.14 and the process of forming direct products $\mathfrak{A} = \mathfrak{B} \times \mathfrak{C}$. We again emphasize our hypotheses that $\mathfrak{A}$ is to be constructed so that $\mathfrak{A}$, $\mathfrak{B}$, $\mathfrak{C}$ are algebras with the same unity quantity, and $\mathfrak{B}$ and $\mathfrak{C}$ are subalgebras of $\mathfrak{A}$. We then prove

Theorem 15. *Let $\mathfrak{A} = \mathfrak{B} \times \mathfrak{C}$ be simple. Then $\mathfrak{B}$ and $\mathfrak{C}$ are simple. If $\mathfrak{A}$ is also normal then $\mathfrak{B}$ and $\mathfrak{C}$ are normal simple.*

For if $\mathfrak{B}_0$ is any non-zero proper ideal of $\mathfrak{B}$ then $\mathfrak{A}_0 = \mathfrak{B}_0\mathfrak{C} < \mathfrak{A}$, $\mathfrak{A}_0\mathfrak{A} = \mathfrak{B}_0\mathfrak{C}\mathfrak{B}\mathfrak{C} = \mathfrak{B}_0\mathfrak{B}\mathfrak{C}^2 \leqq \mathfrak{B}_0\mathfrak{C} = \mathfrak{A}_0$. Similarly $\mathfrak{A}\mathfrak{A}_0 \leqq \mathfrak{A}_0$, $\mathfrak{A}_0$ is a non-zero proper ideal of $\mathfrak{A}$ contrary to our hypothesis that $\mathfrak{A}$ is simple. Similarly $\mathfrak{C}$ is simple. The final statement in the theorem follows from Theorem 3.12.

We next have

Theorem 16. *An algebra $\mathfrak{A}$ is normal simple over $\mathfrak{F}$ if and only if $\mathfrak{A}_\mathfrak{K}$ is normal simple for every scalar extension field $\mathfrak{K}$ of $\mathfrak{F}$.*

For by Theorem 3.14, $\mathfrak{A} \times \mathfrak{A}^{-1} = \mathfrak{M}$ is a total matric algebra. But then $\mathfrak{M}_\mathfrak{K}$ is normal simple for every $\mathfrak{K}$, $\mathfrak{M}_\mathfrak{K} = \mathfrak{A}_\mathfrak{K} \times (\mathfrak{A}^{-1})_\mathfrak{K}$ and our result follows from Theorem 3.15. The converse is trivial. We now derive the consequence

Theorem 17. *An algebra $\mathfrak{A}$ is normal simple over $\mathfrak{F}$ if and only if there exists a scalar extension $\mathfrak{K}$ of finite degree over $\mathfrak{F}$ such that $\mathfrak{A}_\mathfrak{K}$ is a total matric algebra.*

For if $\mathfrak{A}_\mathfrak{K}$ is a total matric algebra its centrum is $\mathfrak{C}_\mathfrak{K}$ where $\mathfrak{C}$ is the centrum of $\mathfrak{A}$. But $\mathfrak{C}_\mathfrak{K}$ has order one over $\mathfrak{K}$, the order of $\mathfrak{C}$ over $\mathfrak{F}$ must be one, $\mathfrak{A}$ is normal. Any non-zero ideal $\mathfrak{B}$ of $\mathfrak{A}$ defines an ideal $\mathfrak{B}_\mathfrak{K} \neq 0$ of $\mathfrak{A}_\mathfrak{K}$ and, since $\mathfrak{A}_\mathfrak{K}$ is simple, $\mathfrak{B}_\mathfrak{K} = \mathfrak{A}_\mathfrak{K}$, $\mathfrak{B}$ has the same order as $\mathfrak{A}$, $\mathfrak{B} = \mathfrak{A}$. Conversely, let $\mathfrak{A}$ be normal simple. Our result is true for algebras of order one and we assume it true for **algebra** of order less than that of $\mathfrak{A}$. If $\mathfrak{A}$ is not a division algebra we apply Theorem 3.9 to write $\mathfrak{A} = \mathfrak{M} \times \mathfrak{D}$ where $\mathfrak{D}$ is a normal division algebra of order less than that of $\mathfrak{A}$. But then $\mathfrak{D}_\mathfrak{K}$ is a total matric algebra; so is $\mathfrak{A}_\mathfrak{K} = \mathfrak{M}_\mathfrak{K} \times \mathfrak{D}_\mathfrak{K}$. Hence let $\mathfrak{A}$ be a normal division algebra. By Theorem 1.9 there exists a scalar extension $\mathfrak{K}_1$ of $\mathfrak{F}$ such that $\mathfrak{A}_{\mathfrak{K}_1}$ is not a division algebra. By Theorem 3.16 and the hypothesis of our induction there exists a scalar extension $\mathfrak{K}$ of $\mathfrak{K}_1$ such that $(\mathfrak{A}_{\mathfrak{K}_1})_\mathfrak{K} = \mathfrak{A}_\mathfrak{K}$ is a total matric algebra. This completes the induction and proves the theorem.

The result just obtained has the following immediate corollary.

Theorem 18. *The order of a normal simple algebra $\mathfrak{A}$ is the square of an integer n called the* **degree** *of $\mathfrak{A}$. If $\mathfrak{B}$ is a normal simple algebra of degree m, then the algebra $\mathfrak{A} \times \mathfrak{B}$ is normal simple of degree nm.*

The degree of an algebra was defined differently in Chapter AX. It will not be necessary for our present treatment to introduce that more complicated concept of degree but we have seen in Section 1.13 that the concepts coincide in the case of normal simple algebras.

We also have

Theorem 19. *Let $\mathfrak{A}$ be a simple algebra of degree n over its centrum $\mathfrak{Z}$ and let $\mathfrak{Z}$ have degree m over $\mathfrak{F}$. Then $\mathfrak{A}$ has order n^2m over $\mathfrak{F}$.*

For $\mathfrak{A}$ has order $t = n^2$ over $\mathfrak{Z}$ and a basis $u_1, \cdots, u_t$ over $\mathfrak{Z} = (v_1, \cdots, v_m)$ over $\mathfrak{F}$. Clearly $\mathfrak{A}$ has the n^2m quantities u_iv_j as a basis over $\mathfrak{F}$.

7. Separable algebras. The concept of a separable field was defined in Section A7.11. We shall generalize this concept by making the

DEFINITION. *An algebra $\mathfrak{A}$ over $\mathfrak{F}$ is called separable over $\mathfrak{F}$ if $\mathfrak{A}_{\mathfrak{K}}$ is semi-simple for every scalar extension $\mathfrak{K}$ of $\mathfrak{F}$.*

The definition implies that every separable algebra is semi-simple. To prove that it includes the definition of a separable field we shall show that *if $\mathfrak{C}$ is a field of degree n over $\mathfrak{F}$ then $\mathfrak{C}_{\mathfrak{K}}$ is semi-simple for all scalar extensions $\mathfrak{K}$ if and only if $\mathfrak{C}$ is separable* in the sense of Section A7.11. For by Theorem 2.34 an inseparable field $\mathfrak{C}$ does not have the property that $\mathfrak{C}_{\mathfrak{K}}$ is semi-simple for all scalar extensions $\mathfrak{K}$. Conversely, if $\mathfrak{C}$ is separable, by Theorem 2.26 there exists a scalar extension $\mathfrak{L}$ of $\mathfrak{F}$ such that $\mathfrak{C}_{\mathfrak{L}}$ is a diagonal algebra over $\mathfrak{L}$. If $\mathfrak{K}$ is any scalar extension of $\mathfrak{F}$ and $\mathfrak{Z}$ is a composite of $\mathfrak{K}$ and $\mathfrak{L}$ the algebra $(\mathfrak{C}_{\mathfrak{K}})_{\mathfrak{Z}} = (\mathfrak{C}_{\mathfrak{L}})_{\mathfrak{Z}}$ is diagonal over $\mathfrak{Z}$ and hence semi-simple, $\mathfrak{C}_{\mathfrak{K}}$ is semi-simple for every $\mathfrak{K}$ as desired.

This proof shows actually that if some scalar extension of an algebra $\mathfrak{A}$ is diagonal, then $\mathfrak{A}$ is separable as well as commutative and semi-simple. But then the reader may readily verify

Theorem 20. *Let $\mathfrak{A}$ be an algebra over $\mathfrak{F}$. Then there exists a scalar extension $\mathfrak{K}$ over $\mathfrak{F}$ such that $\mathfrak{A}_{\mathfrak{K}}$ is a diagonal algebra if and only if $\mathfrak{A}$ is a direct sum of separable fields.*

Let $\mathfrak{C}$ be the centrum of an algebra $\mathfrak{A}$. The nilpotent quantities of $\mathfrak{C}$ are properly nilpotent in $\mathfrak{A}$ and if $\mathfrak{A}$ is separable $\mathfrak{C}$ must be separable. We use this fact and prove

Theorem 21. *A semi-simple algebra $\mathfrak{A}$ is separable if and only if its centrum is separable. Moreover, this is true if and only if the simple components of $\mathfrak{A}$ are separable. Thus $\mathfrak{A}$ is separable if and only if the centrum of each simple component of $\mathfrak{A}$ is a separable field.*

Let $\mathfrak{A} = \mathfrak{B}_1 \oplus \cdots \oplus \mathfrak{B}_t$ with the $\mathfrak{B}_i$ simple algebras. It is clear from the corollary of Theorem 2.21 that $\mathfrak{A}$ is separable if and only if the $\mathfrak{B}_i$ are separable. The centrum of $\mathfrak{A}$ is the direct sum of the centra $\mathfrak{C}_i$ of $\mathfrak{B}_i$ by Theorem 2.24, so that we have reduced the proof of our theorem to the case in which $\mathfrak{A}$ is simple. Then the centrum $\mathfrak{C}$ of $\mathfrak{A}$ is a field and we have proved above that $\mathfrak{C}$ is separable. Conversely, let $\mathfrak{C}$ be a separable field of degree r over $\mathfrak{F}$. By Theorem 2.33 if $\mathfrak{L}$ is the scalar root field of $\mathfrak{C}$ with stem fields $\mathfrak{L}_i$ over $\mathfrak{F}$ then $\mathfrak{A}_{\mathfrak{L}} = \mathfrak{A}_{1_{\mathfrak{L}}} \oplus \cdots \oplus \mathfrak{A}_{r_{\mathfrak{L}}}$, where $\mathfrak{A}_i$ over $\mathfrak{L}_i$ is equivalent over $\mathfrak{F}$ to $\mathfrak{A}$, such that $\mathfrak{L}_i$ corresponds to $\mathfrak{C} = \mathfrak{F}(x)$

and $\mathfrak{L}_i$ thus is the centrum of $\mathfrak{A}_i$. Hence $\mathfrak{A}_i$ is normal simple over $\mathfrak{L}_i$ and there exists a scalar extension $\mathfrak{L}_0$ of finite degree over $\mathfrak{L}$ such that all the algebras $\mathfrak{A}_{i\mathfrak{L}_0}$ are total matric algebras. But then when $\mathfrak{W}$ is a composite of $\mathfrak{L}_0$ and $\mathfrak{K}$ we have $(\mathfrak{A}_{\mathfrak{K}})_{\mathfrak{W}} = (\mathfrak{A}_{\mathfrak{L}_0})_{\mathfrak{W}}$ a direct sum of total matric algebras and thus semi-simple for every $\mathfrak{K}$. It follows that $\mathfrak{A}_{\mathfrak{K}}$ is semi-simple, $\mathfrak{A}$ is separable.

Definition. *A scalar extension $\mathfrak{K}$ of $\mathfrak{F}$ is called a splitting field of $\mathfrak{A}$ if $\mathfrak{A}_{\mathfrak{K}}$ is a direct sum of total matric algebras.*

We now have

Theorem 22. *An algebra $\mathfrak{A}$ over $\mathfrak{F}$ has a splitting field if and only if $\mathfrak{A}$ is separable.*

For, if $\mathfrak{A}$ has a splitting field $\mathfrak{K}$ then $(\mathfrak{A}_{\mathfrak{L}})_{\mathfrak{W}} = (\mathfrak{A}_{\mathfrak{K}})_{\mathfrak{W}}$ is semi-simple for every $\mathfrak{L}$ and composite $\mathfrak{W}$ of $\mathfrak{K}$ and $\mathfrak{L}$, $\mathfrak{A}_{\mathfrak{L}}$ is semi-simple, $\mathfrak{A}$ is separable. The converse is an immediate consequence of our proof of Theorem 3.21.

We also have several immediate consequences of this result and Theorem 3.17.

Corollary I. *If $\mathfrak{A}$ has a splitting field then $\mathfrak{A}$ has a splitting field of finite degree over $\mathfrak{F}$.*

Corollary II. *Let $\mathfrak{A}$ have a splitting field. Then the number of total matric components of $\mathfrak{A}_{\mathfrak{K}}$ is the same for all splitting fields $\mathfrak{K}$ of $\mathfrak{A}$ and is the sum of the degrees over $\mathfrak{F}$ of the centra of the simple components of $\mathfrak{A}$. If $\mathfrak{A}$ is simple the components of $\mathfrak{A}_{\mathfrak{K}}$ are all n-rowed, where n is the degree of $\mathfrak{A}$ over its centrum.*

For, if $\mathfrak{W}$ is a composite of two splitting fields $\mathfrak{K}$ and $\mathfrak{L}$, the number of total matric components in $(\mathfrak{A}_{\mathfrak{K}})_{\mathfrak{W}}$ is the same as that in $\mathfrak{A}_{\mathfrak{K}}$ and also the same as that in $\mathfrak{A}_{\mathfrak{L}}$ since $(\mathfrak{A}_{\mathfrak{L}})_{\mathfrak{W}} = (\mathfrak{A}_{\mathfrak{K}})_{\mathfrak{W}}$. When $\mathfrak{A}$ is simple the components of $\mathfrak{A}_{\mathfrak{K}}$ are n-rowed, and the number of components of $\mathfrak{A}_{\mathfrak{K}}$ is the degree over $\mathfrak{F}$ of the centrum of $\mathfrak{A}$, by the proof of Theorem 3.21. But any separable algebra is a direct sum of simple algebras and our result is evident.

Corollary III. *An algebra $\mathfrak{A}$ with a splitting field $\mathfrak{K}$ is normal simple over $\mathfrak{F}$ if and only if $\mathfrak{A}_{\mathfrak{K}}$ is a total matric algebra.*

8. Structure of algebras with a radical. Let $\mathfrak{A}$ be an algebra of order n over $\mathfrak{F}$ with a radical $\mathfrak{N} \neq 0$. Denote* by $\underline{\mathfrak{A}}$ the difference algebra $\mathfrak{A} - \mathfrak{N}$, by $\underline{a}$ the class of $\underline{\mathfrak{A}}$ with representative the quantity a of $\mathfrak{A}$. Also if $\mathfrak{B} = (a_1, \cdots, a_m)$ is any linear subset of $\mathfrak{A}$ we shall denote by $\underline{\mathfrak{B}}$ the linear subset of $\underline{\mathfrak{A}}$ spanned by the quantities $\underline{a}_1, \cdots, \underline{a}_m$. In particular the quantity $\underline{0}$ is the zero class of $\underline{\mathfrak{A}}$, that is, $\underline{0}$ is the class $\mathfrak{N}$.

We have seen in Section 2.11 that there is a (1-1) correspondence between the subalgebras $\mathfrak{B} > \mathfrak{N}$ of $\mathfrak{A}$ and the corresponding subalgebras $\underline{\mathfrak{B}}$ of $\underline{\mathfrak{A}}$. Let

* Note our usage of $\underline{a}$ instead of our earlier $[a]$. The earlier notation is generally used only when the manipulations with the classes are not as complicated as is the case in the present discussion.

us first interest ourselves in subalgebras $\mathfrak{B}$ of order one of $\underline{\mathfrak{A}}$. Then $\underline{\mathfrak{B}}$ may be a zero algebra $(\underline{u})$ where $\underline{u}^2 = \underline{0}$, u^2 is in $\mathfrak{N}$, u is nilpotent, every quantity of $\mathfrak{B}$ is nilpotent. Hence $\mathfrak{B}$ is a nilpotent subalgebra of $\mathfrak{A}$. Conversely if the radical $\mathfrak{N} = (u_1, \cdots, u_r)$ over $\mathfrak{F}$ and $\mathfrak{B} = (u, u_1, \cdots, u_r)$ is a nilpotent subalgebra of $\mathfrak{A}$ then $\underline{\mathfrak{B}} = (\underline{u})$ is a zero algebra. For $u^2 = \alpha u + v_2$ with α in $\mathfrak{F}$ and v_2 in $\mathfrak{N}$, $u^3 = \alpha u^2 + v_2 u = \alpha^2 u + v_3$, $u^t = \alpha^{t-1}u + v_t = 0$ if and only if $\alpha = 0$, $v_t = 0$, u^2 is in $\mathfrak{N}$, $\underline{u}^2 = \underline{0}$.

There remains the case where $\underline{\mathfrak{B}} = (\underline{u})$, $\underline{u}^2 = \underline{u} \neq \underline{0}$. Since $\underline{\mathfrak{B}}$ is a division algebra (of order one) over $\mathfrak{F}$ the quantity $\underline{u}$ is the only idempotent of $\underline{\mathfrak{B}}$. Now $\mathfrak{B}$ is not a nilpotent algebra and hence contains an idempotent e. Clearly e is not in $\mathfrak{N}$, $\underline{e}$ is idempotent, $\underline{\mathfrak{B}} = (\underline{e})$, $\underline{e} = \underline{u}$. We have proved the case $t = 1$ of

LEMMA 4. *Let $\underline{u}_1, \cdots, \underline{u}_t$ be a set of t pairwise orthogonal idempotents of $\underline{\mathfrak{A}}$. Then $\underline{u}_i = \underline{e}_i$ $(i = 1, \cdots, t)$, where the e_i are pairwise orthogonal idempotents of $\mathfrak{A}$. Moreover if e is any idempotent of $\mathfrak{A}$ such that $\underline{e} = \underline{u}_1 + \cdots + \underline{u}_t$ we may choose the e_i such that $e = e_1 + \cdots + e_t$.*

We make an induction on t and assume our lemma true for $t - 1$ idempotents. By the case $t = 1$ we have $\underline{u}_1 + \cdots + \underline{u}_t = \underline{e}$ for an idempotent e of $\mathfrak{A}$ and we choose e to be any desired idempotent with this property. Now $\underline{e}\underline{u}_i = \underline{u}_i\underline{e} = \underline{u}_i = \underline{e}\underline{u}_i\underline{e}$ so that we may take the u_i in $e\mathfrak{A}e$. The radical of $e\mathfrak{A}e$ is the intersection $e\mathfrak{N}e$ of $e\mathfrak{A}e$ and $\mathfrak{N}$ and it follows that the classes $u_i + e\mathfrak{N}e$ are pairwise orthogonal idempotents of $e\mathfrak{A}e - e\mathfrak{N}e$. The case $t = 1$ then implies that $u_1 + e\mathfrak{N}e = e_1 + e\mathfrak{N}e$ where e_1 is an idempotent of $e\mathfrak{A}e$. Hence $\underline{e}_1 = \underline{u}_1$, $v_2 = e - e_1$ is an idempotent of $e\mathfrak{A}e$ orthogonal to e_1. But $\underline{v}_2 = \underline{u}_2 + \cdots + \underline{u}_t$. By the assumption of our induction there exist idempotents $e_2, \cdots, e_t$ which are pairwise orthogonal and such that $v_2 = e_2 + \cdots + e_t$; the e_i are thus in $v_2\mathfrak{A}v_2$ for $i = 2, \cdots, t$. Clearly $e_1e_i = e_1v_2e_i = 0 = e_ie_1$ for $i = 2, \cdots, t$, and our lemma is proved.

We next prove

LEMMA 5. *Let $\underline{\mathfrak{A}}$ have a total matric subalgebra $\mathfrak{M}_t$ with unity quantity $\underline{u}$ and e be any idempotent of $\mathfrak{A}$ such that $\underline{e} = \underline{u}$. Then there exists a total matric subalgebra $\mathfrak{M}$ of $\mathfrak{A}$ with unity quantity e such that $\underline{\mathfrak{M}} = \mathfrak{M}_t$.*

For let $\mathfrak{M}_t = (\underline{u}_{ij}; i, j, = 1, \cdots, t)$ with the usual multiplication table, e be as above. Then $\underline{e} = \underline{u}_{11} + \cdots + \underline{u}_{tt}$, and by Lemma 3.4 we have $\underline{u}_{ii} = \underline{e}_{ii}$ with the e_{ii} pairwise orthogonal idempotents of $e\mathfrak{A}e$. Also we have $\underline{e}_{ii}\underline{u}_{i1}\underline{e}_{11} = \underline{u}_{i1}$ so that there is no loss of generality if we take the quantities u_{i1} in $e_{ii}\mathfrak{A}e_{11}$ for $i = 2, \cdots, t$, and in fact for $i = 1$ we may take $u_{11} = e_{11}$. Similarly we may take u_{1j} in $e_{11}\mathfrak{A}e_{jj}$ for $j = 2, \cdots, t$. Now $\underline{u}_{1j}\underline{u}_{j1} = \underline{e}_{11}$, $u_{1j}u_{j1} = e_{11} + a_j$ where a_j is in the intersection $e_{11}\mathfrak{N}e_{11}$ of $\mathfrak{N}$ and $e_{11}\mathfrak{A}e_{11}$ and is nilpotent. If $a_j^m = 0$ and $b_j = \sum_{i=1}^{m-1}(-a_j)^i$ then $b_ja_j = -\sum_{i=2}^{m-1}(-a_j)^i$, $a_j + b_ja_j = -b_j$. Clearly b_j is in $e_{11}\mathfrak{A}e_{11}$, $(e_{11} + b_j)(e_{11} + a_j) = e_{11} + b_j + a_j + b_ja_j = e_{11}$. We put $e_{1j} = (e_{11} + b_j)u_{1j}$ in $e_{11}\mathfrak{A}e_{jj}$, $e_{i1} = u_{i1}$ in $e_{ii}\mathfrak{A}e_{11}$, and obtain

$$e_{1j}e_{j1} = e_{11} \qquad (j = 1, \cdots, t).$$

Define $e_{ij} = e_{i1}e_{1j}$ so that since $\underline{e}_{1j} = \underline{u}_{1j}$, $\underline{e}_{i1} = \underline{u}_{i1}$ we have $\underline{e}_{ij} = \underline{u}_{ij}$. Also

$$e_{ij}e_{hk} = 0,\ e_{ij}e_{jk} = e_{i1}e_{1j}e_{j1}e_{1k} = e_{ik} \qquad (j \neq h;\ i, j, h, k = 1, \cdots, t).$$

The quantities e_{ij} are in $e_{ii}\mathfrak{A}e_{jj}$ and the e_{kk} were chosen in $e\mathfrak{A}e$ so that the e_{ij} are in $e\mathfrak{A}e$ which has e as unity quantity. But then $e = e_{11} + \cdots + e_{tt}$ is the unity quantity of $\mathfrak{M} = (e_{ij};\ i, j = 1, \cdots, t)$.

The result above will be applied to give

LEMMA 6. *Let $\underline{\mathfrak{A}}$ contain a direct sum $\mathfrak{M}_{01} \oplus \cdots \oplus \mathfrak{M}_{0s}$ of total matric algebras $\mathfrak{M}_{0i}$. Then $\mathfrak{A}$ contains $\mathfrak{M}_{(1)} \oplus \cdots \oplus \mathfrak{M}_{(s)}$ where $\mathfrak{M}_{(i)}$ is a total matric algebra such that $\underline{\mathfrak{M}}_{(i)} = \mathfrak{M}_{0i}$.*

For we let e_{0i} be the unity quantity of $\mathfrak{M}_{0i}$ and apply Lemma 3.4 to write $e_{0i} = \underline{e}_i$, $i = 1, \cdots, s$ where the e_i are pairwise orthogonal idempotents of $\mathfrak{A}$. By Lemma 3.5, $\mathfrak{A}$ contains total matric algebras $\mathfrak{M}_{(i)}$ such that $\underline{\mathfrak{M}}_{(i)} = \mathfrak{M}_{0i}$, e_i is the unity quantity of $\mathfrak{M}_{(i)}$. The sum of the algebras $\mathfrak{M}_{(1)}, \cdots, \mathfrak{M}_{(s)}$ is their direct sum since their unity quantities are pairwise orthogonal.

Our preliminary lemmas will now be applied to derive the so-called *Principal Theorem* of Wedderburn.

Theorem 23. *Let $\mathfrak{A}$ be an algebra with radical $\mathfrak{N}$ such that $\mathfrak{A} - \mathfrak{N}$ is separable. Then*

$$\mathfrak{A} = \mathfrak{S} + \mathfrak{N},$$

where $\mathfrak{S}$ is zero or an algebra, and hence $\mathfrak{S}$ is separable and equivalent to $\mathfrak{A} - \mathfrak{N}$.

For by Theorem 2.19 if $\mathfrak{S}$ is an algebra it is equivalent to $\mathfrak{A} - \mathfrak{N}$. Note also that it is sufficient to prove that $\mathfrak{A}$ contains a subalgebra $\mathfrak{S}$ equivalent to $\mathfrak{A} - \mathfrak{N}$. For then $\mathfrak{S}$ is semi-simple and its intersection with $\mathfrak{N}$ is zero, the sum of $\mathfrak{S}$ and $\mathfrak{N}$ is a supplementary sum $\mathfrak{S} + \mathfrak{N}$ whose order is the sum of the orders of $\mathfrak{A} - \mathfrak{N}$ and of $\mathfrak{N}$. This is the order of $\mathfrak{A}$, $\mathfrak{A} = \mathfrak{S} + \mathfrak{N}$. We notice finally that our result is true trivially if $\mathfrak{N} = 0$ or $\mathfrak{A} = \mathfrak{N}$. Hence it is true if $\mathfrak{A}$ has order one. We make an induction on the order of $\mathfrak{A}$ and assume the result true for algebras of order less than that of $\mathfrak{A}$.

Assume first that $\mathfrak{N}^{\cdot 2} \neq 0$. If $\mathfrak{N}^{\cdot 2} = \mathfrak{N}$ then $\mathfrak{N} = \mathfrak{N}^{\cdot 2} = \mathfrak{N}^{\cdot 3} = \cdots = \mathfrak{N}^{\cdot r} = 0$, a contradiction. Hence $\mathfrak{N}^{\cdot 2} < \mathfrak{N}$, $\mathfrak{N}_0 = \mathfrak{N} - \mathfrak{N}^{\cdot 2} \neq 0$. By Theorem 2.20 if $\mathfrak{A}_0 = \mathfrak{A} - \mathfrak{N}^{\cdot 2}$ then $\mathfrak{A}_0 - \mathfrak{N}_0$ is equivalent to $\mathfrak{A} - \mathfrak{N}$ and is semi-simple, and $\mathfrak{N}_0$ is the radical of $\mathfrak{A}_0$ by Theorem 3.2. Since the order of $\mathfrak{A}_0$ is less than that of $\mathfrak{A}$ we apply our hypothesis to derive $\mathfrak{A}_0 = \mathfrak{S}_0 + \mathfrak{N}_0$, $\mathfrak{S}_0 \cong \mathfrak{A}_0 - \mathfrak{N}_0$, $\mathfrak{S}_0 \cong \mathfrak{A} - \mathfrak{N}$. By Theorem 2.20 the subalgebra $\mathfrak{S}_0$ of $\mathfrak{A}_0$ has the form $\mathfrak{S}_0 = \mathfrak{B} - \mathfrak{N}^{\cdot 2}$ where $\mathfrak{B}$ is a subalgebra of $\mathfrak{A}$ containing $\mathfrak{N}^{\cdot 2}$. Since $\mathfrak{S}_0$ is semi-simple $\mathfrak{N}^{\cdot 2}$ is the radical of $\mathfrak{B}$ and, since $\mathfrak{N}^{\cdot 2} < \mathfrak{N}$, $\mathfrak{B} < \mathfrak{A}$. But then $\mathfrak{B} = \mathfrak{S} + \mathfrak{N}^{\cdot 2}$, $\mathfrak{S} \cong \mathfrak{S}_0 \cong \mathfrak{A} - \mathfrak{N}$, $\mathfrak{A}$ contains $\mathfrak{S}$, as desired.

There remains the more difficult case, $\mathfrak{N}^{\cdot 2} = 0$, $\mathfrak{N} \neq 0$. By Theorem 3.22 and its Corollary I there exists an algebraic extension $\mathfrak{K} = (\xi_1, \cdots, \xi_r)$ over $\mathfrak{F}$ such that $(\mathfrak{A} - \mathfrak{N})_{\mathfrak{K}}$ is a direct sum of total matric algebras. Moreover we may take $\xi_1 = 1$, the unity quantity of $\mathfrak{F}$. Since $(\mathfrak{A} - \mathfrak{N})_{\mathfrak{K}} = \mathfrak{A}_{\mathfrak{K}} - \mathfrak{N}_{\mathfrak{K}}$ is semi-simple, $\mathfrak{N}_{\mathfrak{K}}$ is the radical of $\mathfrak{A}_{\mathfrak{K}}$ by Theorem 3.2. By Lemma 3.6 $\mathfrak{A}_{\mathfrak{K}}$ contains a subalgebra $\mathfrak{B}_0 \cong (\mathfrak{A} - \mathfrak{N})_{\mathfrak{K}}$. We let $\mathfrak{A} - \mathfrak{N} = (\underline{u}_1, \cdots, \underline{u}_m)$ over $\mathfrak{F}$ with the u_i in $\mathfrak{A}$, $\mathfrak{B}_0 = (v_1, \cdots, v_m)$ over $\mathfrak{K}$ with the v_i in $\mathfrak{A}_{\mathfrak{K}}$, and have

$$u_i = \sum_{j=1}^{m} \alpha_{ij} v_j + w_i \qquad (\alpha_{ij} \text{ in } \mathfrak{K}),$$

with the w_i in $\mathfrak{N}_{\mathfrak{K}}$. Since $\mathfrak{A} - \mathfrak{N}$ is an algebra

$$\underline{u}_i \underline{u}_j = \sum_{k=1}^{m} g_{ijk} \underline{u}_k \qquad (g_{ijk} \text{ in } \mathfrak{F}).$$

Also if $u_i' = u_i - w_i$ then $\underline{u}_i' = \underline{u}_i$ and we have $\underline{u}_i' \underline{u}_j' = \sum_{k=1}^{m} g_{ijk} \underline{u}_k'$. The quantities u_i' are in $\mathfrak{B}_0$ and are linearly independent in $\mathfrak{K}$ since the classes $\underline{u}_i' = \underline{u}_i$ are linearly independent in $\mathfrak{F}$, and hence in $\mathfrak{K}$. Hence $\mathfrak{B}_0 = (u_1', \cdots, u_m')$ over $\mathfrak{K}$. But the intersection of $\mathfrak{B}_0$ and $\mathfrak{N}_{\mathfrak{K}}$ is zero, and $u_i' u_j' - \sum_{k=1}^{m} g_{ijk} u_k'$ is in this intersection. Hence

$$u_i' u_j' = \sum_{k=1}^{m} g_{ijk} u_k' \qquad (i, j = 1, \cdots, m).$$

Write

$$w_i = w_{i1} + w_{i2}\xi_2 + \cdots + w_{ir}\xi_r \qquad (w_{ij} \text{ in } \mathfrak{N}),$$

and put $y_i = u_i - w_{i1}$ in $\mathfrak{A}$. Clearly $\mathfrak{A} = \mathfrak{S} + \mathfrak{N}$ where $\mathfrak{S} = (y_1, \cdots, y_m)$ is a linear set over $\mathfrak{F}$. But $\mathfrak{N}^2 = 0$ and $y_i = u_i' + \sum_{\mu=2}^{r} w_{i\mu}\xi_\mu$,

$$u_i' u_j' - \sum g_{ijk} u_k' = y_i y_j - \sum g_{ijk} y_k + \sum z_\mu \xi_\mu,$$

where the z_μ are in $\mathfrak{A}$. The linear independence of $1, \xi_2, \cdots, \xi_r$ in $\mathfrak{A}$ then implies that the $z_\mu = 0$ and thus the set $\mathfrak{S}$ is an algebra. Hence $\mathfrak{S}$ is equivalent to $\mathfrak{A} - \mathfrak{N}$.

The Wedderburn theorem is actually equivalent, in view of Theorem 2.20, to a result which contains Lemmas 3.4, 3.5, 3.6 as special instances and which we state as

Theorem 24. *Let $\mathfrak{A}$ be an algebra with radical $\mathfrak{N}$ and $\mathfrak{B}_0$ be any separable subalgebra of $\mathfrak{A} - \mathfrak{N}$. Then $\mathfrak{A}$ contains a subalgebra $\mathfrak{S}$ equivalent to $\mathfrak{B}_0$.*

For clearly Theorem 3.23 is the special case $\mathfrak{B}_0 = \mathfrak{A} - \mathfrak{N}$ of Theorem 3.24. Conversely, assume the result of Theorem 3.23. Then by Theorem 2.20 there exists a subalgebra $\mathfrak{B}$ of $\mathfrak{A}$ such that $\mathfrak{B} > \mathfrak{N}$, $\mathfrak{B} - \mathfrak{N} = \mathfrak{B}_0$. Apply Theorem 3.23 to write $\mathfrak{B} = \mathfrak{S} + \mathfrak{N}$; $\mathfrak{S}$ is the desired subalgebra of $\mathfrak{A}$.

The remainder of these LECTURES will be devoted principally to the theory of the structure and other properties of normal simple algebras.

CHAPTER IV

SIMPLE ALGEBRAS

1. The uniqueness theorem. If $\mathfrak{A}$ has a unity quantity 1 and g is any regular quantity of $\mathfrak{A}$ the correspondence

$$S: \qquad a \leftrightarrow a^S = gag^{-1} \qquad (a \text{ in } \mathfrak{A})$$

was seen in Section 1.8 to define what we called an inner automorphism of $\mathfrak{A}$. Every automorphism S of $\mathfrak{A}$ carries any subalgebra $\mathfrak{B}$ of $\mathfrak{A}$ into an equivalent subalgebra $\mathfrak{B}^S$, and when S is as above we shall write $\mathfrak{B}^S = g\mathfrak{B}g^{-1}$. This agrees with our customary notation for linear subsets of $\mathfrak{A}$.

It is clear that if $\mathfrak{A}$ is a simple algebra which is neither a total matric algebra $\mathfrak{M}$ nor a division algebra $\mathfrak{D}$ then the expression of $\mathfrak{A}$ in the form $\mathfrak{M} \times \mathfrak{D}$ is not unique. For both $\mathfrak{M}$ and $\mathfrak{D}$ have order greater than unity and we let d be in $\mathfrak{D}$ and not in $\mathfrak{F}$, $\mathfrak{M} = (e_{ij}\,; i, j = 1, \cdots, m)$, $g = de_{11} + e_{22} + \cdots + e_{mm}$. Then $g^{-1} = d^{-1}e_{11} + e_{22} + \cdots + e_{mm}$, $ge_{12}g^{-1} = de_{12}$ is not in $\mathfrak{M}$, $\mathfrak{A} = (g\mathfrak{M}g^{-1}) \times (g\mathfrak{D}g^{-1})$ where $g\mathfrak{M}g^{-1}$ is equivalent to but distinct from $\mathfrak{M}$. However we shall prove that the expression of $\mathfrak{A}$ in the form $\mathfrak{M} \times \mathfrak{D}$ is unique apart from the type of distinctness above, that is, we shall prove the following *uniqueness* theorem.

Theorem 1. *The expression of a simple algebra $\mathfrak{A}$ in the form $\mathfrak{A} = \mathfrak{M} \times \mathfrak{D}$, where $\mathfrak{M}$ is a total matric algebra and $\mathfrak{D}$ is a division algebra, is unique in the sense that if also $\mathfrak{A} = \mathfrak{M}_1 \times \mathfrak{D}_1$ then there exists a regular quantity g of $\mathfrak{A}$ such that*

$$\mathfrak{M}_1 = g\mathfrak{M}g^{-1} \cong \mathfrak{M}, \qquad \mathfrak{D}_1 = g\mathfrak{D}g^{-1} \cong \mathfrak{D}.$$

For, let $\mathfrak{M} = (e_{ij}\,; i, j = 1, \cdots, m)$, $\mathfrak{M}_1 = (f_{ij}\,; i, j = 1, \cdots, \mu)$. Without loss of generality we may assume that $m \leqq \mu$. Also the unity quantity of $\mathfrak{A}$ is

$$1 = e_{11} + \cdots + e_{mm} = f_{11} + \cdots + f_{\mu\mu}.$$

The f_{ij} are all not zero, and in particular $f_{11} = \sum d_{ij}e_{ij}$, where the d_{ij} are in $\mathfrak{D}$ and at least one $d_{pq} \neq 0$. Write $a = d_{pq}^{-1}e_{1p}f_{11}$, $b = f_{11}e_{q1}$ so that since $f_{11}^2 = f_{11}$

$$ab = d_{pq}^{-1}e_{1p}\sum d_{ij}e_{ij}e_{q1} = d_{pq}^{-1}d_{pq}e_{11} = e_{11}.$$

Now $ba = f_{11}(e_{q1}d_{pq}^{-1}e_{1p})f_{11}$ is a quantity of $f_{11}\mathfrak{A}f_{11} = \mathfrak{D}_1f_{11}$, and is not zero since $e_{11} = e_{11}^2 = a(ba)b$. Also $(ba)^2 = b(ab)a = be_{11}a = f_{11}e_{q1}e_{11}a = ba$, ba is an idempotent of $\mathfrak{D}_1f_{11}$. Since $\mathfrak{D}_1$ is a division algebra, $ba = f_{11}$. We form

$$(1) \qquad h = \sum_{i=1}^{m} e_{i1}af_{1i}, \qquad g = \sum_{j=1}^{m} f_{j1}be_{1j} = \sum_{j=1}^{m} f_{j1}e_{qj},$$

and obtain

$$hg = \sum_{i=1}^{m} e_{i1} a f_{1i} f_{i1} b e_{1i} = \sum_{i=1}^{m} e_{i1} e_{11} e_{1i} = 1 .$$

Hence h is a regular quantity of $\mathfrak{A}$ and $g = h^{-1}$, $gh = 1$. But

$$gh = \sum_{j=1}^{m} f_{j1} b e_{1j} e_{j1} a f_{1j} = \sum_{j=1}^{m} f_{jj} = 1$$

if and only if $m = \mu$. We have proved that $\mathfrak{M}$ and $\mathfrak{M}_1$ are equivalent. In fact

$$g e_{ij} g^{-1} = g e_{ij} h = \sum_{p,q} f_{p1} b e_{1p} e_{ij} e_{q1} a f_{1q} = f_{ij} , \qquad \mathfrak{M}_1 = g \mathfrak{M} g^{-1}.$$

But $\mathfrak{D} = \mathfrak{A}^{\mathfrak{M}}$, $\mathfrak{D}_1 = \mathfrak{A}^{\mathfrak{M}_1}$ and clearly $\mathfrak{A}^{g\mathfrak{M}g^{-1}} = g\mathfrak{D}g^{-1} = \mathfrak{D}_1$ as desired.

A consequence of this proof is the following result.

Theorem 2. *Let e be an idempotent of a simple algebra $\mathfrak{A}$ and write $e = e_1 + \cdots + e_r$ for pairwise orthogonal primitive idempotents e_i. Then r is an integer uniquely determined by e and called its* **rank**. *Moreover, two idempotents are similar in $\mathfrak{A}$ if and only if they have the same rank. The rank of the unity quantity of $\mathfrak{A} = \mathfrak{M} \times \mathfrak{D}$ is the degree of $\mathfrak{M}$, and all other idempotents of $\mathfrak{A}$ have smaller rank.*

For, by Theorem 3.6 the algebra $e\mathfrak{A}e$ is simple and has unity quantity $e = e_1 + \cdots + e_r$ where the e_i are in $e\mathfrak{A}e$. The final statement of Theorem 3.9 applied to the algebra $e\mathfrak{A}e$ and the proof of Theorem 4.1 then imply the invariance of r. Also, $1 - e$ is an idempotent orthogonal to e so that we have $1 = e_1 + \cdots + e_t$, $t \geqq r$, where the e_i are pairwise orthogonal primitive idempotents by Theorem 2.4. Similarly if $f = f_1 + \cdots + f_r$ then $1 = f_1 + \cdots + f_t$ with pairwise orthogonal primitive idempotents f_i. By Theorem 3.9 and the proof given above we have $gf_ig^{-1} = e_i$, $gfg^{-1} = e$. Conversely, if $gfg^{-1} = e = e_1 + \cdots + e_r = gf_1g^{-1} + \cdots + gf_sg^{-1}$ then the quantities gf_ig^{-1} are pairwise orthogonal idempotents which may be seen to be primitive in $\mathfrak{A}$ by the use of either Lemma 2.10 or Theorem 3.7. Then we must have $r = s$, the idempotents e and f have the same rank.

Clearly, we have used the definition that quantities a and b of $\mathfrak{A}$ are *similar in* $\mathfrak{A}$ if $\mathfrak{A}$ contains a regular quantity g such that $b = gag^{-1}$. The case $r = 1$ of the last theorem yields

Theorem 3. *All primitive idempotents of a simple algebra are similar.*

The uniqueness theorem implies

Theorem 4. *Let $\mathfrak{A} = \mathfrak{M} \times \mathfrak{B}$ where $\mathfrak{A}$ and $\mathfrak{M}$ are total matric algebras. Then $\mathfrak{B}$ is a total matric algebra.*

For, by Theorems 3.15 and 3.9 we have $\mathfrak{B} = \mathfrak{M}_1 \times \mathfrak{D}$ with $\mathfrak{D}$ a division algebra, $\mathfrak{M}_1$ a total matric algebra, $\mathfrak{M} \times \mathfrak{M}_1$ a total matric algebra. But then $\mathfrak{A} = (\mathfrak{M} \times \mathfrak{M}_1) \times \mathfrak{D}$. By Theorem 4.1, $\mathfrak{D}$ has order one and $\mathfrak{B} = \mathfrak{M}_1$ is a total matric algebra.

The uniqueness theorem may be readily seen to imply

Theorem 5. *Every automorphism of a normal simple algebra is an inner automorphism.*

For if S is an automorphism of $\mathfrak{A}$ there exists a unique induced automorphism T of the total matric algebra $\mathfrak{M} = \mathfrak{A} \times \mathfrak{A}^{-1}$ such that $aT = aS$, $bT = b$ for every a of $\mathfrak{A}$ and b of $\mathfrak{A}^{-1}$. By Theorem 4.1 every automorphism of $\mathfrak{M}$ is inner and there exists an element t in $\mathfrak{M}$ such that $xT = txt^{-1}$ for every x of $\mathfrak{M}$. Then $tbt^{-1} = b$ for every b of $\mathfrak{A}^{-1}$, t is in the M-commutator of $\mathfrak{A}^{-1}$, t is in $\mathfrak{A}$, $aT = aS = tat^{-1}$ as desired.

Theorem 6. *If the unity quantity of an algebra $\mathfrak{A}$ coincides with that of a normal simple subalgebra $\mathfrak{B}$ then $\mathfrak{A} = \mathfrak{B} \times \mathfrak{C}$.*

Note that by Theorem 3.10, $\mathfrak{C}$ is the uniquely determined $\mathfrak{A}$-commutator of $\mathfrak{B}$. For proof of our theorem we form $\mathfrak{H} = \mathfrak{B}^{-1} \times \mathfrak{A}$ whose unity quantity coincides with that of $\mathfrak{A}$ and hence with that of the subalgebra $\mathfrak{M} = \mathfrak{B}^{-1} \times \mathfrak{B}$ of $\mathfrak{H}$. By Theorem 1.17, $\mathfrak{H} = \mathfrak{M} \times \mathfrak{C}$ where $\mathfrak{C} = \mathfrak{H}^{\mathfrak{M}}$. Now $\mathfrak{B}^{-1} \times \mathfrak{B} = \mathfrak{M}$, $\mathfrak{H} = \mathfrak{B}^{-1} \times \mathfrak{A} = \mathfrak{M} \times \mathfrak{C} = \mathfrak{B}^{-1} \times \mathfrak{B} \times \mathfrak{C}$. However $\mathfrak{A} = \mathfrak{H}^{\mathfrak{B}^{-1}} = \mathfrak{B} \times \mathfrak{C}$ as desired.

3. Elementary properties. We shall proceed to a derivation of certain elementary but useful results on simple algebras which are direct consequences of the uniqueness theorem and the results of Sections 3.4 to 3.6. The first of these is given as

Theorem 7. *Let $\mathfrak{A}$ be simple and $\mathfrak{A} = \mathfrak{M}_t \times \mathfrak{D} = \mathfrak{M}_s \times \mathfrak{C}$ where $\mathfrak{D}$ is a division algebra, $\mathfrak{M}_t$ and $\mathfrak{M}_s$ are total matric algebras of respective degrees t and s. Then s divides t.*

For, by Theorem 3.15, $\mathfrak{C}$ is simple, $\mathfrak{C} = \mathfrak{M}_r \times \mathfrak{D}_1$, $\mathfrak{A} = (\mathfrak{M}_s \times \mathfrak{M}_r) \times \mathfrak{D}_1 = \mathfrak{M}_t \times \mathfrak{D}$. Our uniqueness theorem implies that $\mathfrak{M}_t$ and $\mathfrak{M}_s \times \mathfrak{M}_r = \mathfrak{M}_{rs}$ are equivalent, $t = rs$.

We next obtain

Theorem 8. *Let $\mathfrak{B}$ and $\mathfrak{D}$ be normal division algebras whose direct product $\mathfrak{B} \times \mathfrak{D}$ is a total matric algebra $\mathfrak{M}$. Then $\mathfrak{B}$ and $\mathfrak{D}$ are reciprocal.*

For $\mathfrak{M} \times \mathfrak{D}^{-1} = \mathfrak{B} \times (\mathfrak{D} \times \mathfrak{D}^{-1})$ where $\mathfrak{D} \times \mathfrak{D}^{-1}$ and $\mathfrak{M}$ are total matric algebras, $\mathfrak{B}$ and $\mathfrak{D}^{-1}$ are normal division algebras. Then $\mathfrak{B}$ and $\mathfrak{D}^{-1}$ are equivalent by the uniqueness theorem.

We use the result just obtained to prove

Theorem 9. *Let $\mathfrak{M}$ be a total matric algebra of degree n, $\mathfrak{D}$ be a normal division algebra of degree m. Then $\mathfrak{M}$ has a subalgebra equivalent to $\mathfrak{D}$ and with the same unity quantity as $\mathfrak{M}$ if and only if m^2 divides n.*

For if $n = m^2 q$ we may write $\mu = m^2$, $\mathfrak{M} = \mathfrak{M}_n = \mathfrak{M}_\mu \times \mathfrak{M}_q = \mathfrak{D} \times \mathfrak{D}^{-1} \times \mathfrak{M}_q$ as desired. Conversely if $\mathfrak{M}$ contains $\mathfrak{D}$ with the same unity quantity as $\mathfrak{M}$ we use Theorem 4.6 to write $\mathfrak{M} = \mathfrak{D} \times \mathfrak{B}$ where, by Theorem 3.15, $\mathfrak{B}$ is normal simple. Hence $\mathfrak{B} = \mathfrak{M}_q \times \mathfrak{C}$ where $\mathfrak{M}_q$ is a total matric algebra of degree q and $\mathfrak{C}$ is a normal division algebra. Thus $\mathfrak{M} = \mathfrak{M}_q \times \mathfrak{D} \times \mathfrak{C}$. By Theorem 4.4, $\mathfrak{D} \times \mathfrak{C}$ is a total matric algebra, and $\mathfrak{C}$ is reciprocal to $\mathfrak{D}$ by Theorem 4.8. Then $\mathfrak{D} \times \mathfrak{C} = \mathfrak{M}_\mu$, $n = \mu q$ as desired.

In forming the direct products of normal division algebras it is important to know

Theorem 10. *Let $\mathfrak{B}$ and $\mathfrak{D}$ be normal division algebras of respective degrees m and n. Then if m and n are relatively prime the algebra $\mathfrak{B} \times \mathfrak{D}$ is a division algebra.*

For, we write $\mathfrak{A} = \mathfrak{B} \times \mathfrak{D} = \mathfrak{M} \times \mathfrak{C}$ where $\mathfrak{C}$ is a normal division algebra and $\mathfrak{M}$ is a total matric algebra of degree q. Then $\mathfrak{B}^{-1} \times \mathfrak{A} = (\mathfrak{B}^{-1} \times \mathfrak{B}) \times \mathfrak{D} = \mathfrak{M} \times (\mathfrak{B}^{-1} \times \mathfrak{C}) = (\mathfrak{M} \times \mathfrak{M}_0) \times \mathfrak{G}$ where $\mathfrak{M} \times \mathfrak{M}_0$ and $\mathfrak{B}^{-1} \times \mathfrak{B}$ are total matric algebras, $\mathfrak{G}$ and $\mathfrak{D}$ are division algebras. By Theorem 4.1 $\mathfrak{M} \times \mathfrak{M}_0$ is equivalent to $\mathfrak{B}^{-1} \times \mathfrak{B}$. Hence q divides the degree m^2 of $\mathfrak{B}^{-1} \times \mathfrak{B}$. Similarly q divides n^2 and, since m is prime to n, $q = 1$, $\mathfrak{B} \times \mathfrak{D} = \mathfrak{C}$ is a normal division algebra.

4. Subfields of a total matric algebra. A subalgebra $\mathfrak{Z}$ of an algebra $\mathfrak{A}$ with a unity quantity will be called a *subfield* of $\mathfrak{A}$ if $\mathfrak{Z}$ is a field over $\mathfrak{F}$ and the unity quantity of $\mathfrak{Z}$ is that of $\mathfrak{A}$. We use this definition in considering a subfield $\mathfrak{Z}$ of a total matric algebra $\mathfrak{M}_n$.

There is no loss of generality if we suppose, as in Section 1.9, that $\mathfrak{M}_n$ is the algebra of all linear transformations on a linear set $\mathfrak{L}$ of order n over $\mathfrak{F}$. Write then $\mathfrak{Z} = (I, S_2, \cdots, S_m)$ over $\mathfrak{F}$ for linear transformations S_i. If $x \neq 0$ is in $\mathfrak{L}$ then

(2) $$x\alpha_1 + x^{S_2}\alpha_2 + \cdots + x^{S_m}\alpha_m = x^S,$$

where $S = I\alpha_1 + \cdots + S_m\alpha_m$, S is the zero quantity of $\mathfrak{Z}$ if and only if the α_i are all zero. When $S \neq 0$ it is in the field $\mathfrak{Z}$ and is non-singular, $x^S \neq 0$. Hence $x, x^{S_2}, \cdots, x^{S_m}$ are linearly independent in $\mathfrak{F}$. We let q be the largest integer such that

(3) $$x_1, x_1^{S_2}, \cdots, x_1^{S_m}, x_2, x_2^{S_2}, \cdots, x_2^{S_m}, \cdots, x_q, x_q^{S_2}, \cdots, x_q^{S_m}$$

are linearly independent in $\mathfrak{F}$, where $x_1, \cdots, x_n$ is a suitably chosen permutation of a basis $u_1, \cdots, u_n$ of $\mathfrak{L}$. Then the quantities (3) span a linear subset $\mathfrak{L}_0$ of $\mathfrak{L}$ such that every quantity of $\mathfrak{L}_0$ is expressible in the form $x = x_1^{z_1} + \cdots + x_q^{z_q}$ for z_i uniquely determined in $\mathfrak{Z}$. Moreover if u is any one of the x_i with subscript greater than q we have

$$x_1^{z_1} + \cdots + x_q^{z_q} + u^z = 0$$

for the z_i, z not all zero and in $\mathfrak{Z}$. It follows immediately that $z \neq 0$, $u = -(x_1^{z_1 z^{-1}} + \cdots + x_q^{z_q z^{-1}})$ is in $\mathfrak{L}_0$. Hence $\mathfrak{L} = \mathfrak{L}_0$. There are mq quantities in the basis (3) of $\mathfrak{L}$ and thus $n = mq$.

Define scalar multiplication on $\mathfrak{L}\mathfrak{Z}$ to $\mathfrak{L}$ by

$$zx = xz = x^z \tag{4}$$

for every x of $\mathfrak{L}$, and see that when z is in $\mathfrak{F}$ this scalar product is the same as that in $\mathfrak{L}$ over $\mathfrak{F}$. *Then $\mathfrak{L}$ may be regarded as a linear set of order q over $\mathfrak{Z}$* with a basis $x_1, \cdots, x_q$. A linear transformation S on $\mathfrak{L}$ over $\mathfrak{F}$ is commutative with every z of $\mathfrak{Z}$ if and only if

$$(x^z)^S = (xz)^S = (x^S)^z = x^S z$$

for every x of $\mathfrak{L}$. But then S is a linear transformation on $\mathfrak{L}$ of order q over $\mathfrak{Z}$. Conversely any linear transformation on $\mathfrak{L}$ over $\mathfrak{Z}$ is a linear transformation on $\mathfrak{L}$ over $\mathfrak{F}$ commutative with every z of $\mathfrak{Z}$, and hence $Sz = zS$ for every z of $\mathfrak{Z}$ if and only if S is in the total matric algebra $\mathfrak{M}_q$ of all linear transformations on $\mathfrak{L}$ over $\mathfrak{Z}$. We now have

Theorem 11. *A field $\mathfrak{Z}_0$ of degree m over $\mathfrak{F}$ is equivalent to a subfield $\mathfrak{Z}$ of a total matric algebra $\mathfrak{M}$ of degree n over $\mathfrak{F}$ if and only if $n = mq$. The $\mathfrak{M}$-commutator of a subfield $\mathfrak{Z}$ of $\mathfrak{M}$ is a total matric algebra $\mathfrak{M}_q$ over $\mathfrak{Z}$.*

For, our argument above shows that if $\mathfrak{Z}_0$ is equivalent to $\mathfrak{Z}$ then $n = mq$, $\mathfrak{M}^{\mathfrak{Z}} = \mathfrak{M}_q$ over $\mathfrak{Z}$. Conversely, let $n = mq$ so that $\mathfrak{M}_n = \mathfrak{M}_m \times \mathfrak{M}_q$. By Theorem 1.3 we may represent $\mathfrak{Z}_0$ as a subfield $\mathfrak{Z}$ of $\mathfrak{M}_m$. But then $\mathfrak{Z}$ is a subfield of $\mathfrak{M}_n$.

5. Simple subalgebras. The result of Theorem 4.11 may be generalized to the case of subfields of a normal simple algebra and provides an important tool for their study. We do this and first prove

Theorem 12. *Let $\mathfrak{Z}$ be a subfield of degree m over $\mathfrak{F}$ of a normal simple algebra $\mathfrak{A}$ of degree n over $\mathfrak{F}$. Then m divides n, $n = mq$, and the $\mathfrak{A}$-commutator of $\mathfrak{Z}$ is a normal simple algebra of degree q over $\mathfrak{Z}$.*

We form $\mathfrak{A}^{-1} \times \mathfrak{A} = \mathfrak{M}$. Then $\mathfrak{Z}$ is a subfield of $\mathfrak{M}$ and $\mathfrak{M}^{\mathfrak{Z}} \geqq \mathfrak{A}^{-1}_{\mathfrak{Z}}$. By Theorem 3.16 $\mathfrak{A}^{-1}_{\mathfrak{Z}}$ is normal simple over $\mathfrak{Z}$, $\mathfrak{M}^{\mathfrak{Z}}$ is a total matric algebra over $\mathfrak{Z}$, and $\mathfrak{M}^{\mathfrak{Z}} = \mathfrak{A}^{-1}_{\mathfrak{Z}} \times \mathfrak{C}$ where $\mathfrak{C}$ is normal simple over $\mathfrak{Z}$ by Theorem 3.15. Clearly $\mathfrak{C} \leqq \mathfrak{M}^{\mathfrak{A}^{-1}}$. However $\mathfrak{A} = \mathfrak{M}^{\mathfrak{A}^{-1}}$, $\mathfrak{C} \leqq \mathfrak{A}$, and since $\mathfrak{C} \leqq \mathfrak{M}^{\mathfrak{Z}}$ we have $\mathfrak{C} \leqq \mathfrak{A}^{\mathfrak{Z}}$. Now $\mathfrak{C}$ is the $\mathfrak{M}^{\mathfrak{Z}}$-commutator of $\mathfrak{A}^{-1}_{\mathfrak{Z}}$. This commutator contains $\mathfrak{A}^{\mathfrak{Z}}$ since $\mathfrak{A}^{\mathfrak{Z}} \leqq \mathfrak{M}^{\mathfrak{Z}}$, $\mathfrak{A}^{\mathfrak{Z}} \leqq \mathfrak{A} = \mathfrak{M}^{\mathfrak{A}^{-1}}$. It follows that $\mathfrak{A}^{\mathfrak{Z}} = \mathfrak{C}$ is normal simple over $\mathfrak{Z}$. The algebra $\mathfrak{M}^{\mathfrak{Z}}$ is a total matric algebra of degree ρ over $\mathfrak{Z}$ where, by Theorem 4.11, $m\rho = n^2$. But $\mathfrak{M}^{\mathfrak{Z}} = \mathfrak{A}^{-1} \times \mathfrak{A}^{\mathfrak{Z}}$ so that $\rho = nq$, $mnq = n^2$, $n = mq$ as desired.

As a corollary of the result above and Theorem 4.6 we may prove the generalization given in

Theorem 13. *The simple subalgebras of a normal simple algebra $\mathfrak{A}$ over $\mathfrak{F}$ which have the same unity quantity as $\mathfrak{A}$ occur in pairs $\mathfrak{B}$, $\mathfrak{C}$ such that*

$$\mathfrak{C} = \mathfrak{A}^{\mathfrak{B}}, \qquad \mathfrak{B} = \mathfrak{A}^{\mathfrak{C}},$$

and the order of $\mathfrak{A}$ *over* $\mathfrak{F}$ *is the product of the order of* $\mathfrak{B}$ *by that of* $\mathfrak{C}$. *Moreover, the algebras* $\mathfrak{B}$ *and* $\mathfrak{C}$ *have the same centrum* $\mathfrak{Z}$ *and in fact if* $\mathfrak{B}$ *and* $\mathfrak{C}$ *are expressed as algebras over* $\mathfrak{Z}$ *then*

$$\mathfrak{A}^{\mathfrak{Z}} = \mathfrak{B} \times \mathfrak{C}.$$

For, let $\mathfrak{A}$ be any normal simple algebra over $\mathfrak{F}$ and $\mathfrak{B}$ be any simple subalgebra of $\mathfrak{A}$ such that the unity quantities of $\mathfrak{A}$ and $\mathfrak{B}$ coincide. Then the centrum of $\mathfrak{B}$ is a subfield $\mathfrak{Z}$ of $\mathfrak{A}$. Let the degree of $\mathfrak{A}$ over $\mathfrak{F}$ be n so that its order is n^2 and $n = mq$ where m is the degree of $\mathfrak{Z}$ over $\mathfrak{F}$. The $\mathfrak{A}$-commutator $\mathfrak{H}$ of $\mathfrak{Z}$ is normal simple of degree q over $\mathfrak{Z}$ by Theorem 4.12. It contains $\mathfrak{B}$ as a normal simple subalgebra of degree t over $\mathfrak{Z}$, and hence $\mathfrak{H} = \mathfrak{B} \times \mathfrak{C}$ over $\mathfrak{Z}$, $q = ts$, $\mathfrak{C}$ is normal simple of degree s over $\mathfrak{Z}$. The order of $\mathfrak{B}$ over $\mathfrak{F}$ is t^2m, the order of $\mathfrak{C}$ over $\mathfrak{F}$ is s^2m and the product of their orders is $s^2t^2m^2 = q^2m^2 = n^2$ as desired. Now the $\mathfrak{A}$-commutator of $\mathfrak{B}$ contains $\mathfrak{C}$ and is evidently contained in the $\mathfrak{A}$-commutator $\mathfrak{H}$ of $\mathfrak{Z}$. But then the $\mathfrak{A}$-commutator of $\mathfrak{B}$ is contained in the $\mathfrak{H}$-commutator of $\mathfrak{B}$ which is $\mathfrak{C}$. This proves $\mathfrak{C} = \mathfrak{A}^{\mathfrak{B}}$. Similarly $\mathfrak{B} = \mathfrak{A}^{\mathfrak{C}}$.

6. Extensions of equivalences. In Section 1.8 we observed that every automorphism S of an algebra $\mathfrak{A}$ carries each subalgebra $\mathfrak{B}$ of $\mathfrak{A}$ into an equivalent subalgebra $\mathfrak{B}^S$. There are certain cases where the converse is true. We define the concept explicitly in the following

DEFINITION. *Let* $\mathfrak{B}$ *and* $\mathfrak{C}$ *be subalgebras of an algebra* $\mathfrak{A}$. *Then we shall say that an equivalence over* $\mathfrak{F}$ *of* $\mathfrak{B}$ *and* $\mathfrak{C}$,

$$S: \qquad b \leftrightarrow c = b^S \qquad (b \text{ in } \mathfrak{B},\ b^S \text{ in } \mathfrak{C}),$$

may be extended to an automorphism of $\mathfrak{A}$ *if there exists an automorphism*

$$T: \qquad a \leftrightarrow a^T \qquad (a,\ a^T \text{ in } \mathfrak{A})$$

such that for every b *of* $\mathfrak{B}$ *the corresponding* $c = b^S$ *is the quantity* b^T.

We shall customarily use the same symbol S to designate the automorphism of $\mathfrak{A}$ which induces the equivalence S of $\mathfrak{B}$ and $\mathfrak{B}^S$ and which we consider as an extension of S. Moreover we shall say that S has been extended to be an automorphism of $\mathfrak{A}$. We repeat that according to the convention which we are always assuming all of the concepts above are relative to a fixed field $\mathfrak{F}$.

If $\mathfrak{A}$ is normal simple, every automorphism of $\mathfrak{A}$ is inner. Hence an equivalence S may be extended to an automorphism of $\mathfrak{A}$ if and only if $b^S = gbg^{-1}$ for b in $\mathfrak{B}$ and a fixed regular g of $\mathfrak{A}$.

The case in which $\mathfrak{A}$ is normal simple and $\mathfrak{B}$ is a simple subalgebra of $\mathfrak{A}$ will be the only case to interest us. We shall treat it completely in

Theorem 14. *Let* $\mathfrak{A}$ *be a normal simple algebra whose unity quantity coincides with that of two simple subalgebras,* $\mathfrak{B}$ *and* $\mathfrak{B}_0$ *of* $\mathfrak{A}$. *Then any equivalence of* $\mathfrak{B}$ *and* $\mathfrak{B}_0$ *may be extended to an automorphism of* $\mathfrak{A}$.

We shall prove the result first for the case in which $\mathfrak{B}$ is normal simple. Then by Theorem 4.6, $\mathfrak{A} = \mathfrak{B} \times \mathfrak{C} = \mathfrak{B}_0 \times \mathfrak{C}_0$. If $\mathfrak{B}^{-1}$ is reciprocal to $\mathfrak{B}$ it is also reciprocal to $\mathfrak{B}_0$ and $\mathfrak{B}^{-1} \times \mathfrak{B} = \mathfrak{M}$, $\mathfrak{B}^{-1} \times \mathfrak{B}_0 = \mathfrak{M}_0$ for total matric algebras $\mathfrak{M}$ and $\mathfrak{M}_0$ of the same degree. Write $\mathfrak{C} = \mathfrak{M}_1 \times \mathfrak{D}$, $\mathfrak{C}_0 = \mathfrak{M}_{01} \times \mathfrak{D}_0$ for total matric algebras $\mathfrak{M}_0$, $\mathfrak{M}_{01}$, and normal division algebras $\mathfrak{D}$, $\mathfrak{D}_0$. This is possible by Theorem 3.15. Then

$$\mathfrak{B}^{-1} \times \mathfrak{A} = (\mathfrak{M} \times \mathfrak{M}_1) \times \mathfrak{D} = (\mathfrak{M}_0 \times \mathfrak{M}_{01}) \times \mathfrak{D}_0 .$$

It follows that $\mathfrak{D}$ and $\mathfrak{D}_0$ are equivalent and so are $\mathfrak{M}_1$, $\mathfrak{M}_{01}$. Thus $\mathfrak{C}$ and $\mathfrak{C}_0$ are equivalent under a correspondence $c \leftrightarrow c_0$. But if $b \leftrightarrow b_0$ is any equivalence of $\mathfrak{B}$ and $\mathfrak{B}_0$ the correspondence carrying each b into b_0 and each c into c_0 is an automorphism S of $\mathfrak{A} = \mathfrak{B} \times \mathfrak{C} = \mathfrak{B}_0 \times \mathfrak{C}_0$ such that $b_0 = b^S$. This completes the proof for the case where $\mathfrak{B}$ is normal simple.

If $\mathfrak{A}$ is a total matric algebra and $f(\lambda)$ is an irreducible polynomial of $\mathfrak{F}(\lambda)$ such that $f(x) = 0$ for an x in $\mathfrak{A}$ then all non-trivial invariant factors of $\lambda - x$ are identical with $f(\lambda)$, x_0 in $\mathfrak{A}$ is similar to x if and only if $f(x_0) = 0$. This theorem on matrices is the case $\mathfrak{B} = \mathfrak{F}(x)$ a field, $\mathfrak{A}$ a total matric algebra, of our theorem. We may use our two results to prove

Lemma 2. *The theorem is true for any field* $\mathfrak{B} = \mathfrak{F}(x)$.

For we form $\mathfrak{M} = \mathfrak{A} \times \mathfrak{A}^{-1}$ and have $\mathfrak{B}$ a subfield $\mathfrak{F}(x)$ of $\mathfrak{M}$. By our proof if $\mathfrak{B}_0 = \mathfrak{F}(x_0)$ is equivalent to $\mathfrak{B}$ under a correspondence $\beta(x) \leftrightarrow \beta(x_0)$, there exists an automorphism T of $\mathfrak{M}$ such that $x_0 = x^T$. It is clear that $\mathfrak{M}^{\mathfrak{B}} = \mathfrak{A}^{\mathfrak{B}} \times \mathfrak{A}^{-1}$, $\mathfrak{M}^{\mathfrak{B}_0} = \mathfrak{A}^{\mathfrak{B}_0} \times \mathfrak{A}^{-1}$. Apply T and obtain $(\mathfrak{M}^{\mathfrak{B}})^T = \mathfrak{M}^{\mathfrak{B}_0} = (\mathfrak{A}^{\mathfrak{B}})^T \times (\mathfrak{A}^{-1})^T_{\mathfrak{B}_0}$. However $(\mathfrak{A}^{-1})^T_{\mathfrak{B}_0}$ is equivalent to $\mathfrak{A}^{-1}_{\mathfrak{B}_0}$ and by the normal simple case of our theorem there exists an automorphism U of $\mathfrak{M}^{\mathfrak{B}_0}$ over $\mathfrak{B}_0$ such that $(\mathfrak{A}^{-1})^{TU} = \mathfrak{A}^{-1}$. Since U is inner it may be considered to be an automorphism of $\mathfrak{M}$ and then if $S = TU$ we have $x_0^U = x_0$, $x^S = x^{TU} = x_0$, $(\mathfrak{A} \times \mathfrak{A}^{-1})^S = \mathfrak{A} \times \mathfrak{A}^{-1} = \mathfrak{A}^S \times \mathfrak{A}^{-1}$ so that $\mathfrak{A}^S = \mathfrak{A}$, S is the desired automorphism of $\mathfrak{A}$.

We now complete our proof. The result is trivial if the order r of $\mathfrak{B}$ is unity and we make an induction on r assuming the theorem true for subalgebras of order less than r. We have proved the result true when $\mathfrak{B}$ is normal and so may assume that the centrum of $\mathfrak{B}$ has order greater than unity. If $\mathfrak{B}$ is a field with a proper subfield or if $\mathfrak{B}$ is not a field there exists a subfield $\mathfrak{Z}$ of the centrum of $\mathfrak{B}$ of positive degree $m < r$. By our hypothesis there exists an automorphism T of $\mathfrak{A}$ such that if the equivalence $b \leftrightarrow b_0$ of $\mathfrak{B}$ and $\mathfrak{B}_0$ carries each z of $\mathfrak{Z}$ into z_0 of $\mathfrak{Z}_0$ then $z_0 = z^T$. But $\mathfrak{B}^T$ over $\mathfrak{Z}_0$ is equivalent over $\mathfrak{Z}_0$ to $\mathfrak{B}_0$ over $\mathfrak{Z}_0$ under the correspondence $b^T \leftrightarrow b_0$. By Theorem 4.12 the algebra $\mathfrak{A}^{\mathfrak{Z}_0}$ is normal simple and contains the equivalent normal simple algebras $\mathfrak{B}^T$ and $\mathfrak{B}_0$ over $\mathfrak{Z}_0$. Hence there exists an inner automorphism U_0 of $\mathfrak{A}^{\mathfrak{Z}_0}$ such that $b^{TU_0} = b_0 = gb^Tg^{-1}$. Clearly g is regular in $\mathfrak{A}$ and if U is the automorphism $a \leftrightarrow gag^{-1}$ we have $S = TU$, $b^S = b_0$ as desired. There remains the case where $\mathfrak{B}$ is a field with no proper subfield. But then $\mathfrak{B} = \mathfrak{F}(x)$ for any x in $\mathfrak{B}$ and not in $\mathfrak{F}$, and this is the case treated in our lemma. Our induction is complete and the theorem proved.

The theorem we have just proved has numerous applications in the remainder of our theory.

As a corollary of Theorem 4.14 we have

Theorem 15. *Let $\mathfrak{B}$ and $\mathfrak{B}_0$ be equivalent simple subalgebras of a normal simple algebra $\mathfrak{A}$ whose unity quantity coincides with those of $\mathfrak{B}$, $\mathfrak{B}_0$. Then the $\mathfrak{A}$-commutator of $\mathfrak{B}$ is equivalent over $\mathfrak{F}$ to that of $\mathfrak{B}_0$.*

For, by Theorem 4.14 and 4.5 there exists a regular quantity g of $\mathfrak{A}$ such that $b \leftrightarrow b_0 = gbg^{-1}$ is the equivalence between $\mathfrak{B}$ and $\mathfrak{B}_0$. If a is such that $ab = ba$ for every b of $\mathfrak{B}$ then $gag^{-1}b_0 = b_0gag^{-1}$ for each corresponding b_0 of $\mathfrak{B}_0$, $g\mathfrak{A}^{\mathfrak{B}}g^{-1} \leqq \mathfrak{A}^{\mathfrak{B}_0}$. Similarly $g^{-1}\mathfrak{A}^{\mathfrak{B}_0}g \leqq \mathfrak{A}^{\mathfrak{B}}$ so that $\mathfrak{A}^{\mathfrak{B}}$ and $\mathfrak{A}^{\mathfrak{B}_0}$ have the same orders, $\mathfrak{A}^{\mathfrak{B}_0} = g\mathfrak{A}^{\mathfrak{B}}g^{-1}$. Hence $\mathfrak{A}^{\mathfrak{B}}$ and $\mathfrak{A}^{\mathfrak{B}_0}$ are equivalent.

The argument above is simply an amplification of the statement that an inner automorphism of $\mathfrak{A}$ carrying $\mathfrak{B}$ into $\mathfrak{B}_0$ must carry $\mathfrak{A}^{\mathfrak{B}}$ into $\mathfrak{A}^{\mathfrak{B}_0}$. We apply this result to prove

Theorem 16. *Let $\mathfrak{A}$ be a normal simple algebra of degree $n = mq$ over $\mathfrak{F}$ and $\mathfrak{Z}$ be a scalar extension of degree m over $\mathfrak{F}$ equivalent to a subfield $\mathfrak{Y}$ of $\mathfrak{A}$. Then*

$$\mathfrak{A}_{\mathfrak{Z}} = \mathfrak{M} \times \mathfrak{B}$$

where $\mathfrak{M}$ is a total matric algebra of degree m and $\mathfrak{B}$ is a normal simple algebra over $\mathfrak{Z}$ equivalent over $\mathfrak{F}$ to the $\mathfrak{A}$-commutator of $\mathfrak{Y}$.

For we form $\mathfrak{H} = \mathfrak{M}_m \times \mathfrak{A}$ and may assume that $\mathfrak{Z}$ is a subfield of $\mathfrak{M}_m$ by Theorem 4.11. By Theorem 4.12 $\mathfrak{A}_{\mathfrak{Z}}$ is clearly the $\mathfrak{H}$-commutator of $\mathfrak{Z}$ and is equivalent to $\mathfrak{H}^{\mathfrak{Y}} = \mathfrak{M}_m \times \mathfrak{A}^{\mathfrak{Y}}$ by Theorem 4.15. This is our desired result.

7. The existence of maximal subfields of normal division algebras. The theory of the subfields of an algebra is complicated by the possibility that a field may be inseparable. In our consideration of the subfields of division algebras we shall thus require in particular a consideration of the case where the characteristic of $\mathfrak{F}$ is a divisor of the degree of the algebra to be considered.

If $\mathfrak{A}$ is a normal simple algebra of degree n over $\mathfrak{F}$ and $\mathfrak{Z}$ is a subfield of $\mathfrak{A}$ of degree m over $\mathfrak{F}$ then, by Theorem 4.12, m divides n. Hence the maximum possible degree of a subfield of $\mathfrak{A}$ is n. We thus make the

DEFINITION. *A subfield $\mathfrak{Z}$ of a normal simple algebra $\mathfrak{A}$ is called maximal if its degree is the degree of $\mathfrak{A}$.*

Let us note that Theorem 4.12 states that if $\mathfrak{Z}$ is a maximal subfield of $\mathfrak{A}$ then $\mathfrak{A}^{\mathfrak{Z}} = \mathfrak{Z}$, that is, the only quantities of $\mathfrak{A}$ commutative with every quantity of $\mathfrak{Z}$ are the quantities in $\mathfrak{Z}$. We remark also that, while the degree of $\mathfrak{Z}$ has been defined to be its order as an algebra over $\mathfrak{F}$, the order of $\mathfrak{A}$ is the square of its degree.

We shall prove the existence of a separable maximal subfield of any normal division algebra $\mathfrak{D}$. Our proof will be an application of the theory of the commutator subalgebra $\mathfrak{A}^{\mathfrak{Z}}$ of a subfield $\mathfrak{Z}$ of $\mathfrak{D}$. Note that we are discussing all commutative subalgebras of $\mathfrak{D}$. For, every such algebra is a field.

The necessity of considering inseparable fields is evidenced when we begin our theory with

Theorem 17. *Let $\mathfrak{A}$ be a normal simple algebra of degree p over a field $\mathfrak{F}$ of characteristic p and let y in $\mathfrak{A}$ have $\phi(\lambda) = \lambda^p - \gamma, \gamma \neq 0$ in $\mathfrak{F}$, as its minimum function. Then there exists a quantity x in $\mathfrak{A}$ such that*

$$xy = y(x + 1). \tag{5}$$

For, if $\mathfrak{A}$ is a total matric algebra we may regard y as a p-rowed square matrix. Since $\phi(\lambda)$ has degree p it is the characteristic function and the only non-trivial invariant factor of y, every y_0 of $\mathfrak{A}$ with $\phi(\lambda)$ as minimum function is similar to y. Hence we may take $\mathfrak{A} = \mathfrak{M}_p = (e_{ij}\ ;\ i, j = 1, \cdots, p)$ such that $y = e_{21} + e_{32} + \cdots + e_{p,p-1} + \gamma e_{1p}$. By direct multiplication $e_{i+1,i+1}\, y = ye_{ii}$, and $x = e_{11} + 2e_{22} + \cdots + (p - 1)e_{p-1,p-1}$ is the quantity of our theorem. If $\mathfrak{A}$ is not a total matric algebra it must be a normal division algebra $\mathfrak{D}$. For, the degree of $\mathfrak{A} = \mathfrak{M} \times \mathfrak{D}$ is the prime p. Let $\mathfrak{K} = (1, \xi_2, \cdots, \xi_t)$ over $\mathfrak{F}$ be a splitting field of $\mathfrak{D}$ so that $x_0 y = y(x_0 + 1)$ for x_0 in $\mathfrak{D}_{\mathfrak{K}}$. Write $x_0 = x + x_2\xi_2 + \cdots + x_t\xi_t$ with x and the x_i in $\mathfrak{D}$ and have

$$x_0 y - y(x_0 + 1) = [xy - y(x + 1)] + \sum_{i=2}^{t} (x_i y - y x_i)\xi_i = 0$$

only if $xy = y(x + 1)$ as desired.

We apply this result in the case in which $\mathfrak{A}$ is a division algebra $\mathfrak{D}$. If $\mathfrak{Z}$ is a subfield of $\mathfrak{D}$ then either $\mathfrak{Z}$ is separable or inseparable over $\mathfrak{F}$. In the latter case the hypothesis that $\mathfrak{D}$ has prime degree p implies that $\mathfrak{Z} = \mathfrak{F}(y)$, y as in Theorem 4.17. But then the subfield $\mathfrak{F}(x)$ of that theorem cannot be inseparable. For, otherwise $x^p = \lambda$ is in $\mathfrak{F}$, $(x + 1)^p = \lambda + 1 = (y^{-1}xy)^p = \lambda$ which is impossible. We use this property to prove our principal result:

Theorem 18. *Every normal division algebra $\mathfrak{D}$ has a separable maximal subfield. Every subfield $\mathfrak{Y}$ of $\mathfrak{D}$ is contained in a maximal subfield which may be taken to be separable when $\mathfrak{Y}$ is separable.*

First let $\mathfrak{D}$ have prime degree p. Then every quantity of $\mathfrak{D}$ not in $\mathfrak{F}$ generates a maximal subfield of $\mathfrak{D}$ which is either separable or, by the argument above, may be replaced by a maximal subfield which is separable. Hence our theorem is true for $\mathfrak{D}$ of prime degree. We assume that the degree n of $\mathfrak{D}$ is not a prime and make an induction on n.

Clearly $\mathfrak{D}$ contains subfields $\mathfrak{Y}$ of degree m greater than unity. If $m < n$, the algebra $\mathfrak{D}^{\mathfrak{Y}}$ is a normal division algebra of degree q over $\mathfrak{Y}$, $n = mq$, by Theorem 4.12. By our hypothesis $\mathfrak{D}^{\mathfrak{Y}}$ contains a field $\mathfrak{Z}$ which is separable of degree q over $\mathfrak{Y}$. The degree of $\mathfrak{Z}$ over $\mathfrak{F}$ is then $mq = n$. We have proved that every subfield $\mathfrak{Y}$ of $\mathfrak{D}$ is contained in a maximal subfield $\mathfrak{Z}$ of $\mathfrak{D}$. Moreover, by Theorem 2.30, $\mathfrak{Z}$ is separable if $\mathfrak{Y}$ is separable. It remains only to show that if $\mathfrak{Y}$ is inseparable it may be replaced by a separable subfield of degree greater than unity. By our proof the maximal subfield $\mathfrak{Z}$ over $\mathfrak{Y}$ obtained above is separable of degree $q > 1$ over $\mathfrak{Y}$. The maximal separable subfield $\mathfrak{Z}_0$ of $\mathfrak{Z}$ over $\mathfrak{F}$ is not $\mathfrak{F}$ by Theorem 2.29 and is the desired field. This completes our induction.

We have proved the existence of maximal subfields of normal division algebras. Such a proof cannot be made in general, for normal simple algebras. For example, if $\mathfrak{F}$ is algebraically closed no algebraic extensions of $\mathfrak{F}$ exist, yet there exist normal simple algebras of any degree over $\mathfrak{F}$, and indeed all such algebras are easily seen to be total matric algebras. We shall consequently not interest ourselves further in the discussion of the existence of maximal subfields of normal simple algebras over an arbitrary field $\mathfrak{F}$. The question will be studied in later chapters wherein $\mathfrak{F}$ will be restricted.

8. The class group. Let $\mathfrak{A}$ be a normal simple algebra so that, by Theorem 3.13, $\mathfrak{A}$ is expressible in the form

$$\mathfrak{A} = \mathfrak{M} \times \mathfrak{D} \tag{6}$$

where $\mathfrak{M}$ is a total matric algebra and $\mathfrak{D}$ is a normal division algebra. The expression (6) is unique apart from an inner automorphism of $\mathfrak{A}$ so that in particular the degree m of $\mathfrak{D}$ is a unique integer. We shall call m the *index** of $\mathfrak{A}$. If the degree of $\mathfrak{A}$ is n and the degree of $\mathfrak{M}$ is q then $n = mq$. Hence the order of $\mathfrak{M}$ is uniquely determined by the degree and index of $\mathfrak{A}$. Moreover the structure of $\mathfrak{A}$ is uniquely determined when $\mathfrak{D}$ and the degree of $\mathfrak{A}$ are given.

If $\mathfrak{A}_1 = \mathfrak{M}_1 \times \mathfrak{D}_1$ with $\mathfrak{D}_1$ equivalent to $\mathfrak{D}$ we call $\mathfrak{A}_1$ and $\mathfrak{A}$ *similar algebras*, say that $\mathfrak{A}_1$ is similar to $\mathfrak{A}$, and write

$$\mathfrak{A}_1 \sim \mathfrak{A}.$$

Thus we have $\mathfrak{A} \sim \mathfrak{D}$. The relative structure of any two similar algebras is then determined by their degrees and common index.

The relation of similarity is an equivalence relation in the sense of page A5 and may be used to classify the set of all normal simple algebras over $\mathfrak{F}$ into classes $(\mathfrak{A})$. Each *class* $(\mathfrak{A})$ consists of all normal simple algebras similar to $\mathfrak{A}$. By Theorem 4.1 each class contains a normal division algebra $\mathfrak{D}$ uniquely determined in the sense of equivalence. Let us call the degree of $\mathfrak{D}$, which is then the index of every algebra in $(\mathfrak{A})$, the *index* of $(\mathfrak{A})$, and let $\mathfrak{G}$ be the set of all such classes.

The *product* $(\mathfrak{A})(\mathfrak{B})$ of two classes will be defined as the class $(\mathfrak{A} \times \mathfrak{B})$. Evidently the operation so defined is a commutative and associative operation on $\mathfrak{G}\mathfrak{G}$ to $\mathfrak{G}$. We then have immediately

* This invariant of $\mathfrak{A}$ is sometimes called the *Schur* index. The index of an algebra as defined in Chapter I is a trivial concept when applied to normal simple algebras and should not be confused with the present one.

Theorem 19. *The set $\mathfrak{G}$ is an abelian group with respect to the operation defined above. Its identity element is the class $\mathfrak{E}$ of all total matric algebras over $\mathfrak{F}$ and the inverse of every $(\mathfrak{A})$ is the class $(\mathfrak{A}^{-1})$ defined by the algebra $\mathfrak{A}^{-1}$ reciprocal over $\mathfrak{F}$ to $\mathfrak{A}$.*

The result above is clearly merely a restatement of theorems we have already derived. We shall complete the discussion in Chapter V by proving that every element of the group $\mathfrak{G}$ has finite order.

The division algebra $\mathfrak{D}$ which is a representative of the identity class of the group $\mathfrak{G}$ has order one over $\mathfrak{F}$ and hence any total matric algebra $\mathfrak{M}$ is similar to $\mathfrak{F}$. But $\mathfrak{F} = (1)$ over $\mathfrak{F}$ and it is desirable to express the fact that the structure of a total matric algebra is essentially independent of the field of reference. Hence it has become conventional to write

$$\mathfrak{A} \sim 1$$

to indicate that $\mathfrak{A}$ is a total matric algebra. We shall adopt this notation and use it frequently in the remainder of these LECTURES. To repeat, $\mathfrak{A} \sim 1$ implies that $\mathfrak{A}_{\mathfrak{K}} \sim 1$ for every scalar extension $\mathfrak{K}$ of $\mathfrak{F}$.

It will be convenient to note the following elementary result.

COROLLARY. *Let $\mathfrak{A}$ and $\mathfrak{B}$ be normal simple algebras such that the degree of $\mathfrak{B}$ divides that of $\mathfrak{A}$, and $\mathfrak{A} \sim \mathfrak{B}$. Then $\mathfrak{A} \cong \mathfrak{M} \times \mathfrak{B}$ where $\mathfrak{M}$ is a total matric algebra.*

This result states that while $\mathfrak{A} \sim \mathfrak{B}$ merely means that $\mathfrak{M}_s \times \mathfrak{A} \cong \mathfrak{M}_t \times \mathfrak{B}$ for total matric algebras $\mathfrak{M}_s$ and $\mathfrak{M}_t$ it actually means that if the proper relation exists between the degrees of $\mathfrak{A}$ and $\mathfrak{B}$ then we have the more explicit result, $\mathfrak{A} \cong \mathfrak{M} \times \mathfrak{B}$.

9. Index reduction factor. We shall discuss the effect on the index of a normal simple algebra $\mathfrak{A}$ when the base field $\mathfrak{F}$ is extended to a scalar extension $\mathfrak{K}$ of finite degree over $\mathfrak{F}$. The result is evidently the same for any $\mathfrak{B} \sim \mathfrak{A}$. We shall prove

Theorem 20. *Let r be the degree of $\mathfrak{K}$ over $\mathfrak{F}$, m and μ be the respective indices of $\mathfrak{A}$, $\mathfrak{A}_{\mathfrak{K}}$. Then $m = \mu q$ where q is a divisor of r called the* ***index reduction factor*** *of $\mathfrak{A}$* ***relative to*** *$\mathfrak{K}$.*

For by Theorem 1.3 the field $\mathfrak{K}$ may be taken to be a subfield of an $\mathfrak{M}_r \sim 1$ and of degree r. There is no loss of generality if we restrict our attention to the case in which $\mathfrak{A} = \mathfrak{D}$ is a normal division algebra. Form $\mathfrak{H} = \mathfrak{M}_r \times \mathfrak{D} \geqq \mathfrak{K} \times \mathfrak{D} = \mathfrak{D}_{\mathfrak{K}}$. By Theorem 3.16, $\mathfrak{D}_{\mathfrak{K}}$ is normal simple over $\mathfrak{K}$ and hence

$$\mathfrak{D}_{\mathfrak{K}} = \mathfrak{M}_q \times \mathfrak{B},$$

where q is the degree of $\mathfrak{M}_q \sim 1$, μ is the degree of the normal division algebra $\mathfrak{B}$ over $\mathfrak{K}$, $m = \mu q$. But $\mathfrak{M}_q$ over $\mathfrak{K}$ has the form $\mathfrak{M} \times \mathfrak{K}$ over $\mathfrak{F}$ where $\mathfrak{M} \sim 1$

over $\mathfrak{F}$ is a subalgebra of $\mathfrak{H}$ over $\mathfrak{F}$ and, by Theorem 4.6, $\mathfrak{H} = \mathfrak{M} \times \mathfrak{H}^{\mathfrak{M}} = \mathfrak{M}_r \times \mathfrak{D}$. By Theorem 4.7, q divides r.

We also have the

COROLLARY. *Let $\mathfrak{A}$ be a normal simple algebra whose index is prime to the degree of a scalar extension $\mathfrak{K}$ of finite degree over $\mathfrak{F}$. Then $\mathfrak{A}_{\mathfrak{K}}$ has the same index as $\mathfrak{A}$. Hence if $\mathfrak{A}$ is a division algebra so is $\mathfrak{A}_{\mathfrak{K}}$.*

If $\mathfrak{K}$ is a splitting field of $\mathfrak{A}$ so that $\mu = 1$, then $q = m$ divides r. We state this result as

Theorem 21. *The degree of a splitting field of a normal simple algebra $\mathfrak{A}$ is divisible by the index of $\mathfrak{A}$.*

10. Representation of fields by normal simple algebras. The proof of Theorem 4.20 has some further consequences. We let $r = qs$ in that proof, $\mathfrak{M}_r = \mathfrak{M}_q \times \mathfrak{M}_s$ and have $\mathfrak{Q} = \mathfrak{H}^{\mathfrak{M}_q} = \mathfrak{M}_s \times \mathfrak{D}_0$ where $\mathfrak{D}_0$ is equivalent to $\mathfrak{D}$. The degree of $\mathfrak{Q}$ is $sm = s\mu q = r\mu$, and $\mathfrak{Q}$ contains the algebra $\mathfrak{B}$ of degree μ over $\mathfrak{K}$, where $\mathfrak{K}$ is of degree r over $\mathfrak{F}$. But $\mathfrak{B} \leqq \mathfrak{Q}^{\mathfrak{K}}$ of degree μ over $\mathfrak{K}$ by Theorem 4.12, $\mathfrak{B} = \mathfrak{Q}^{\mathfrak{K}}$. We have proved

Theorem 22. *Let $\mathfrak{D}$ be a normal division algebra, $\mathfrak{K}$ be a field of degree r over $\mathfrak{F}$, and q be the index reduction factor of $\mathfrak{D}$ relative to $\mathfrak{K}$ so that*

$$(7) \qquad r = qs, \qquad \mathfrak{D}_{\mathfrak{K}} \sim \mathfrak{B},$$

where $\mathfrak{B}$ is a normal division algebra over $\mathfrak{K}$. Then $\mathfrak{K}$ is equivalent to a subfield $\mathfrak{K}_0$ of $\mathfrak{Q} = \mathfrak{M}_s \times \mathfrak{D}$ and the $\mathfrak{Q}$-commutator of $\mathfrak{K}_0$ is equivalent over $\mathfrak{F}$ to $\mathfrak{B}$.

The case in which $\mathfrak{K}$ is a splitting field of $\mathfrak{D}$ is of special interest. In this case q is the degree of $\mathfrak{D}$ and $\mathfrak{M}_s \times \mathfrak{D}$ of degree $qs = r$ has $\mathfrak{K}_0$ as maximal subfield. We thus have the

COROLLARY. *Every splitting field of degree ms of a normal division algebra $\mathfrak{D}$ of degree m is equivalent to a maximal subfield of $\mathfrak{M}_s \times \mathfrak{D}$.*

We may amplify the result of Theorem 4.22 and prove

Theorem 23. *The algebra $\mathfrak{Q} = \mathfrak{M}_s \times \mathfrak{D}$ of Theorem 4.22 is the algebra of least degree similar to $\mathfrak{D}$ and with a subfield equivalent to $\mathfrak{K}$. In fact $\mathfrak{A} \sim \mathfrak{D}$ has a subfield equivalent to $\mathfrak{K}$ if and only if $\mathfrak{A} = \mathfrak{M}_t \times \mathfrak{Q}$.*

For, we have seen that $\mathfrak{Q}$ does contain a subfield $\mathfrak{K}_0$ equivalent to $\mathfrak{K}$. Let $\mathfrak{A} = \mathfrak{M}_f \times \mathfrak{D}$ contain a subfield $\mathfrak{K}_1$ equivalent to $\mathfrak{K}$ and hence let $\mathfrak{K}_1$ be a subfield of $\mathfrak{H} = \mathfrak{M}_s \times \mathfrak{A} = \mathfrak{M}_f \times (\mathfrak{M}_s \times \mathfrak{D})$. Then $\mathfrak{K}_1$ is equivalent to the subfield $\mathfrak{K}_0$ of $\mathfrak{M}_s \times \mathfrak{D}$ and, by Theorem 4.14, there exists an inner automorphism of $\mathfrak{H}$ carrying $\mathfrak{K}_1$ to $\mathfrak{K}_0$. This carries $\mathfrak{H}^{\mathfrak{K}_1}$ to $\mathfrak{H}^{\mathfrak{K}_0}$, that is, $\mathfrak{H}^{\mathfrak{K}_1} \cong \mathfrak{H}^{\mathfrak{K}_0}$ over $\mathfrak{F}$. But $\mathfrak{H}^{\mathfrak{K}_0} = \mathfrak{M}_f \times \mathfrak{B}$, $\mathfrak{H}^{\mathfrak{K}_1} = \mathfrak{M}_s \times \mathfrak{A}^{\mathfrak{K}_1}$. Since $\mathfrak{B}$ is a normal division algebra over $\mathfrak{K}_0$ we apply Theorem 4.7 and have $f = st$ as desired.

We also have the useful result obtained as the case $s = 1$ of Theorems 4.22, 4.23.

Theorem 24. *The index reduction factor of $\mathfrak{D}$ relative to $\mathfrak{K}$ is equal to the degree of $\mathfrak{K}$ if and only if $\mathfrak{K}$ is equivalent to a subfield of $\mathfrak{D}$.*

We may think of the algebras $\mathfrak{A}$ containing subfields $\mathfrak{K}_0 \cong \mathfrak{K}$ as representation algebras of $\mathfrak{K}$ by $\mathfrak{D}$. For actually $\mathfrak{A}$ does provide a representation of $\mathfrak{K}$ as a set of matrices with elements in $\mathfrak{D}$. As we have seen, $\mathfrak{Q}$ is then a *least representation algebra.* Moreover we have

Theorem 25. *A representation algebra $\mathfrak{A} = \mathfrak{M} \times \mathfrak{D}$ of $\mathfrak{K}$ is a least representation algebra if and only if $\mathfrak{A}^{\mathfrak{K}_0}$ is a division algebra.*

For, in general we have shown that $\mathfrak{A}^{\mathfrak{K}_0} = \mathfrak{M}_t \times \mathfrak{B}$, and $\mathfrak{A} = \mathfrak{Q}$ if and only if $t = 1$.

The result just proved may be stated in another form more useful for certain applications.

Theorem 26. *Let $\mathfrak{K}$ be a subfield of degree r of a normal simple algebra $\mathfrak{A}$ such that the $\mathfrak{A}$-commutator algebra $\mathfrak{A}^{\mathfrak{K}}$ is a division algebra. Then the degree μ of $\mathfrak{A}^{\mathfrak{K}}$ over $\mathfrak{K}$ divides the index m of $\mathfrak{A}$ and $m = \mu q$ where q divides r.*

Note that in fact $\mathfrak{A}$ has degree μr, $r = sq$, $\mu r = sm$, and $\mathfrak{A} = \mathfrak{M}_s \times \mathfrak{D}$ is clearly a least representation algebra of $\mathfrak{K}$ by $\mathfrak{D}$.

The result above may be applied in an inductive construction of division algebras. For example, we shall construct an algebra $\mathfrak{A}$ in Chapter XI such that $\mathfrak{K}$ has prime degree $r = p$ over $\mathfrak{F}$ and $\mathfrak{A}^{\mathfrak{K}}$ is a division algebra. Then either $\mathfrak{A}$ is a division algebra or $\mathfrak{A}$ has degree pm, index m, where m is the degree of $\mathfrak{A}^{\mathfrak{K}}$ over $\mathfrak{K}$. We shall then show that this provides a criterion that $\mathfrak{A}$ be a division algebra.

11. Splitting fields of an algebra. The study of normal simple algebras with particular types of maximal subfields will be made in later chapters. The problem will consequently arise as to whether a given algebra has a field with given structure as maximal subfield. This study will be considerably simplified by the use of the next theorem. We may say that $\mathfrak{K}$ splits $\mathfrak{A}$ if $\mathfrak{K}$ is a splitting field of $\mathfrak{A}$ and prove

Theorem 27. *A field $\mathfrak{K}$ of degree n over $\mathfrak{F}$ is equivalent to a (maximal) subfield of a normal simple algebra $\mathfrak{A}$ of degree n over $\mathfrak{F}$ if and only if $\mathfrak{A}_{\mathfrak{K}} \sim 1$, that is, $\mathfrak{K}$ splits $\mathfrak{A}$.*

For, if $\mathfrak{K}$ splits $\mathfrak{A} = \mathfrak{M} \times \mathfrak{D}$, then $\mathfrak{K}$ splits $\mathfrak{D}$ and we apply the corollary of Theorem 4.22 to see that $\mathfrak{K}$ is equivalent to a maximal subfield of $\mathfrak{A}$. The converse is the case $m = n$ of Theorem 4.16.

A proof of this latter fact which is independent of Theorem 4.16 and is very elegant may be given and we shall present it. Let $\mathfrak{K}$ be equivalent to a subfield of degree n of $\mathfrak{A}$ so that this field may be taken to be a maximal subfield of $\mathfrak{A}^{-1}$. We form $\mathfrak{M} = \mathfrak{A} \times \mathfrak{A}^{-1}$ and use Theorem 4.11 to see that $\mathfrak{M}^{\mathfrak{K}}$ is a total matric algebra of degree n over $\mathfrak{K}$. But $\mathfrak{M}^{\mathfrak{K}} \geqq \mathfrak{A}_{\mathfrak{K}}$ of degree n over $\mathfrak{K}$, $\mathfrak{M}_{\mathfrak{K}} = \mathfrak{A}_{\mathfrak{K}} \sim 1$ as desired.

The result just obtained is of exceedingly great importance in the study of the structure of normal division algebras. It is applied many times in the literature on that subject, and is so well known that it is frequently applied without explicit reference being made to it.

By Theorem 4.18 we have the following result.

COROLLARY. *Every normal simple algebra has a separable splitting field.*

12. Finite simple algebras. We shall apply our results on normal simple algebras to obtain a determination of all such algebras having only a finite number of quantities. These are then all normal simple algebras over a finite field. We first prove

LEMMA 3. *Let $\mathfrak{H}$ be a proper subgroup of a finite group $\mathfrak{G}$. Then there exists a quantity of $\mathfrak{G}$ not in any subgroup of $\mathfrak{G}$ conjugate to $\mathfrak{H}$.*

For let $\mathfrak{G}$ have order $\nu = \mu\rho$ where μ is the order of $\mathfrak{H}$. Then $\mathfrak{G} = \mathfrak{H} + \mathfrak{H}S_2 + \cdots + \mathfrak{H}S_\rho$, where the S_i are in $\mathfrak{G}$ and a complete set of subgroups conjugate to $\mathfrak{H}$ is given by $\mathfrak{H}$, $S_2^{-1}\mathfrak{H}S_2, \cdots, S_\rho^{-1}\mathfrak{H}S_\rho$. Each of these groups contains precisely μ quantities and they all have the identity of $\mathfrak{G}$ in common. Hence the maximum number of distinct quantities in these conjugate groups is $\mu\rho - (\rho - 1)$ which is less than the number $\mu\rho = \nu$ of distinct quantities of $\mathfrak{G}$ since $\rho > 1$.

We now prove the determination result given by

Theorem 28. *Every finite simple algebra is a total matric algebra over its centrum.*

For, let $\mathfrak{D}$ be a normal division algebra of degree n over a finite field $\mathfrak{F}$. If $n > 1$ we apply Theorem 4.18 to obtain a field $\mathfrak{Z} < \mathfrak{D}$ and of degree n over $\mathfrak{F}$. If y is any non-zero quantity of $\mathfrak{D}$, the field $\mathfrak{F}(y)$ is contained in a subfield $\mathfrak{Z}_0$ of degree n over $\mathfrak{F}$ of $\mathfrak{D}$. But by Section A7.10 all finite fields of degree n over $\mathfrak{F}$ are equivalent. Thus by Theorem 4.14 there exists an inner automorphism of $\mathfrak{D}$ carrying $\mathfrak{Z}_0$ into $\mathfrak{Z}$. We let $\mathfrak{G}$ be the set of all non-zero quantities of $\mathfrak{D}$, a finite multiplicative group. The set $\mathfrak{H}$ of all non-zero quantities of $\mathfrak{Z}$ is a proper subgroup of $\mathfrak{G}$ since $\mathfrak{Z} < \mathfrak{D}$. But our proof above shows that every quantity of $\mathfrak{G}$ is contained in a group $\mathfrak{H}_0$, of all non-zero quantities of $\mathfrak{Z}_0$, $\mathfrak{H}_0$ is conjugate to $\mathfrak{H}$. This contradicts Lemma 4.3 and proves that $n = 1$.

Our theorem may be interpreted to state that the only finite division algebras are finite fields, all finite simple algebras are total matric algebras.

13. Applications of the Galois theory. The fundamental theorems of the Galois theory of fields may be applied to the Sylow theorems on groups as stated in Section A6.7 and yield certain theorems on normal fields. We shall use these results. Their verification is left to the reader.

LEMMA 4. *Let $\mathfrak{N}$ be a normal field of degree $m = gp^r$ over $\mathfrak{F}$ where p is a prime and not a divisor of g. Then there exists a subfield $\mathfrak{K}$ of $\mathfrak{N}$, called a Sylow subfield of $\mathfrak{N}$ for the prime p, such that $\mathfrak{K}$ has degree g over $\mathfrak{F}$. Any two such subfields are equivalent over $\mathfrak{F}$.*

LEMMA 5. *A normal field $\mathfrak{N}$ of degree p^e over $\mathfrak{F}$ is metacyclic.*

The result above will be applied to an arbitrary $\mathfrak{N}$ considered as a field over its Sylow subfield $\mathfrak{K}$. It states that

$$\mathfrak{N} = \mathfrak{N}_r > \mathfrak{N}_{r-1} > \cdots > \mathfrak{N}_1 > \mathfrak{K} = \mathfrak{N}_0 \tag{8}$$

where $\mathfrak{N}_i$ is cyclic of degree p over $\mathfrak{N}_{i-1}$ $(i = 1, \cdots, r)$.

The final Sylow theorem is

LEMMA 6. *Every subfield $\mathfrak{N}_s$ of degree $p^s g$ of a normal field $\mathfrak{N}$ of degree $p^r g$ over $\mathfrak{F}$ is a member of a sequence of fields* (8).

We may apply these results with our Theorem 4.18 to obtain

Theorem 29. *Let $\mathfrak{A}$ be a normal simple algebra of index $m = p^e q$, p a prime not dividing q. Then there exists a field $\mathfrak{K}$ of degree prime to p such that $\mathfrak{A}_\mathfrak{K}$ has index p^e.*

For, by Theorems 4.18, 4.27 there exists a separable splitting field of degree m of $\mathfrak{A}$. Every separable splitting field is contained in a normal* field $\mathfrak{N}$ which also splits $\mathfrak{A}$. The degree of $\mathfrak{N}$ is $p^r g$ where the prime p does not divide g and, by Theorem 4.21, $p^e q$ divides $p^r g$. Hence q divides g, $r \geqq e$. By Lemma 4.4, $\mathfrak{N}$ has a subfield $\mathfrak{K}$ of degree g over $\mathfrak{F}$ so that $\mathfrak{N}$ has degree p^r over $\mathfrak{K}$. Then by Theorem 4.20 the index reduction factor of $\mathfrak{A}$ relative to $\mathfrak{K}$ is prime to p and hence $\mathfrak{A}_\mathfrak{K}$ has index $p^e t$. But $\mathfrak{N}$ splits $\mathfrak{A}$, $\mathfrak{N}$ of degree p^r over $\mathfrak{K}$ splits $\mathfrak{A}_\mathfrak{K}$ so that $t = 1$, $\mathfrak{A}_\mathfrak{K}$ has index p^e as desired.

We apply the Sylow theory further and derive

Theorem 30. *Let p be any prime divisor of the index of a normal simple algebra $\mathfrak{A}$ over $\mathfrak{F}$. Then there exists a field $\mathfrak{L}$ of finite degree over $\mathfrak{F}$ such that the index of $\mathfrak{A}_\mathfrak{L}$ is p and $\mathfrak{A}_\mathfrak{L}$ has a splitting field which is cyclic of degree p over $\mathfrak{L}$.*

For, we choose $\mathfrak{N}$ and $\mathfrak{K}$ as in the proof of Theorem 4.29 so that $\mathfrak{A}_\mathfrak{K}$ has index p^e and $\mathfrak{N}$ is the metacyclic field of (8). Let j be the largest integer such that $\mathfrak{N}_j$ does not split $\mathfrak{A}$. We put $\mathfrak{L} = \mathfrak{N}_j$ and see that $\mathfrak{N}_{j+1}$ is cyclic of degree p over $\mathfrak{L}$ and splits $\mathfrak{A}$. Now $\mathfrak{A}_\mathfrak{L}$ is not a total matric algebra by hypothesis. The index reduction factor of $\mathfrak{A}_\mathfrak{L}$ relative to $\mathfrak{N}_{j+1}$ divides p and is not unity and hence must be p. It follows that $\mathfrak{A}_\mathfrak{L}$ has index p as desired.

* Note that we are using the definition of a normal field $\mathfrak{N}$ which has as a part the hypothesis that $\mathfrak{N}$ is separable.

The result just proved may also be seen to be a consequence of

Theorem 31. *Let $\mathfrak{A}$ be normal simple of index $m = p^e q$ where p is a prime not dividing q. Then there exists a field $\mathfrak{K}$ of degree prime to p such that $\mathfrak{A}_{\mathfrak{K}} \sim \mathfrak{D}$, where $\mathfrak{D}$ is a normal division algebra of degree p^e over $\mathfrak{K}$ with a separable maximal subfield $\mathfrak{Z}_e$ such that*

$$(9) \qquad \mathfrak{Z}_e > \mathfrak{Z}_{e-1} > \cdots > \mathfrak{Z}_1 > \mathfrak{Z}_0 \cong \mathfrak{K},$$

with $\mathfrak{Z}_i$ cyclic of degree p over $\mathfrak{Z}_{i-1}$ $(i = 1, \cdots, e)$.

For, let $\mathfrak{K}_0$ be the field of degree g of Theorem 4.29. Then $\mathfrak{A}_{\mathfrak{K}_0}$ has index p^e and hence a separable splitting field $\mathfrak{W}$ of degree p^e over $\mathfrak{K}_0$. We let $\mathfrak{N}$ be its normal root field of degree $p^t\mu$ over $\mathfrak{K}_0$ where μ is prime to p, $t \geqq e$. The corresponding Sylow subfield $\mathfrak{K}$ over $\mathfrak{K}_0$ of $\mathfrak{N}$ has degree μ over $\mathfrak{K}_0$ and $\mathfrak{A}_{\mathfrak{K}}$ has index p^e, $\mathfrak{A}_{\mathfrak{K}} \sim \mathfrak{D}$ where $\mathfrak{D}$ is a normal division algebra over $\mathfrak{K}$. Moreover, the degree of $\mathfrak{K}$ over $\mathfrak{F}$ is μg prime to p. Now $\mathfrak{W}$ is equivalent to a separable maximal subfield of the normal division algebra $\mathfrak{D}_0 \sim \mathfrak{A}_{\mathfrak{K}_0}$ and, since $(\mathfrak{D}_0)_{\mathfrak{K}} = \mathfrak{D}$ is a division algebra, $\mathfrak{W}_{\mathfrak{K}}$ is a field of degree p^e over $\mathfrak{K}$ and is equivalent to a separable maximal subfield $\mathfrak{Z}_e$ of $\mathfrak{D}$ over $\mathfrak{K}$. By the Sylow Lemma 4.6 the field $\mathfrak{Z}_e \cong \mathfrak{W}_{\mathfrak{K}}$ of degree $p^e\mu$ over $\mathfrak{K}_0$ is contained in a sequence (8) and hence $\mathfrak{Z}_e$ has the form (9) of our theorem.

To derive Theorem 4.30 as a consequence of 4.31 we choose the field $\mathfrak{L}$ of the former theorem to contain the field $\mathfrak{K}$ of the latter and be equivalent over $\mathfrak{K}$ to $\mathfrak{Z}_{e-1}$ over $\mathfrak{K}$. Then by Theorem 4.16, $\mathfrak{D}_{\mathfrak{L}} \sim \mathfrak{B}$, where $\mathfrak{B}$ is equivalent to the $\mathfrak{D}$-commutator of $\mathfrak{Z}_{e-1}$. This latter algebra, by Theorem 4.12, is a normal division algebra of degree p over $\mathfrak{Z}_{e-1}$ and clearly has the cyclic maximal subfield $\mathfrak{Z}_e$. Hence $\mathfrak{A}_{\mathfrak{L}} \sim \mathfrak{D}_{\mathfrak{L}} \sim \mathfrak{B}$ of degree and index p and with the desired cyclic splitting field. Observe however that we have now taken $\mathfrak{L}$ of degree p^{e-1} over $\mathfrak{K}$ so that $\mathfrak{L}$ has degree $p^{e-1}r$ over $\mathfrak{F}$ with r prime to p. We state this additional result as

Theorem 32. *The field $\mathfrak{L}$ of Theorem 4.30 may be chosen to have degree $p^{e-1}r$ over $\mathfrak{F}$, where r is prime to p, and p^e is the exact power of p dividing the index of $\mathfrak{A}$.*

The explicit restriction on the degree of $\mathfrak{L}$ obtained above is of importance in the proof we shall give of the finiteness of the orders, as group elements, of the classes of normal simple algebras. We pass to this latter theory, called the theory of exponents.

CHAPTER V

CROSSED PRODUCTS AND EXPONENTS

1. Connections of the theories. Some of the principal discoveries about the structure of normal division algebras have been obtained by the use of the theory of the exponents of such algebras. The proofs we shall give of the theorems on exponents are inherently connected with certain elementary properties of those normal simple algebras called cyclic algebras. It would be possible to derive directly the results on cyclic algebras which we require and so to obtain the exponent theory. However cyclic algebras are algebras of a type which is but a special instance of the more general type of algebra called a crossed product and the cyclic algebra theorems needed for our theory of exponents are obtainable by specialization of analogous theorems on crossed products. It is equally as important to obtain the theory of crossed products as to obtain these other theories and our derivation will actually turn out to be not very much more complicated and longer than the corresponding independent study of cyclic algebras. We shall therefore order our work so as first to study crossed products and then the consequent results on cyclic algebras, after which we shall apply our theory to obtain quickly the theory of exponents.

2. Equivalence of algebra-group pairs. The known types of normal simple algebras have been constructed by means of definitions relating their structure to that of certain types of fields. Each of the algebras we shall construct will be seen to contain a subfield equivalent to the defining field. But the equivalence of subfield and defining field will be a strengthened type of equivalence which we shall define explicitly in order to insure clarity of exposition.

DEFINITION. *Let $\mathfrak{G}$ and $\mathfrak{G}_0$ be respective groups of automorphisms over $\mathfrak{F}$ of algebras $\mathfrak{N}$ and $\mathfrak{N}_0$ over $\mathfrak{F}$ and $\mathfrak{G}$ be equivalent to $\mathfrak{G}_0$ under a correspondence $S \leftrightarrow S_0$ for S in $\mathfrak{G}$. Then we shall call the pairs $\mathfrak{N}$, $\mathfrak{G}$ and $\mathfrak{N}_0$, $\mathfrak{G}_0$ equivalent if there exists an equivalence $z \leftrightarrow z_0$ of $\mathfrak{N}$ and $\mathfrak{N}_0$ such that $(z^S)_0 = (z_0)^{S_0}$ for every S of $\mathfrak{G}$ and z of $\mathfrak{N}$.*

Observe that the concept above is really one restricting our notation. Thus when we say that $\mathfrak{N}$, $\mathfrak{G}$ and $\mathfrak{N}_0$, $\mathfrak{G}_0$ are equivalent we insist that when we write S_0 in $\mathfrak{G}_0$ we shall mean an automorphism corresponding to S in $\mathfrak{G}$. When $\mathfrak{N}$ is a normal field over $\mathfrak{F}$ so that $\mathfrak{G} = (S_1, \cdots, S_n)$ will be taken to be the group of all automorphisms of $\mathfrak{N}$ over $\mathfrak{F}$ then we are assuming that the notation $\mathfrak{G}_0 = (S_{01}, \cdots, S_{0n})$ shall mean that $S_i \leftrightarrow S_{0i}$ is an equivalence of $\mathfrak{G}$ and $\mathfrak{G}_0$.

Confusion sometimes arises in case we do not explicitly write down $S_1, \cdots, S_n$ and we shall consider a particularly illuminating example. We let $\mathfrak{G} = [S]$ be a cyclic group and shall henceforth designate the pair $\mathfrak{N}$, $[S]$ by $\mathfrak{N}$, S. Assume that $\mathfrak{N}$ is a cyclic field of degree $n > 1$ over $\mathfrak{F}$ with generating automorphism S so that $\mathfrak{N} = \mathfrak{F}(x)$ where x is a root of an irreducible polynomial

$$f(\lambda) = (\lambda - x)(\lambda - x^S)(\lambda - x^{S^2}) \cdots (\lambda - x^{S^{n-1}})$$

with coefficients in $\mathfrak{F}$. The quantities z of $\mathfrak{N}$ are uniquely expressible in the form $z = z(x) = \alpha_0 + \alpha_1 x + \cdots + \alpha_{n-1}x^{n-1}$ with the α_i in $\mathfrak{F}$, and in particular $x^S = \theta(x)$ for such a polynomial in x, $x^{S^i} = \theta^i(x)$ is the ith iterative of $\theta(x)$. We now call $\mathfrak{N}$, S and $\mathfrak{N}_0$, S_0 equivalent pairs if there exists a correspondence $z \leftrightarrow z_0$ of $\mathfrak{N}$ and $\mathfrak{N}_0$ such that $(z^S)_0 = (z_0)^{S_0}$, that is, $z(x) \leftrightarrow z(x_0)$ such that $[z(x_0)]^{S_0} = z[\theta(x_0)]$. Thus our definition of the equivalence of $\mathfrak{N}$, S and $\mathfrak{N}_0$, S_0 *prescribes* the correspondence $S \leftrightarrow S_0$.

We now let t be any positive integer less than n and prime to n. Then $S_0 = S^t$ is a generating automorphism of $\mathfrak{N}$ over $\mathfrak{F}$ and we have the pair $\mathfrak{N}$, S_0. If it were equivalent to $\mathfrak{N}$, S we would have a correspondence with $x \leftrightarrow x_0$, a root in $\mathfrak{N}$ of $f(\lambda)$, and hence $x_0 = x^{S^r}$. Then $(x^S)_0 = (x^S)^{S^r} = (x_0)^{S_0} = (x^{S^r})^{S^t}$ if and only if $t = 1$. It is clear that all of the pairs $\mathfrak{N}$, S^t are inequivalent for t taking on the distinct positive integral values less than n and prime to n.

3. Crossed products. An algebra $\mathfrak{A}$ over $\mathfrak{F}$ will be called a *crossed product* if $\mathfrak{A}$ is normal simple of degree n over $\mathfrak{F}$ and $\mathfrak{A}$ has a subfield $\mathfrak{N}$ of degree n over $\mathfrak{F}$ which is normal over $\mathfrak{F}$. Thus by Theorem 4.27 a normal simple algebra $\mathfrak{A}$ of degree n over $\mathfrak{F}$ is a crossed product if and only if there exists a scalar extension $\mathfrak{N}_0$ of degree n over $\mathfrak{F}$ which is normal over $\mathfrak{F}$ and splits $\mathfrak{A}$.

The importance of crossed products is due not merely to the fact that up to the present they are the only normal simple algebras which have actually been constructed but also to

Theorem 1. *Every normal simple algebra is similar to a crossed product.*

For by the corollary to Theorem 4.27 there exist separable splitting fields $\mathfrak{K}$ of $\mathfrak{A}$. But every such $\mathfrak{K}$ is contained in a normal field $\mathfrak{N}_0$ over $\mathfrak{F}$. We may take the degree n of $\mathfrak{K}$ to be the degree of the normal division algebra $\mathfrak{D} \sim \mathfrak{A}$, the degree of $\mathfrak{N}_0$ is tn, $\mathfrak{A} \sim \mathfrak{B} = \mathfrak{M}_t \times \mathfrak{D}$ of degree tn with $\mathfrak{N}_0$ as splitting field. Hence $\mathfrak{B}$ is a crossed product.

We shall now study the structure of crossed products relative to a fixed normal maximal subfield $\mathfrak{N}$ with automorphism group $\mathfrak{G} = (S_1, \cdots, S_n)$. We shall prove

Theorem 2. *A crossed product* $\mathfrak{A}$ *has the form*

$$\mathfrak{A} = u_{S_1}\mathfrak{N} + \cdots + u_{S_n}\mathfrak{N}, \tag{1}$$

where the u_S *are quantities of* $\mathfrak{A}$ *defined for each* S *of* $\mathfrak{G}$ *such that*

$$zu_S = u_S z^S \qquad (z \textit{ in } \mathfrak{N}). \tag{2}$$

There exist n^2 *quantities* $g_{S,T} \neq 0$ *of* $\mathfrak{N}$ *defined for every* S *and* T *of* $\mathfrak{G}$ *such that*

$$u_S u_T = u_{ST} g_{S,T}. \tag{3}$$

Moreover, the $g_{S,T}$ *satisfy the equations*

(4) $$g_{S,TR}\,g_{T,R} = g_{ST,R}(g_{S,T})^R$$

for every S, T, R *of* $\mathfrak{G}$.

For $\mathfrak{A}$ contains $\mathfrak{N} = \mathfrak{N}^S$ for every S of the automorphism group $\mathfrak{G}$ of $\mathfrak{N}$. By Theorem 4.14 there exist regular quantities u_S in $\mathfrak{A}$ such that (2) holds. Now $zu_Su_T = u_Sz^Su_T = u_Su_Tz^{ST}$, $zu_{ST} = u_{ST}z^{ST}$. It follows that

$$(u_{ST})^{-1}z = z^{ST}(u_{ST})^{-1},\ zu_Su_T(u_{ST})^{-1} = u_Su_Tz^{ST}(u_{ST})^{-1} = u_Su_T(u_{ST})^{-1}z.$$

By Theorem 4.12 $h_{S,T} = u_Su_T(u_{ST})^{-1}$ is in $\mathfrak{N}$, $u_Su_T = h_{S,T}u_{ST} = u_{ST}g_{S,T}$, where $g_{S,T} = (h_{S,T})^{ST}$ is in $\mathfrak{N}$. The algebra $\mathfrak{A}$ is associative, and the computation of $(u_Su_T)u_R = u_S(u_Tu_R)$ by the use of (3) together with the regularity of u_{STR} gives (4).

It remains to prove that $(u_{S_1}\mathfrak{N}, \cdots, u_{S_n}\mathfrak{N})$ is a supplementary sum. Since u_{S_1} is regular we may assume that the S_i have been so ordered that $\mathfrak{A}_r = u_{S_1}\mathfrak{N} + \cdots + u_{S_r}\mathfrak{N}$ is a supplementary sum for $1 \leqq r \leqq n$ but that if $r < n$, $(\mathfrak{A}_r, u_S\mathfrak{N})$ is not supplementary for every $S \neq S_i$ $(i = 1, \cdots, r)$ in $\mathfrak{G}$. Then $u_S\mathfrak{N}$ contains a quantity $u_Sz_0 \neq 0$ in $\mathfrak{A}_r$. Since $z_0 \neq 0$ has an inverse in $\mathfrak{N}$, and since $\mathfrak{A}_r = \mathfrak{A}_r\mathfrak{N}$ the quantity u_S is in $\mathfrak{A}_r$, $u_S = u_{S_1}z_1 + \cdots + u_{S_r}z_r$ for z_i in $\mathfrak{N}$. Hence

$$zu_S - u_Sz^S = \sum_{i=1}^{r} u_{S_i}z_i(z^{S_i} - z^S) = 0.$$

Since $\mathfrak{A}_r$ is supplementary this implies that $z_i(z^{S_i} - z^S) = 0$. If z is a generating quantity of $\mathfrak{N}$ over $\mathfrak{F}$ we have $z^{S_i} \neq z^S$ for every $i = 1, \cdots, r$, $z_i = 0$, $u^S = 0$ contrary to proof. Hence $r = n$ and we have (1).

Note that this proof applies to the subset $\mathfrak{B} = (u_{S_1}\mathfrak{N}, \cdots, u_{S_n}\mathfrak{N})$ of any algebra $\mathfrak{A}$ with a normal subfield $\mathfrak{N}$ such that $\mathfrak{B}$ has the multiplication table (2), (3), (4), of a crossed product. It implies that $\mathfrak{B} = u_{S_1}\mathfrak{N} + \cdots + u_{S_n}\mathfrak{N}$ is a crossed product subalgebra of $\mathfrak{A}$. We use this property in numerous proofs.

4. Factor sets. The sets of quantities $g_{S,T}$ of $\mathfrak{N}$ arise naturally in the study of crossed products and we shall consider some of the properties of such sets. We thus let $\mathfrak{N}$ be a normal field of degree n over $\mathfrak{F}$ and designate the quantities of its automorphism group by $S, T, \cdots$. We then make the

DEFINITION. *Let* $\mathfrak{g} = \{g_{S,T}\}$ *be a set of non-zero quantities* $g_{S,T}$ *of* $\mathfrak{N}$ *such that* $g_{S,T}$ *is uniquely defined for every* S *and* T *of* $\mathfrak{G}$. *Then if* (4) *is satisfied for every* S, T, R *of* $\mathfrak{G}$, *we call* $\mathfrak{g}$ *a factor set of* $\mathfrak{N}$.

The following result may be verified by a direct computation which is left for the reader.

LEMMA 1. *Let* $\mathfrak{g} = \{g_{S,T}\}$ *be a factor set and let there exist quantities* $c_S \neq 0$ *in* $\mathfrak{N}$ *for every* S *of* $\mathfrak{G}$ *such that*

(5) $$h_{S,T} = c_Tc_S^T(c_{ST})^{-1}g_{S,T}.$$

Then $\mathfrak{h} = \{h_{S,T}\}$ *is a factor set of* $\mathfrak{N}$ *said to be* ***associated*** *with* $\mathfrak{g}$.

It is trivial to verify that the relation of association is a formal equivalence relation in the sense that such relations were defined in Section A1.3. We shall introduce the notation

$$g \sim h \tag{6}$$

to indicate that g and h are associated.

There are two further results which have verifications of a direct computational character which we also leave to the reader.

LEMMA 2. *Let* $g = \{g_{S,T}\}$ and $h = \{h_{S,T}\}$ *be factor sets. Then*

$$k = g \cdot h = \{k_{S,T}\}, \qquad k_{S,T} = g_{S,T}h_{S,T} \tag{7}$$

for every S and T of $\mathfrak{G}$ is a factor set called the **product** *of g and h.*

LEMMA 3. *The set $i = \{i_{S,T}\}$, $i_{S,T} = 1$ for every S and T of $\mathfrak{G}$, is a factor set called the* **unit** *set. It has the property $g \cdot i = g$ for every factor set g.*

Note that the operation of multiplication of factor sets is an associative commutative operation with i as identity quantity. Moreover we easily verify that $g^{-1} = \{(g_{S,T})^{-1}\}$ is a factor set such $g \cdot g^{-1} = i$.

If g is a factor set of $\mathfrak{N}$ then the definition of g really depends also upon the group $\mathfrak{G}$. If $\mathfrak{N}_0$, $\mathfrak{G}_0$ is a pair equivalent to $\mathfrak{N}$, $\mathfrak{G}$ under correspondences $z \leftrightarrow z_0$, $S \leftrightarrow S_0$, $(z^S)_0 = (z_0)^{S_0}$, then a factor set $g = \{g_{S,T}\}$ of $\mathfrak{N}$ corresponds to a factor set $\{g_{S_0,T_0}\}$ of $\mathfrak{N}_0$ such that $(g_{S,T})_0 = g_{S_0,T_0}$ for every S and T of $\mathfrak{G}$ and corresponding S_0, T_0 of $\mathfrak{G}_0$. We shall later consider a factor set g of $\mathfrak{N}$, a factor set h of an equivalent field again indicated by $\mathfrak{N}$, and the product $g \cdot h$. Here we are actually considering three equivalent group-field pairs with a factor set $g = \{g_{S,T}\}$ defined in the first pair $\mathfrak{N}$, $\mathfrak{G}$, a factor set $h = \{h_{S_0,T_0}\}$ defined in $\mathfrak{N}_0$, $\mathfrak{G}_0$, and the factor set $\{g_{S_1,T_1} \cdot h_{S_1,T_1}\}$ which we call the product of g and h but which is really the product of the factor set in $\mathfrak{N}_1$, $\mathfrak{G}_1$ corresponding to g by that corresponding to h.

5. Construction of crossed products. The algebra considered in Theorem 5.2 was assumed to be a normal simple algebra with $\mathfrak{N}$ as maximal subfield and the relations (1)–(4) were obtained as consequent properties. However for every $\mathfrak{N}$ and every factor set $g = \{g_{S,T}\}$ of $\mathfrak{N}$ the expression in (1) together with the relations (2), (3) will be seen to define an algebra $\mathfrak{A}$ which is normal simple. We shall prove

Theorem 3. *Let g be a factor set of a normal field $\mathfrak{N}$ of degree n over $\mathfrak{F}$ with group $\mathfrak{G}$. Then there exists an algebra*

$$\mathfrak{A} = (\mathfrak{N}, g) \tag{8}$$

called the crossed product of $\mathfrak{N}$ by g which has the form (1) *and the multiplication table given by* (2), (3). *All such algebras defined by the same $\mathfrak{N}$ and g are equivalent, and are normal simple of degree n over $\mathfrak{F}$. Each has its unity quantity given by*

$$u_I (g_{I,I})^{-1}, \tag{9}$$

with I the identity automorphism, and has $\mathfrak{N}$ as a maximal subfield.

In the proof of our theorem we shall show that $\mathfrak{A}$ contains a normal field $\mathfrak{N}_0$ such that $\mathfrak{N}_0$, $\mathfrak{G}_0$ is equivalent to $\mathfrak{N}$, $\mathfrak{G}$, and if $\mathfrak{g}_0 = \{g_{S_0,T_0}\}$ is the factor set of $\mathfrak{N}_0$ corresponding to $\mathfrak{g}$ then

$$(10) \qquad \mathfrak{A} = u_{S_{01}}\mathfrak{N}_0 + \cdots + u_{S_{0n}}\mathfrak{N}_0$$

$$(11) \qquad z_0u_{S_0} = u_{S_0}z^{S_0}, \qquad u_{S_0}u_{T_0} = u_{S_0T_0}g_{S_0,T_0}.$$

By Theorem A1.9 we may then construct an equivalent $\mathfrak{A}$ as stated in the theorem. To make our proof we shall actually construct $\mathfrak{A}$ as a subalgebra of the algebra $\mathfrak{M} \times \mathfrak{N}$ where $\mathfrak{M}$ is a total matric algebra of degree n over $\mathfrak{F}$. Then $\mathfrak{A}$ will be associative without further proof.

Order the elements of the group $\mathfrak{G}$ of $\mathfrak{N}$ over $\mathfrak{F}$ in some fixed order. Then $\mathfrak{G} = (S_1, \cdots, S_n)$, and $\mathfrak{M}$ has a basis e_{S_i,S_j} $(i, j = 1, \cdots, n)$ such that

$$(12) \qquad e_{S,T}e_{T,U} = e_{S,U}, \qquad e_{S,T}e_{R,U} = 0$$

for all automorphisms S, T, R, U of $\mathfrak{G}$ such that $T \neq R$. In particular write $e_{T,T} = e_T$ and form the sum

$$(13) \qquad z_0 = \sum_T z^T e_T \qquad (z \text{ in } \mathfrak{N})$$

taken over all T of $\mathfrak{G}$. The correspondence $z \leftrightarrow z_0$ is an equivalence over $\mathfrak{F}$ of $\mathfrak{N}$ and the set $\mathfrak{N}_0$ of all the quantities z_0. Moreover $\mathfrak{N}_0$ is a subfield over $\mathfrak{F}$ of $\mathfrak{M}_{\mathfrak{N}} = \mathfrak{M} \times \mathfrak{N}$. We observe that (5.13) is another form of equation (2.13) and by the considerations leading to Theorem 2.27 we have $\mathfrak{N}_0\,\mathfrak{N} = \mathfrak{N}_0 \times \mathfrak{N}$,

$$(14) \qquad \mathfrak{N}_{0\mathfrak{N}} = \mathfrak{N}_0 \times \mathfrak{N} = (e_{S_1}, \cdots, e_{S_n}) \text{ over } \mathfrak{N}.$$

The field $\mathfrak{N}_0$ is normal of degree n over $\mathfrak{F}$ and its automorphism group consists of all the automorphisms

$$S_0: \qquad z_0 \leftrightarrow z_0^{S_0} = \sum_T (z^S)^T e_T.$$

Clearly

$$(15) \qquad S \leftrightarrow S_0$$

is then an equivalence of $\mathfrak{G}$ and $\mathfrak{G}_0$ such that the pairs $\mathfrak{N}$, $\mathfrak{G}$ and $\mathfrak{N}_0$, $\mathfrak{G}_0$ are equivalent.

Define n quantities

$$(16) \qquad u_{S_0} = \sum_U g_{S,U}e_{SU,U}.$$

Then u_{S_0} is defined for every S of $\mathfrak{G}$ and corresponding S_0 of $\mathfrak{G}_0$. We compute $z_0u_{S_0} = \sum_U z^{SU}g_{S,U}e_{SU,U}$, and obtain

$$(17) \qquad u_{S_0}z_0^{S_0} = u_{S_0}\sum_T z^{ST}e_T = \sum_U z^{SU}g_{S,U}e_{SU,U} = z_0u_{S_0}.$$

This proves that (2) holds in the sense of the remarks we have made. Also if we write $u_{T_0} = \sum_R g_{T,R}e_{TR,R}$ then the product $e_{SU,U}e_{TR,R} = 0$ unless $U = TR$, and we have

$$(18) \qquad u_{S_0}u_{T_0} = \sum_R g_{S,TR}g_{T,R}e_{S(TR),R}.$$

Also

$$(19) \qquad \begin{aligned} u_{S_0T_0}g_{S_0,T_0} &= \Big(\sum_R g_{ST,R}e_{(ST)R,R}\Big)\Big(\sum_V (g_{S,T})^V e_V\Big) \\ &= \sum_R g_{ST,R}(g_{S,T})^R e_{(ST)R,R}. \end{aligned}$$

But $\mathfrak{g}$ is a factor set and (4) holds. Comparing (18) and (19) and using $(ST)R = S(TR)$ we have $u_{S_0}u_{T_0} = u_{S_0T_0}g_{S_0,T_0}$ as desired.

We now define

$$(20) \qquad \mathfrak{A} = (u_{S_{01}}\mathfrak{N}_0, \cdots, u_{S_{0n}}\mathfrak{N}_0)$$

so that $\mathfrak{A}$ is a linear subset over $\mathfrak{F}$ of $\mathfrak{M} \times \mathfrak{N}$. As we have seen $u_{S_0}\mathfrak{N}_0u_{T_0}\mathfrak{N}_0 = u_{S_0T_0}\mathfrak{N}_0$, so that $\mathfrak{A}$ is clearly a subalgebra of $\mathfrak{M} \times \mathfrak{N}$. The quantity (9) has the form

$$u_{I_0}(g_{I_0,I_0})^{-1} = \Big(\sum_R g_{I,R}e_R\Big)\Big(\sum_T (g_{I,I}^T)^{-1}e_T\Big) = \sum_R g_{I,R}(g_{I,I}^R)^{-1}e_R.$$

By (4) with $S = T = I$ we have $g_{I,R}g_{I,R} = g_{I,R}g_{I,I}^R$, $(g_{I,I}^R)^{-1}g_{I,R} = 1$, $u_{I_0}(g_{I_0,I_0})^{-1} = \sum_R e_R = 1$ is the unity quantity of both $\mathfrak{M}$ and $\mathfrak{A}$. Then $u_{I_0}\mathfrak{N}_0 = \mathfrak{N}_0$, $\mathfrak{A}$ contains $\mathfrak{N}_0$. Hence $\mathfrak{A}\mathfrak{N} \geqq \mathfrak{N}_0\mathfrak{N}$ and, by (14), $\mathfrak{A}\mathfrak{N}$ contains all the $e_{S,S}$. Thus

$$\mathfrak{A}\mathfrak{N} \geqq g_{S,U}^{-1}e_{T,SU}u_{S_0} = e_{T,U}$$

for every T and U of $\mathfrak{G}$, $\mathfrak{A}\mathfrak{N} \geqq \mathfrak{M}$ and hence $\mathfrak{A}\mathfrak{N} = \mathfrak{M} \times \mathfrak{N}$, $\mathfrak{A}\mathfrak{N} = \mathfrak{A} \times \mathfrak{N} = \mathfrak{A}_{\mathfrak{N}} \sim 1$. By Theorem 3.17, $\mathfrak{A}$ is normal simple of degree n over $\mathfrak{F}$. Clearly any two algebras satisfying (1), (2), (3) are equivalent. We have proved our theorem. Note that

$$(21) \qquad u_{S_0}u_{S_0^{-1}} = u_{I_0}g_{S_0,S_0^{-1}},$$

so that u_{S_0} is regular for every S.

We shall use the notation $(\mathfrak{N}, \mathfrak{g})$ for crossed products not necessarily containing $\mathfrak{N}$ as subfield but merely an $\mathfrak{N}_0$ with $\mathfrak{N}_0$, $\mathfrak{G}_0$ equivalent to $\mathfrak{N}$, $\mathfrak{G}$. This usage is to be particularly noticed in the discussion of two or more crossed products where it may be desirable to have the defining fields distinct. Observe the notations introduced and indicating this in the later proofs, particularly those of of Theorems 5.5 and 5.6. We now study some further properties of crossed products.

If in $\mathfrak{A}$ we replace u_S by $v_S = u_Sc_S$ with $c_S \neq 0$ in $\mathfrak{N}$ we have

$$(22) \qquad v_Sv_T = u_Su_Tc_S^T c_T = u_{ST}g_{S,T}c_S^T c_T = v_{ST}h_{S,T},$$

where $h_{S,T}$ is given by (5). Clearly if $\mathfrak{A} = u_{S_1}\mathfrak{N} + \cdots + u_{S_n}\mathfrak{N}$ then also $\mathfrak{A} = v_{S_1}\mathfrak{N} + \cdots + v_{S_n}\mathfrak{N}$ for every $v_S = u_Sc_S$. Hence we have proved

Theorem 4. *Every crossed product* $\mathfrak{A} = (\mathfrak{N}, \mathfrak{g})$ *also has the form* $\mathfrak{A} = (\mathfrak{N}, \mathfrak{h})$ *for every* $\mathfrak{h}$ *associated with* $\mathfrak{g}$.

We may, however, actually derive the deeper theorem stated as.

Theorem 5. *Two crossed products* $\mathfrak{A} = (\mathfrak{N}, \mathfrak{g})$ *and* $\mathfrak{B} = (\mathfrak{N}, \mathfrak{h})$ *are equivalent if and only if* $\mathfrak{g}$ *and* $\mathfrak{h}$ *are associated.*

The equivalence of $\mathfrak{A} = u_{S_1}\mathfrak{N} + \cdots + u_{S_n}\mathfrak{N}$ and $\mathfrak{B} = u_{S_{01}}\mathfrak{N}_0 + \cdots + u_{S_{0n}}\mathfrak{N}_0$ carries $\mathfrak{N}_0$ into a subfield $\mathfrak{N}_1$ of $\mathfrak{A}$ equivalent to $\mathfrak{N}$ in the sense of Section 5.2. By Theorem 4.14 there is an inner automorphism of $\mathfrak{A}$ carrying $\mathfrak{N}_1$ into $\mathfrak{N}$. The product of our equivalence of $\mathfrak{A}$ and $\mathfrak{B}$ by this inner automorphism is an equivalence W of $\mathfrak{A}$ and $\mathfrak{B}$ carrying $\mathfrak{N}_0$ into $\mathfrak{N}$. Now W must carry $u_{S_{0i}}$ of $\mathfrak{B}$ into a quantity v_S of $\mathfrak{A}$ such that $zv_S = v_S z^S$ for every z of $\mathfrak{N}$, $v_S v_T = v_{ST} h_{S,T}$. We are actually trying to show that the factor set $\mathfrak{g}$ of $\mathfrak{A}$ and the set $\mathfrak{h} = \{h_{S,T}\}$ now defined are associated. Since $zu_S = u_S z^S$ for every z of $\mathfrak{N}$ we have $v_S u_S^{-1}$ in the $\mathfrak{A}$-commutator $\mathfrak{N}$ of $\mathfrak{N}$, $v_S = u_S c_S$ for $c_S \neq 0$ in $\mathfrak{N}$, $\mathfrak{g} \sim \mathfrak{h}$ as desired. The converse is merely an interpretation of Theorem 5.4.

6. Direct products of crossed products. We shall consider the structure of the direct product of two crossed products defined by the same $\mathfrak{N}$. The result will be stated first as

Theorem 6. *The direct product*

$$(\mathfrak{N}, \mathfrak{g}) \times (\mathfrak{N}, \mathfrak{h}) = \mathfrak{M} \times (\mathfrak{N}, \mathfrak{g}\cdot\mathfrak{h}),$$

where $\mathfrak{M}$ *is a total matric algebra.*

For proof write $\mathfrak{A} = u_{S_1}\mathfrak{N} + \cdots + u_{S_n}\mathfrak{N}$, $\mathfrak{B} = u_{S_{01}}\mathfrak{N}_0 + \cdots + u_{S_{0n}}\mathfrak{N}_0$ where $u_S u_T = u_{ST} g_{S,T}$, $u_{S_0} u_{T_0} = u_{S_0 T_0} h_{S_0,T_0}$. The algebra $\mathfrak{A} \times \mathfrak{B}$ contains $\mathfrak{B}_{\mathfrak{N}} = \mathfrak{B} \times \mathfrak{N} = \mathfrak{M} \times \mathfrak{N}$, where the n^2 quantities $e_{S,T}$ of the proof of Theorem 5.3 are a basis of the total matric algebra $\mathfrak{M}$. By Theorem 4.6 we have

$$\mathfrak{H} = \mathfrak{A} \times \mathfrak{B} = \mathfrak{M} \times \mathfrak{C}$$

where $\mathfrak{C}$ is the $\mathfrak{H}$-commutator of $\mathfrak{M}$. Then $\mathfrak{N}$ is in $\mathfrak{C}$, $\mathfrak{C} = (\mathfrak{N}, \mathfrak{k})$ with $\mathfrak{k}$ to be determined. The quantity $e = e_{I,I}$ is a primitive idempotent of $\mathfrak{M}$ with the properties

$$e^S = e^{S_0^{-1}} \tag{23}$$

for every S of $\mathfrak{G}$ by Theorem 2.27. Also in Section 2.15 we observed that $\mathfrak{N}$ and $\mathfrak{N}_0$ are equivalent under a correspondence $z \leftrightarrow z_0$ such that $ez = ez_0$ for every z of $\mathfrak{N}$. Observe in addition that (23) implies

$$e = e^{SS_0}, \qquad ey_S = y_S e = ey_S e \tag{24}$$

for every S of $\mathfrak{G}$ where $y_S = u_S u_{S_0} = u_{S_0} u_S$. We use these properties to determine $\mathfrak{k}$.

The algebra $e\mathfrak{H}e = e\mathfrak{C}e = e\mathfrak{C}$ is evidently equivalent to $\mathfrak{C}$ under a correspondence $eae = ece = ec \leftrightarrow c$ with c in $\mathfrak{C}$. Clearly this equivalence carries the subfield $\mathfrak{N}$ of $\mathfrak{C}$ into the field $e\mathfrak{N}$ with unity quantity e and such that $ez \leftrightarrow ez^S$ is the automorphism corresponding to S on $\mathfrak{N}$. As we have seen, ey_S is in $e\mathfrak{C}$ and has the property

$$ez \cdot ey_S = ez \cdot y_S = ey_S z^S = ey_S \cdot ez^S, \; ey_S \cdot ey_T = eu_S u_T u_{S_0} u_{T_0} = ey_{ST} \cdot eg_{S,T} \cdot eh_{S_0,T_0} = ey_{ST} \cdot eg_{S,T} h_{S,T},$$

where $h_{S,T}$ in $\mathfrak{N}$ corresponds to h_{S_0,T_0} in $\mathfrak{N}_0$. It follows that $e\mathfrak{C}e = (e\mathfrak{N}, e_{\mathfrak{g}} \cdot \mathfrak{k}) \cong \mathfrak{C}$ as desired.

The result above may also be stated in the alternative form

Theorem 6′. $(\mathfrak{N}, \mathfrak{g}) \times (\mathfrak{N}, \mathfrak{k}) = (\mathfrak{N}, \iota) \times (\mathfrak{N}, \mathfrak{g} \cdot \mathfrak{k})$.

For we may easily prove* the important

Theorem 7. *A crossed product* $(\mathfrak{N}, \mathfrak{g}) \sim 1$ *if and only if* $\mathfrak{g} \sim \iota$.

In view of Theorem 5.5 this result is equivalent to the statement that $(\mathfrak{N}, \iota) \sim 1$. To prove this we let $\mathfrak{W}$ be a field equivalent to $\mathfrak{N}$ and consider a certain set of linear transformations on $\mathfrak{W}$, a linear set of order n over $\mathfrak{F}$. The first of these are the transformations

$$U_S: \qquad w \leftrightarrow w^S \qquad (w \text{ in } \mathfrak{W}),$$

where S ranges over all automorphisms of the group $\mathfrak{G}$ of $\mathfrak{N}$ and we have used the same notation S for the corresponding automorphism of $\mathfrak{W}$. That U_S is a linear transformation on $\mathfrak{W}$ is true since $(w_1 + w_2)^S = w_1^S + w_2^S$, $(w\alpha)^S = w^S\alpha$ for every w_1, w_2, w of $\mathfrak{W}$ and α of $\mathfrak{F}$. The correspondences

$$Z_z: \qquad w \leftrightarrow wz = w^{Z_z} \qquad (z \text{ in } \mathfrak{W})$$

are the linear transformations forming the regular representation $\mathfrak{N}$ of $\mathfrak{W}$ as a field of linear transformations equivalent to $\mathfrak{W}$ under the correspondence $z \leftrightarrow Z_z$. But then Z_{z^S} is the transformation in $\mathfrak{N}$ corresponding to the result of applying S in $\mathfrak{N}$ to Z_z, that is, $Z_{z^S} = (Z_z)^S$. Hence

$$w^{Z_z U_S} = (wz)^S = w^S z^S = (w^{U_S})^{(Z_z)^S}, \tag{25}$$

that is, $ZU_S = U_S Z^S$ for every Z of $\mathfrak{N}$. Finally

$$w^{U_S U_T} = (w^S)^T = w^{ST} = w^{U_{ST}}. \tag{26}$$

It follows that the set $(U_{S_1}\mathfrak{N}, \cdots, U_{S_n}\mathfrak{N})$ forms an algebra of linear transformations which is a crossed product $(\mathfrak{N}, \iota)$ by Theorem 5.3. Its order is n^2 and it is a subalgebra of the total matric algebra $\mathfrak{M}_n$ of all linear transformations on $\mathfrak{W}$. Hence $(\mathfrak{N}, \iota) = \mathfrak{M}_n$ as desired.

7. Scalar extensions of crossed products. Let $\mathfrak{A} = (\mathfrak{N}, \mathfrak{g})$ of degree n and $\mathfrak{Y}$ be a subfield of degree m over $\mathfrak{F}$ of $\mathfrak{N}$. Then the group $\mathfrak{H}$ of all automorphisms

* The proof we give is very elegant. It is due to N. Jacobson, who also suggested the proof of Theorem 4.11.

of $\mathfrak{N}$ over $\mathfrak{F}$ which leave $\mathfrak{Y}$ unaltered is the group of $\mathfrak{N}$ over $\mathfrak{Y}$. If S is in $\mathfrak{H}$ then $yu_S = u_S y^S = u_S y$, so that u_S is in the $\mathfrak{A}$-commutator $\mathfrak{B}$ of $\mathfrak{Y}$. The degree of $\mathfrak{N}$ over $\mathfrak{Y}$ is q and this is the order of $\mathfrak{H} = (T_1, \cdots, T_q)$, $\mathfrak{B}$ contains $\mathfrak{N}$ and $u_{T_1}, \cdots, u_{T_q}$. It follows from Theorem 5.3 and Theorem 4.12 that $\mathfrak{B} = u_{T_1}\mathfrak{N} + \cdots + u_{T_q}\mathfrak{N} = (\mathfrak{N}, \bar{g})$ over $\mathfrak{Y}$, where

$$\bar{g} = \{g_{S,T}\} \qquad (S, T \text{ in } \mathfrak{H}). \tag{27}$$

If $\mathfrak{Y}_0$ is now a scalar extension of $\mathfrak{F}$ equivalent over $\mathfrak{F}$ to $\mathfrak{Y}$ we apply Theorem 4.16 and have $\mathfrak{A}_{\mathfrak{y}_0} = \mathfrak{M}_m \times \mathfrak{B}_0$, where $\mathfrak{B}_0$ over $\mathfrak{Y}_0$ is equivalent over $\mathfrak{F}$ to $\mathfrak{B}$ over $\mathfrak{Y}$.

We now let $\mathfrak{K}$ be any scalar extension of $\mathfrak{F}$, $\mathfrak{N}$ be normal over $\mathfrak{F}$, and let $\mathfrak{Y}_0$ be the maximal subfield over $\mathfrak{F}$ of $\mathfrak{K}$ equivalent to a subfield $\mathfrak{Y}$ over $\mathfrak{F}$ of $\mathfrak{N}$. By Theorem A8.12 the field $(\mathfrak{N}, \mathfrak{K})$ which is the composite over $\mathfrak{K}$ of $\mathfrak{N}$ and $\mathfrak{K}$ has the group $\mathfrak{H}_0$ equivalent to the group $\mathfrak{H}$ above, and is actually obtained by forming $\mathfrak{N}_0$ over $\mathfrak{Y}_0$ equivalent to $\mathfrak{N}$ over $\mathfrak{Y}$, and then forming the extension $\mathfrak{N}_0 \times \mathfrak{K} = (\mathfrak{N}_0)_{\mathfrak{K}}$. Since $\mathfrak{A}_{\mathfrak{K}} = (\mathfrak{A}_{\mathfrak{Y}_0})_{\mathfrak{K}}$ the argument above implies that $\mathfrak{A}_{\mathfrak{K}} = \mathfrak{M}_m \times (\mathfrak{N}_0, \bar{g})_{\mathfrak{K}}$ where $(\mathfrak{N}_0, \bar{g})$ is an algebra over $\mathfrak{Y}_0$. But the composite of $\mathfrak{N}$ and $\mathfrak{K}$ over $\mathfrak{Y}_0$ is their direct product $\bar{\mathfrak{N}} = (\mathfrak{N}_0)_{\mathfrak{K}}$ and is normal over $\mathfrak{K}$, $(\mathfrak{N}_0, \bar{g})_{\mathfrak{K}} = (\bar{\mathfrak{N}}, \bar{g})$. We state the result as

Theorem 8. *Let $\mathfrak{A} = (\mathfrak{N}, g)$ and $\mathfrak{K}$ be a scalar extension of $\mathfrak{F}$. Then if $\bar{g}$ is the subset of g corresponding to the automorphisms of the group $\mathfrak{H}$ of the composite $\bar{\mathfrak{N}}$ over $\mathfrak{K}$ of $\mathfrak{N}$ and $\mathfrak{K}$, we have*

$$\mathfrak{A}_{\mathfrak{K}} = \mathfrak{M}_m \times (\bar{\mathfrak{N}}, \bar{g}).$$

As a corollary of the above and Theorem 5.7 we see that a field $\mathfrak{K}$ is a splitting field of $\mathfrak{A} = (\mathfrak{N}, g)$ if and only if $\bar{g}$ is associated in $\bar{\mathfrak{N}}$ with the unit factor set. This result is sometimes useful in considering the structure of a given normal simple algebra $\mathfrak{A}$. Observe that in our results we do not assume that $\mathfrak{K}$ has finite degree over $\mathfrak{F}$. We shall actually study a case later where $\mathfrak{K}$ is a p-adic extension of an algebraic number field $\mathfrak{F}$.

8. Normalizations of crossed products. We have seen that a crossed product $\mathfrak{A} = (\mathfrak{N}, g) = (\mathfrak{N}, h)$ for every h associated with g. The operation of replacing any given factor set g by an associated factor set h is then a tool by means of which certain simple normalizations of $\mathfrak{A}$ may be accomplished.

In particular it is generally desirable to have $u_I = 1$, the unity quantity of $\mathfrak{A}$. This is equivalent to replacing $g = \{g_{S,T}\}$ by an associated $h = \{h_{S,T}\}$ such that $h_{I,S} = h_{S,I} = 1$ for every S of $\mathfrak{G}$. But in (9) we showed that $u = u_I g_{I,I^{-1}}$ is the unity quantity of $\mathfrak{A}$, and in the proof of Theorem 5.4 we showed that the replacement of u_I by u replaces g by an associated h. Hence we may always take u_I to be the unity quantity of $\mathfrak{A}$.

The normalization so accomplished is retained when we pass to an associated factor set if and only if $c_I = 1$ in (5). Moreover it is clear that if $g_{S,I} = g_{S,I} = 1$ and $h_{S,I} = h_{I,S} = 1$ for every S of $\mathfrak{G}$ then the product $g \cdot h = \{g_{S,T} h_{S,T}\}$ also has the property that u_I is the unity quantity of the corresponding $\mathfrak{A}$. Hence our complete theory of factor sets and crossed products could have been developed with the restriction that u_I be the unity quantity of $\mathfrak{A}$.

If $\mathfrak{G} = [S]$ is a cyclic group we have $y = u_S$, $zy = yz^S$ for every z of $\mathfrak{N}$. But then $zy^r = y^r z^{S^r}$ so that $y^r = u_{S^r} c_{S^r}$ for c_{S^r} in $\mathfrak{N}$, and there is no loss of generality if we replace every u_{S^r} by y^r. This simply replaces the factor set $\mathfrak{g}$ by an associated factor set. When we make this replacement we have

$$\mathfrak{A} = \mathfrak{N} + y\mathfrak{N} + \cdots + y^{n-1}\mathfrak{N},$$

with $zy^r = y^r z^{S^r}$ as above. Now $zy^n = y^n z$ for every z of $\mathfrak{Z}$ so that $y^n = \gamma \neq 0$ in $\mathfrak{N}$. But $y^n y = yy^n$, $\gamma = \gamma^S$ must be in $\mathfrak{F}$. We now make the

DEFINITION. *An algebra $\mathfrak{A}$ over $\mathfrak{F}$ is called a cyclic algebra if it is normal simple of degree n over $\mathfrak{F}$ and has a cyclic splitting field (or cyclic subfield) of degree n over $\mathfrak{F}$.*

Thus cyclic algebras are crossed products defined by cyclic fields and our crossed product theorems combined with the normalization above then yield

Theorem 9. *Let $\mathfrak{Z}$ be a cyclic field with automorphism group $\mathfrak{G} = [S]$. Then if $\mathfrak{A}$ is a cyclic algebra with $\mathfrak{Z}$ as maximal subfield we have*

$$\mathfrak{A} = \mathfrak{Z} + y\mathfrak{Z} + \cdots + y^{n-1}\mathfrak{Z}, \tag{28}$$

$$y^n = \gamma \neq 0 \text{ in } \mathfrak{F}, \qquad zy = yz^S \tag{29}$$

for every z of $\mathfrak{Z}$. Conversely, if $\mathfrak{Z}$ and S are as above and $\gamma \neq 0$ is in $\mathfrak{F}$, there exists a cyclic algebra, designated by

$$\mathfrak{A} = (\mathfrak{Z}, S, \gamma),$$

which satisfies (28), (29) *and is unique in the sense of equivalence.*

9. Elementary properties of cyclic algebras. We shall proceed to a derivation of those properties of cyclic algebras which we shall use in our exposition of the theory of exponents. Our results are corollaries of the theorems on crossed products and we thus first represent a cyclic algebra as a crossed product.

The quantities T, U of a cyclic group $[S]$ are the powers $T = S^j$, $U = S^k$ for $j, k = 0, 1, \cdots, n-1$. Then by Theorem 5.9 every cyclic algebra

$$\mathfrak{A} = (\mathfrak{Z}, S, \gamma) = (\mathfrak{Z}, \mathfrak{g}) \tag{30}$$

where $\mathfrak{g} = \{g_{T,U}\}$, $g_{T,U} = 1$ if $j + k < n$, $g_{T,U} = \gamma$ if $j + k \geqq n$. For, $y^j y^k = y^{j+k}$ or γy^{j+k-n} in the respective cases. If $\mathfrak{B} = (\mathfrak{Z}, \mathfrak{k}) = (\mathfrak{Z}, S, \delta)$ with $\mathfrak{k}$ also determined by Theorem 5.9 we see that the product $\mathfrak{g} \cdot \mathfrak{k} = \mathfrak{l} = \{k_{T,U}\}$ has $k_{T,U} = 1$ or $\gamma\delta$ in the respective cases. But then we apply Theorems 5.7, 5.6 to obtain the corresponding

Theorem 10. *A cyclic algebra $(\mathfrak{Z}, S, 1)$ is a total matric algebra.*

Theorem 11. *The direct product of two cyclic algebras defined by the same cyclic field has the form*

$$(31) \qquad (\mathfrak{Z}, S, \gamma) \times (\mathfrak{Z}, S, \delta) = (\mathfrak{Z}, S, 1) \times (\mathfrak{Z}, S, \gamma\delta).$$

Using an immediate induction we obtain

Theorem 12. *The direct power*

$$(32) \qquad (\mathfrak{Z}, S, \gamma)^r \sim (\mathfrak{Z}, S, \gamma^r).$$

The condition of Theorem 5.5 may be used to prove

Theorem 13. *Two cyclic algebras* $\mathfrak{A} = (\mathfrak{Z}, S, \gamma)$ *and* $\mathfrak{B} = (\mathfrak{Z}, S, \delta)$ *are equivalent if and only if*

$$(33) \qquad \delta = \gamma \cdot N_{\mathfrak{Z}|\mathfrak{F}}(z),$$

where $N_{\mathfrak{Z}|\mathfrak{F}}(z) = zz^S \cdots z^{S^{n-1}}$ *is the norm of a quantity* z *of* $\mathfrak{Z}$.

For we have seen that $\mathfrak{A} \cong \mathfrak{B}$ only if there exists a quantity $z = c_S \neq 0$ in $\mathfrak{Z}$ such that $u_S c_S = yz = y_0$ has the property $y_0^n = \delta$. Here $y^n = \gamma$, $y_0^n = (yz)^n = \gamma N_{\mathfrak{Z}|\mathfrak{F}}(z) = \delta$. Conversely, if $\delta = \gamma N_{\mathfrak{Z}|\mathfrak{F}}(z)$ then we replace y of (29) by $y_0 = yz$ and have $\mathfrak{A} = (\mathfrak{Z}, S, \delta) \cong \mathfrak{B}$.

We combine this result with Theorem 5.10 and have

Theorem 14. *A cyclic algebra* $(\mathfrak{Z}, S, \gamma) \sim 1$ *if and only if* γ *is the norm* $N_{\mathfrak{Z}|\mathfrak{F}}(z)$ *of a quantity of* $\mathfrak{Z}$.

We put $n = r$ in Theorem 5.12 and use $\gamma^n = N_{\mathfrak{Z}|\mathfrak{F}}(\gamma)$ together with Theorem 5.14 to obtain

Lemma 4. *The direct nth power of a cyclic algebra of degree n is* $\mathfrak{M} \sim 1$.

We also have

Lemma 5. *Let* $\mathfrak{D}$ *be a cyclic division algebra of prime degree p. Then* $\mathfrak{D}^t \sim 1$ *if and only if p divides t.*

For $\mathfrak{D} = (\mathfrak{Z}, S, \gamma)$, $\mathfrak{D}^t \sim (\mathfrak{Z}, S, \gamma^t) \sim (\mathfrak{Z}, S, 1)$ if and only if $\gamma^t = N_{\mathfrak{Z}|\mathfrak{F}}(z)$. If p does not divide t it is prime to t, $ap + bt = 1$, $\gamma = (\gamma^a)^p[N_{\mathfrak{Z}|\mathfrak{F}}(z)]^b = N_{\mathfrak{Z}|\mathfrak{F}}(z^b\gamma^a)$ contrary to our hypothesis that $\mathfrak{D}$ is a division algebra. The converse is trivial.

10. The exponent of a normal simple algebra. If $\mathfrak{D}$ and $\mathfrak{D}_0$ are equivalent normal division algebras the direct product $\mathfrak{D}_0 \times \mathfrak{D}$ contains $\mathfrak{D}_0 \times \mathfrak{Z} = (\mathfrak{D}_0)_{\mathfrak{Z}}$ for every maximal subfield $\mathfrak{Z}$ of $\mathfrak{D}$. Such fields exist by Theorem 4.18, and by Theorem 4.27 $(\mathfrak{D}_0)_{\mathfrak{Z}} \sim 1$, $(\mathfrak{D}_0)_{\mathfrak{Z}} = \mathfrak{M} \times \mathfrak{Z}$ where $\mathfrak{M}$ over $\mathfrak{F}$ has the same degree as $\mathfrak{D}$ and $\mathfrak{M} \sim 1$. By Theorem 4.6

$$\mathfrak{D}_0 \times \mathfrak{D} = \mathfrak{M} \times \mathfrak{B}$$

where $\mathfrak{B}$ is normal simple and of the same degree as $\mathfrak{D}$, $\mathfrak{B}$ contains $\mathfrak{Z}$. By an evident induction the direct power $\mathfrak{D}$ has the property

$$(34) \qquad \mathfrak{D}^t = \mathfrak{M}^{t-1} \times \mathfrak{B}_t,$$

where $\mathfrak{B}_t$ is normal simple of the same degree as $\mathfrak{D}$ and contains a field $\mathfrak{Z}_t$

equivalent over $\mathfrak{F}$ to $\mathfrak{Z}$. Clearly if $\mathfrak{A} \sim \mathfrak{D}$ then $\mathfrak{A}^t = (\mathfrak{M}^{(0)})^{t-1} \times \mathfrak{A}_t$ where $\mathfrak{M}^{(0)} \sim 1$ and $\mathfrak{A}_t$ has the same degree as $\mathfrak{A}$. Hence we have the first part of

Theorem 15. *The index of the direct power $\mathfrak{A}^t$ of a normal simple algebra $\mathfrak{A}$ divides the index of $\mathfrak{A}$. Moreover every splitting field of $\mathfrak{A}$ splits $\mathfrak{A}^t$.*

The final result above is an immediate consequence of the property

$$(35) \qquad (\mathfrak{A}^t)_{\mathfrak{K}} = (\mathfrak{A}_{\mathfrak{K}})^t$$

for every scalar extension $\mathfrak{K}$ of $\mathfrak{F}$ so that if $\mathfrak{A}_{\mathfrak{K}} \sim 1$, $(\mathfrak{A}^t)_{\mathfrak{K}} \sim 1$. We shall use (35) several times in this section. Apply it to Lemma 5.5 to obtain

LEMMA 6. *The tth direct power of a normal division algebra $\mathfrak{D}$ of prime degree p over $\mathfrak{F}$ is ~ 1 if and only if p divides t.*

For by Theorem 4.31 there exists a scalar extension $\mathfrak{K}$ of degree r prime to p such that $\mathfrak{D}_{\mathfrak{K}}$ is a cyclic division algebra. By Lemma 5.5 the index of $(\mathfrak{D}_{\mathfrak{K}})^t$ is one if and only if p divides t. However the index of $\mathfrak{D}^t$ divides p by Theorem 5.15 and $\mathfrak{K}$ has degree prime to p, so that $(\mathfrak{D}^t)_{\mathfrak{K}} = (\mathfrak{D}_{\mathfrak{K}})^t$ has the same index as $\mathfrak{D}^t$ by the corollary of the Theorem 4.20. This proves the lemma.

We use this result to derive

LEMMA 7. *Let $\mathfrak{A}$ be a normal simple algebra of index m and p be a prime factor of m. Then the index of $\mathfrak{A}^p$ divides mp^{-1}.*

Let $m = p^e q$, where p does not divide q, and $\mathfrak{L}$ be the field of Theorems 4.30, 4.32 such that $\mathfrak{A}_{\mathfrak{L}}$ has index p, $\mathfrak{L}$ has degree $p^{e-1}r$ over $\mathfrak{F}$ with r prime to p. Then $(\mathfrak{A}^p)_{\mathfrak{L}} = (\mathfrak{A}_{\mathfrak{L}})^p \sim 1$, and the index of $\mathfrak{A}^p$ divides the degree of $\mathfrak{L}$. By Theorem 5.15 it divides the integer $p^e q$, as well as $p^{e-1}r$, and hence divides $p^{e-1}q$.

An immediate induction based on this lemma yields

Theorem 16. *Let $\mathfrak{A}$ be a normal simple algebra of index $m = rs$. Then $\mathfrak{A}^r$ has index a divisor of s and, in particular, $\mathfrak{A}^m \sim 1$. Furthermore $\mathfrak{A}^{rt}$ has index a divisor of s for every integer t.*

The last statement follows from Theorem 5.15. We now prove

Theorem 17. *The order of a class $(\mathfrak{A})$ of normal simple algebras as an element of the class group is a finite integer ρ called the* **exponent** *of $\mathfrak{A}$. It divides the index m of $\mathfrak{A}$ and is divisible by every prime factor p of m.*

For clearly ρ is the least integer such that $\mathfrak{A}^\rho \sim 1$. Hence $1 \leqq \rho \leqq m$. Write $m = \rho a + b$ with $0 \leqq b < \rho$, so that if $b = 0$ we have the result desired. If $b > 0$ the power $\mathfrak{A}^m = (\mathfrak{A}^\rho)^a \times \mathfrak{A}^b \sim \mathfrak{A}^b \sim 1$ contrary to our definition of ρ. Hence ρ divides m. If p is a prime factor of m we define $\mathfrak{L}$ as in Theorem 4.30 and form $(\mathfrak{A}_{\mathfrak{L}})^\rho = (\mathfrak{A}^\rho)_{\mathfrak{L}} \sim 1$. But $\mathfrak{A}_{\mathfrak{L}}$ has index p and by Lemma 5.6 p must divide ρ.

The finiteness of ρ implies that the group generated by each class is a finite cyclic group and that $(\mathfrak{A}^{\rho-1})$ is the inverse of $(\mathfrak{A})$. If $\mathfrak{A} \sim \mathfrak{D}$ and $\mathfrak{A}^{\rho-1} \sim \mathfrak{B}$

for normal division algebras $\mathfrak{B}$ and $\mathfrak{D}$ we have $\mathfrak{D} \times \mathfrak{B} \sim \mathfrak{A}^\rho \sim 1$. By Theorem 4.8, $\mathfrak{B}$ is reciprocal to $\mathfrak{D}$. Hence $(\mathfrak{A}^{\rho-1}) = (\mathfrak{A}^{-1})$. Thus the notation $\mathfrak{A}^{-1}$ is given a new justification.

We may finally use our results to derive the following very important structure theorem.

Theorem 18. *Let $\mathfrak{D}$ be a normal division algebra of degree n so that we may write $n = p_1^{e_1} \cdots p_t^{e_t}$ for the p_i distinct primes. Then*

$$\mathfrak{D} = \mathfrak{D}_1 \times \cdots \times \mathfrak{D}_t, \tag{36}$$

for normal division algebras $\mathfrak{D}_i$ of degree $p_i^{e_i}$ uniquely determined in the sense of equivalence.

Note that we showed in Theorem 4.10 that conversely every direct product (36) of normal division algebras $\mathfrak{D}_i$ of distinct prime-power degrees $p_i^{e_i}$ is a normal division algebra.

For proof let ρ be the exponent of $\mathfrak{D}$ so that $\rho = \rho_1 \cdots \rho_t$ with $\rho_i = p_i^{f_i}$, $0 < f_i \leqq e_i$ by Theorem 5.17. By Theorem A6.5

$$(\mathfrak{D}) = (\mathfrak{D}_1) \cdots (\mathfrak{D}_t) \tag{37}$$

where $(\mathfrak{D}_i)$ generates a cyclic group of order ρ_i, $\mathfrak{D}_i$ has exponent ρ_i. But then by Theorem 5.17, $\mathfrak{D}_i$ may be taken to be a normal division algebra of degree $p_i^{g_i}$. Now by (37) the class $(\mathfrak{D}) = (\mathfrak{D}_1 \times \cdots \times \mathfrak{D}_t)$ and we apply Theorem 4.10 to see that $\mathfrak{D}_1 \times \cdots \times \mathfrak{D}_t$ is a normal division algebra. Since the division algebra of any class $(\mathfrak{D})$ is unique in the sense of equivalence we have (36). We now assume that also $\mathfrak{D} = \mathfrak{D}_{01} \times \cdots \times \mathfrak{D}_{0t}$ so that $(\mathfrak{D}) = (\mathfrak{D}_{01})(\mathfrak{D}_{02}) \cdots (\mathfrak{D}_{0t})$. By the finite group Theorem A6.5 we have $(\mathfrak{D}_{0i}) = (\mathfrak{D}_i)$ and, by the uniqueness property, $\mathfrak{D}_{0i} \cong \mathfrak{D}_i$.

Our result evidently implies also

Theorem 19. *The exponent of $\mathfrak{D}$ of* (36) *is the product of the exponents of factors $\mathfrak{D}_i$ of distinct prime-power degrees.*

A normal simple algebra $\mathfrak{A}$ is called *primary* if it does not have the form $\mathfrak{A} = \mathfrak{B} \times \mathfrak{C}$. We may use Theorem 5.18 to obtain

Theorem 20. *The degree of a primary normal simple algebra is a power of a prime.*

For if $\mathfrak{A} = \mathfrak{M} \times \mathfrak{D}$ is primary we must either have $\mathfrak{A} = \mathfrak{D}$ or $\mathfrak{A} = \mathfrak{M}$. If $\mathfrak{A} = \mathfrak{M}$ has composite degree $n = pq$ we have $\mathfrak{A} = \mathfrak{M}_p \times \mathfrak{M}_q$, a contradiction. Otherwise $\mathfrak{A} = \mathfrak{D}$ is primary by Theorem 5.18 only if $n = p^e$, p a prime.

In closing we observe that then a primary normal simple algebra is either a p-rowed total matric algebra or else a normal division algebra of degree p^e for a prime p. The converse of Theorem 5.20 is not necessarily true as was shown in [29].

CHAPTER VI

CYCLIC SEMI-FIELDS

1. Groups of automorphisms of algebras. Probably the most interesting special type of normal simple algebra is the cyclic algebra. This is due partly to the fact that such algebras are the earliest known normal division algebras, and partly to the result, obtained in Chapter IX, that every normal simple algebra over an algebraic number field is a cyclic algebra. Some of the elementary properties of such algebras are closely connected with and were obtained as corollaries of more general theorems on crossed products. Others will be related to and expressed in terms of a theory of certain commutative algebras which we shall call cyclic semi-fields and shall study in the present chapter.

The initial stage in our discussion will be a consideration of certain general properties of algebras $\mathfrak{Z}$ with a unity quantity and a group $\mathfrak{G}$ of automorphisms of $\mathfrak{Z}$ over $\mathfrak{F}$. We have already seen that every S of $\mathfrak{G}$ carries any one of its subalgebras $\mathfrak{B}$ into an equivalent algebra $\mathfrak{B}^S$ and that S induces an automorphism $b \leftrightarrow b^S$ in $\mathfrak{B}$ if $\mathfrak{B} = \mathfrak{B}^S$. We now make the

Definition. *Let $\mathfrak{G}$ be a group of automorphisms over $\mathfrak{F}$ of an algebra $\mathfrak{Z}$ over $\mathfrak{F}$. If $\mathfrak{Y}$ is any subalgebra of $\mathfrak{Z}$ such that $\mathfrak{Y} = \mathfrak{Y}^S$ for every S of $\mathfrak{G}$ then $\mathfrak{Y}$ is called $\mathfrak{G}$-reducible if $\mathfrak{Y} = \mathfrak{B} \oplus \mathfrak{C}$ with $\mathfrak{B} = \mathfrak{B}^S \neq 0$, $\mathfrak{C} = \mathfrak{C}^S \neq 0$, for every S of $\mathfrak{G}$. Otherwise $\mathfrak{Y}$ is called $\mathfrak{G}$-irreducible.*

We shall use the terminology *S-reducible* and *S-irreducible* in the case where $\mathfrak{G} = [S]$ is the cyclic group generated by S. Observe that the concept of irreducibility of Section 2.12 coincides with that of I-irreducibility, I the identity automorphism of $\mathfrak{Z}$. We now prove

Theorem 1. *An algebra $\mathfrak{Z}$ with a unity quantity is $\mathfrak{G}$-irreducible if and only if there exists an irreducible component $\mathfrak{Y}_1$ of $\mathfrak{Z}$ and distinct automorphisms $S_1 = I, S_2, \cdots, S_t$ of $\mathfrak{G}$ such that*

$$(1) \qquad \mathfrak{Z} = \mathfrak{Y}_1^{S_1} \oplus \mathfrak{Y}_1^{S_2} \oplus \cdots \oplus \mathfrak{Y}_1^{S_t}.$$

Every $\mathfrak{G}$-reducible algebra is expressible uniquely, apart from the order of its components, as a direct sum of $\mathfrak{G}$-irreducible algebras.

For by Theorem 2.22 every $\mathfrak{Z}$ is uniquely expressible as a direct sum $\mathfrak{Z} = \mathfrak{Y}_1 \oplus \cdots \oplus \mathfrak{Y}_q$ of irreducible algebras $\mathfrak{Y}_i$. Then also if S is any automorphism of $\mathfrak{G}$ we have $\mathfrak{Z} = \mathfrak{Z}^S = \mathfrak{Y}_1^S \oplus \cdots \oplus \mathfrak{Y}_q^S$ with each $\mathfrak{Y}_i^S$ equivalent to $\mathfrak{Y}_i$ and hence irreducible. It follows that $\mathfrak{Z}^S = \mathfrak{Z}$ if and only if $\mathfrak{Y}_i^S$ is a $\mathfrak{Y}_j$ for every i and S of $\mathfrak{G}$. If an algebra $\mathfrak{Z}$ of the form (1) were $\mathfrak{G}$-reducible we would have $\mathfrak{Z} = \mathfrak{B} \oplus \mathfrak{C}$ with $\mathfrak{Y}_1$ a component of $\mathfrak{B}$, $\mathfrak{B} = \mathfrak{B}^{S_i} \geqq \mathfrak{Y}^{S_i}$ for every i, $\mathfrak{B} \geqq \mathfrak{Z}$, a contradiction. Hence every $\mathfrak{Z}$ of the form (1) is $\mathfrak{G}$-irreducible. Conversely, let $\mathfrak{Z}$ be $\mathfrak{G}$-irreducible. Then $\mathfrak{Y}_1, \mathfrak{Y}_1^{S_2}, \cdots, \mathfrak{Y}_1^{S_t}$ are distinct for a finite set of S_k of $\mathfrak{G}$ and $\mathfrak{Y}_1^S$ is one of the $\mathfrak{Y}_1^{S_k}$ for every S of $\mathfrak{G}$. Clearly $\mathfrak{B} = \mathfrak{Y}_1 \oplus \mathfrak{Y}_1^{S_2} \oplus \cdots \oplus \mathfrak{Y}_1^{S_t}$ is $\mathfrak{G}$-irreducible, $\mathfrak{Z} = \mathfrak{B} \oplus \mathfrak{C}$. If $\mathfrak{Y}_i$ is a component of $\mathfrak{C}$ then $\mathfrak{Y}_i^S = \mathfrak{Y}_1^{S_k}$ implies that $\mathfrak{Y}_i = \mathfrak{Y}_1^{S_k S^{-1}}$ which is impossible. Hence $\mathfrak{C} = \mathfrak{C}^S$ which is contrary to our hypothesis unless $\mathfrak{C} = 0$, $\mathfrak{Z} = \mathfrak{B}$ has the form (1) as desired. We have evidently shown also that if $\mathfrak{B}_1$ and $\mathfrak{B}_2$ are $\mathfrak{G}$-irreducible components of $\mathfrak{Z}$ they are either equal or have no irreducible component in common. This yields the final conclusion in our theorem.

If $\mathfrak{G} = [S]$ the automorphisms S_i of (1) are all powers of S. But $\mathfrak{Y}_1^{S^i} = \mathfrak{Y}_1^{S^j}$ for $i \geqq j$ if and only if $\mathfrak{Y}_1 = \mathfrak{Y}_1^{S^{i-j}}$. Hence the integer t of (1) is the least t for which $\mathfrak{Y}_1^{S^t} = \mathfrak{Y}_1$ and we have

Theorem 2. *An algebra $\mathfrak{Z}$ with a unity quantity is S-irreducible if and only if*

$$\mathfrak{Z} = \mathfrak{Y}_1 \oplus \mathfrak{Y}_1^S \oplus \cdots \oplus \mathfrak{Y}_1^{S^{t-1}}, \tag{2}$$

where $\mathfrak{Y}_1$ is an irreducible algebra such that $\mathfrak{Y}_1^{S^t} = \mathfrak{Y}_1$.

2. Notational hypotheses. Diagonal algebras will occur with extreme frequency in our study and we shall save numerous repetitions by the adoption of certain conventions about notation. We shall use the letters $\mathfrak{E}$, $\mathfrak{E}_0$, $\mathfrak{E}_1$ repeatedly for diagonal algebras and shall frequently indicate this fact by writing

$$\mathfrak{E} \dot{\sim} 1.$$

Whenever $\mathfrak{E} \dot{\sim} 1$ it has a basis $e_1, \cdots, e_t$ consisting of pairwise orthogonal idempotent quantities. We shall not use any other type of basis and shall therefore assume that whenever we write $\mathfrak{E} = (e_1, \cdots, e_t)$ for $\mathfrak{E}$ diagonal the basal quantities indicated are pairwise orthogonal idempotents.

Every diagonal algebra $\mathfrak{E} = (e_1, \cdots, e_t)$ has an automorphism S of order t over $\mathfrak{F}$ given by

$$\alpha_1 e_1 + \cdots + \alpha_t e_t \leftrightarrow \alpha_1 e_2 + \cdots + \alpha_{t-1} e_t + \alpha_t e_1 \qquad (\alpha_i \text{ in } \mathfrak{F}). \tag{3}$$

Clearly $\mathfrak{E}$ is the S-irreducible algebra (2) with $\mathfrak{Y}_1 = (e_1)$, and hence $\mathfrak{E} = (e_1, e_1^S, \cdots, e_1^{S^{t-1}})$. We shall also use this notation frequently and shall always mean that the $e_1^{S^{i-1}}$ are distinct pairwise orthogonal idempotents for $i = 1, \cdots, t$ such that $e_1^{S^t} = e_1$. We make a similar hypothesis about $\mathfrak{E} = (e_1, e_1^{S_2}, \cdots, e_1^{S_t})$ for the S_i in a group $\mathfrak{G}$.

We shall consider direct products $\mathfrak{E} \times \mathfrak{Y}$ of a diagonal algebra $\mathfrak{E}$ and an algebra $\mathfrak{Y}$. Observe then that if $I = T_1, \cdots, T_t$ are automorphisms of $\mathfrak{Y}$ and y is in $\mathfrak{Y}$ the quantity $y_0 = ye_1 + y^{T_2}e_2 + \cdots + y^{T_t}e_t$ may be regarded as a diagonal matrix with elements in $\mathfrak{Y}$, and the usual rules of combination of diagonal matrices indicate clearly that the algebra $\mathfrak{Y}_0$ of all such quantities is equivalent to $\mathfrak{Y}$ under the correspondence $y_0 \leftrightarrow y$. It is also equivalent to $\mathfrak{Y}e_1$. If $\mathfrak{Y}$ is any irreducible algebra then

(4) $$\mathfrak{Z} = \mathfrak{E} \times \mathfrak{Y} = e_1\mathfrak{Y} \oplus \cdots \oplus e_t\mathfrak{Y},$$

and the algebras $e_i\mathfrak{Y}$ are the irreducible components of $\mathfrak{Z}$. Note that then $\mathfrak{Z}$ is S-irreducible with respect to the automorphism (3) where the α_i are now in $\mathfrak{Y}$. In this connection we shall prove

Theorem 3. *Let $\mathfrak{G}$ be a group of automorphisms of an algebra $\mathfrak{Z}$ with a unity quantity. Then $\mathfrak{Z}$ is $\mathfrak{G}$-irreducible if and only if $\mathfrak{Z} = \mathfrak{E} \times \mathfrak{Y}$ where $\mathfrak{Y}$ is irreducible and $\mathfrak{E} \dot{\sim} 1$ is $\mathfrak{G}$-irreducible. The algebra $\mathfrak{E}$ is uniquely determined and $\mathfrak{Y}$ is unique in the sense of equivalence.*

For $\mathfrak{Z} = \mathfrak{Y}_1 \oplus \mathfrak{Y}_1^{S_2} \oplus \cdots \oplus \mathfrak{Y}_1^{S_t}$ and if e_1 is the unity quantity of $\mathfrak{Y}_1$ that of $\mathfrak{Y}_1^{S_i}$ is $e_1^{S_i} = e_i$. Hence $\mathfrak{E} = (e_1, \cdots, e_t) \dot{\sim} 1$ and has the same unity quantity $1 = e_1 + \cdots + e_t$ as $\mathfrak{Z}$. Evidently $\mathfrak{E} = (e_1, e_1^{S_2}, \cdots, e_1^{S_t})$ is $\mathfrak{G}$-irreducible. If y_1 ranges over all quantities of $\mathfrak{Y}_1$ the set $\mathfrak{Y}$ of all quantities $y = y_1 + y_1^{S_2} + \cdots + y_1^{S_t}$ is an algebra equivalent to $\mathfrak{Y}_1$ and is irreducible. A simple argument shows that every quantity of $\mathfrak{Z}$ is uniquely expressible as a quantity of $\mathfrak{E}\mathfrak{Y}$ so that $\mathfrak{Z} = \mathfrak{E} \times \mathfrak{Y}$. Conversely, if $\mathfrak{Z} = \mathfrak{E} \times \mathfrak{Y}$, where $\mathfrak{Y}$ is irreducible and $\mathfrak{E} = (e_1, e_1^{S_2}, \cdots, e_1^{S_t}) \dot{\sim} 1$, then $\mathfrak{Z} = \mathfrak{Y}_1 \oplus \cdots \oplus \mathfrak{Y}_t$ with $\mathfrak{Y}_i = \mathfrak{Y}e_1^{S_i}$ irreducible and equivalent to $\mathfrak{Y}$. But $\mathfrak{Z} = \mathfrak{Z}^{S_i}$ and hence $\mathfrak{Y}_1^{S_i} = \mathfrak{Y}_j$ for some j. The unity quantity of $\mathfrak{Y}_1^{S_i}$ is $e_1^{S_i}$ in $\mathfrak{Y}_i$ and hence $\mathfrak{Y}_i = \mathfrak{Y}_1^{S_i}$, $\mathfrak{Z} = \mathfrak{Y}_1 \oplus \mathfrak{Y}_1^{S_2} \oplus \cdots \oplus \mathfrak{Y}_1^{S_t}$ as desired.

If also $\mathfrak{Z} = \mathfrak{E}_0 \times \mathfrak{Y}_0$ where $\mathfrak{E}_0 \dot{\sim} 1$ and $\mathfrak{Y}_0$ is irreducible, we have $\mathfrak{E}_0 = (f_1, \cdots, f_s)$, the idempotents f_j are the unity quantities of the irreducible components $\mathfrak{Y}_0 f_j$ of $\mathfrak{Z}$. They then must coincide with the e_i in some order, $\mathfrak{E}_0 = \mathfrak{E}$. Finally $\mathfrak{Y} \cong e_i\mathfrak{Y} \cong f_j\mathfrak{Y}_0 \cong \mathfrak{Y}_0$ as desired.*

We shall henceforth restrict our attention to the case where the group $\mathfrak{G}$ is the cyclic group generated by a single automorphism S over $\mathfrak{F}$ of $\mathfrak{Z}$. Our results imply that if $\mathfrak{Z}$ is S-irreducible then

(5) $$\mathfrak{Z} = \mathfrak{E} \times \mathfrak{Y}, \qquad \mathfrak{E} = (e_1, \cdots, e_t) = \mathfrak{E}^S, e_i = e_1 s^{i-1},$$

and S^t generates an automorphism T over $\mathfrak{F}$ of $\mathfrak{Z}$. Then $\mathfrak{Y}$ has the same unity quantity e as both $\mathfrak{Z}$ and $\mathfrak{E}$. If also $\mathfrak{Z} = \mathfrak{E} \times \mathfrak{Y}_0$, where $\mathfrak{Y}_0$ has the same unity quantity as $\mathfrak{Z}$, then every element y_0 of $\mathfrak{Y}_0$ is uniquely expressible in the form $y_0 = y_1e_1 + y_2e_2 + \cdots + y_te_t$, where the coefficients y_i are in $\mathfrak{Y}$. We have already shown that $\mathfrak{Y}e_i = \mathfrak{Y}_0e_i$ is the set of all y_ie_i for y_i in $\mathfrak{Y}$, and so the mapping $y_1 \to y_i$ is an automorphism S_i over $\mathfrak{F}$ of $\mathfrak{Y}$. Thus we see that the expression of $\mathfrak{Z}$ in the direct product form $\mathfrak{Z} = \mathfrak{E} \times \mathfrak{Y}_0$ occurs only if there are automorphisms $S_1, S_2, \cdots, S_t$ over $\mathfrak{F}$ of $\mathfrak{Y}$ (with S_1 the identity), such that an element y_0 of $\mathfrak{Z}$ is in $\mathfrak{Y}_0$ if and only if it is expressible in the form $y_0 = ye_1 + y^{S_2}e_2 + \cdots + y^{S_t}e_t$, for y in $\mathfrak{Y}$. Conversely, if $S_1 = I, S_2, \cdots, S_t$ are any automorphisms over $\mathfrak{F}$ of $\mathfrak{Y}$, and we define $\mathfrak{Y}_0$ to be the set of all $y_0 = y^{S_1}e_1 + y^{S_2}e_2 + \cdots + y^{S_t}e_t$ for y in $\mathfrak{Y}$, it should be evident that $\mathfrak{Z} = \mathfrak{E} \times \mathfrak{Y}_0$.

* However we need not have $\mathfrak{Y} = \mathfrak{Y}_0$. For example, let $t = 2$, $\mathfrak{E} = (e_1, e_2) \dot{\sim} 1$, $\mathfrak{Z} = \mathfrak{E} \times \mathfrak{Y}$ where $\mathfrak{Y} = (1, i)$ is a separable quadratic field over $\mathfrak{F}$. Then $\mathfrak{Y}$ has a non-identical automorphism T and $\mathfrak{Z} = \mathfrak{E} \times \mathfrak{Y}_0$ with $\mathfrak{Y}_0 = (1, i_0) \neq \mathfrak{Y}$ for $i_0 = e_1 i + e_2 i^T$.

We now form $\mathfrak{Y}_{00} = \mathfrak{Y}_0^{S^t}$. It consists of all $y_{00} = y_0^{S^t} = y^T e_1 + y^{T_2} e_2 + \cdots + y^{T_t} e_t$, where $T_i = S_i T$. The mapping S^t then induces an *isomorphism* of $\mathfrak{Y}_0$ onto $\mathfrak{Y}_{00}$. It is an automorphism if and only if $\mathfrak{Y}_{00} = \mathfrak{Y}_0$, and this clearly occurs if and only if $S_i T = T S_i$ for $i = 1, \cdots, t$. We see then that S^t induces an automorphism T_0 over $\mathfrak{F}$ of *all possible* $\mathfrak{Y}_0$ if and only if the automorphism T is in the center of the automorphism group of $\mathfrak{Y}$ over $\mathfrak{F}$.

If μ is the order of T then that of S is clearly μt. We may, in fact, prove the following result.

Theorem 4. *Let $\mathfrak{Z} = \mathfrak{E} \times \mathfrak{Y}$ where $\mathfrak{E} \sim 1$ has order t and $\mathfrak{Y}$ is an irreducible algebra with an automorphism T of order μ. Then there exists an automorphism S of order μt of $\mathfrak{Z}$ such that S^t induces T in $\mathfrak{Y}$, and both $\mathfrak{Z}$ and $\mathfrak{E}$ are S-irreducible.*

For $\mathfrak{E} = (e_1, \cdots, e_t)$ and every quantity of $\mathfrak{Z}$ is uniquely expressible in the form

$$z = y_1 e_1 + \cdots + y_t e_t \qquad (y_i \text{ in } \mathfrak{Y}). \tag{6}$$

Let T be an automorphism of order μ in $\mathfrak{Y}$ and define

$$z^S = y_1 e_2 + \cdots + y_{t-1} e_t + y_t^T e_1. \tag{7}$$

An elementary computation shows that $z \leftrightarrow z^S$ is the desired automorphism. The construction of $\mathfrak{Y}$ in the proof of Theorem 6.3 now implies the following result.

Corollary. *If $\mathfrak{Z}$ is S-irreducible, the algebra $\mathfrak{Y}$ of Theorem* 6.3 *may be chosen so that the automorphism S is given by* (6) *and* (7), *where $\mathfrak{E}$ has order t and T is the automorphism induced in $\mathfrak{Y}$ by S^t.*

3. Semi-fields. A commutative semi-simple algebra $\mathfrak{Z}$ is a direct sum of a finite number t of simple commutative algebras $\mathfrak{Y}_i$. These algebras are then fields over $\mathfrak{F}$. We shall call $\mathfrak{Z}$ a *semi-field* if its components $\mathfrak{Y}_i$ are all equivalent over $\mathfrak{F}$.

Let $y_1 \leftrightarrow (y_1)^{S_i}$ be the defining equivalences of $\mathfrak{Y}_1$ and $\mathfrak{Y}_i$. Then the induced equivalences $y_1^{S_i} \leftrightarrow y_1^{S_{i+1}}$ of $\mathfrak{Y}_i$ and $\mathfrak{Y}_{i+1}$ define an automorphism S of $\mathfrak{Z}$ carrying each $y_1 + \cdots + y_t$ of $\mathfrak{Z}$ into $y_1^{S_2} + y_2^{S_2^{-1}S_3} + \cdots + y_t^{S_t^{-1}S_1}$. Evidently $\mathfrak{Y}_i = \mathfrak{Y}_1^{S^{i-1}}$, $\mathfrak{Y}_1^{S^t} = \mathfrak{Y}_1$, S^t induces the identity automorphism in $\mathfrak{Y}_1$. Then $\mathfrak{Z}$ is S-irreducible and by Theorem 6.3 we have

Theorem 5. *Every semi-field is expressible as the direct product $\mathfrak{Z} = \mathfrak{E} \times \mathfrak{Y}$ of a field $\mathfrak{Y}$ and a uniquely determined diagonal algebra $\mathfrak{E}$. The field $\mathfrak{Y}$ is unique in the sense of equivalence.*

We may thus regard any semi-field $\mathfrak{Z}$ as the set of all t-rowed diagonal matrices with elements in a field $\mathfrak{Y}$. It is clear that the order of $\mathfrak{E}$ is the number of irreducible components of $\mathfrak{Z}$, and that these components are all equivalent to $\mathfrak{Y}$. The corollary of Theorem 6.3 clearly may be applied to the present case so that if $\mathfrak{Z}$ is S-irreducible, $\mathfrak{Y}$ may be chosen so that S is given by (6), (7). This choice will always be made in what follows.

If $\mathfrak{E} = (e_1, \cdots, e_t) \dot{\sim} 1$ and $\mathfrak{E}_0 = (f_1, \cdots, f_r) \dot{\sim} 1$ the algebra $\mathfrak{E} \times \mathfrak{E}_0 = (e_1f_1, \cdots, e_tf_r) \dot{\sim} 1$. This result together with $\mathfrak{E}_0 \times (\mathfrak{E} \times \mathfrak{Y}) = (\mathfrak{E}_0 \times \mathfrak{E}) \times \mathfrak{Y}$ implies

Theorem 6. *The direct product of a diagonal algebra and a semi-field $\mathfrak{Z}$ is a semi-field which is a diagonal algebra over $\mathfrak{F}$ if and only if $\mathfrak{Z}$ is a diagonal algebra over $\mathfrak{F}$.*

In a discussion made later we shall be interested in subalgebras of direct products of semi-fields. We shall then require a result obtainable as a consequence of the

LEMMA. *Every subalgebra $\mathfrak{D}$ of a diagonal algebra $\mathfrak{E}$ is a diagonal algebra.*

For proof observe that $\mathfrak{E}$ contains no nilpotent quantities and thus every $\mathfrak{D}$ is a commutative semi-simple algebra. Hence $\mathfrak{D} = \mathfrak{D}_1 \oplus \cdots \oplus \mathfrak{D}_u$ where $\mathfrak{D}_i$ is a field over $\mathfrak{F}$ with unity quantity f_i. If y is in $\mathfrak{D}_i$ then $y = a_1e_1 + \cdots + a_te_t$ with the a_i in $\mathfrak{F}$, $\phi(y) = (y - a_1) \cdots (y - a_t) = 0$. The minimum function $g(\lambda)$ of y over $\mathfrak{F}$ is irreducible in $\mathfrak{F}$ since $\mathfrak{D}_i$ is a field and divides $\phi(\lambda)$, $g(\lambda)$ is linear, y is in (f_i) over $\mathfrak{F}$, $\mathfrak{D}_i = (f_i)$ over $\mathfrak{F}$, $\mathfrak{D}$ is diagonal.

We now obtain

Theorem 7. *Let $\mathfrak{Z}$ and $\mathfrak{C}$ be separable semi-fields. Then any subalgebra $\mathfrak{B}$ of $\mathfrak{Z} \times \mathfrak{C}$ is a commutative separable algebra.*

For by Theorems 3.21 and 3.20 there exists a field $\mathfrak{K}$ such that $\mathfrak{Z}_{\mathfrak{K}} \dot{\sim} 1$, $\mathfrak{C}_{\mathfrak{K}} \dot{\sim} 1$, and by Theorem 6.6, we have $(\mathfrak{Z} \times \mathfrak{C})_{\mathfrak{K}} \sim 1$. Hence by our lemma the subalgebra $\mathfrak{B}_{\mathfrak{K}} \sim 1$, $\mathfrak{B}$ is separable.

4. Diagonal direct factors. We have seen that if a semi-field $\mathfrak{Z}$ is S-irreducible then $\mathfrak{Z} = \mathfrak{E} \times \mathfrak{Y}$ where $\mathfrak{Y}$ is a field with an automorphism T induced by S^t, $\mathfrak{E} = (e_1, e_1^S, \cdots, e_1^{S^{t-1}})$. We shall generalize this property of $\mathfrak{E}$ relative to S and shall prove

Theorem 8. *Let $\mathfrak{Z}$ be S-irreducible so that $\mathfrak{Z} = \mathfrak{E} \times \mathfrak{Y}$ where $\mathfrak{E} \dot{\sim} 1$ has order t, and let $\mathfrak{Z}$ have a subalgebra $\mathfrak{E}_0 = (f, f^S, \cdots, f^{S^{r-1}}) \dot{\sim} 1$ with the same unity quantity as $\mathfrak{Z}$. Then $t = r\tau$, $\mathfrak{E}_0$ is a uniquely determined subalgebra of $\mathfrak{E} = \mathfrak{E}_0 \times \mathfrak{E}_1$, $\mathfrak{Z}_0 = \mathfrak{E}_1 \times \mathfrak{Y}$ is a semi-field which is T-irreducible with respect to the automorphism T induced in $\mathfrak{Z}_0$ by S^r. Conversely, if $t = r\tau$ then $\mathfrak{E} = \mathfrak{E}_0 \times \mathfrak{E}_1$, $\mathfrak{Z} = \mathfrak{E}_0 \times \mathfrak{Z}_0$ with $\mathfrak{E}_0$ as indicated above.*

Write $\mathfrak{E} = (e_1, \cdots, e_t) \dot{\sim} 1$ with $e_i = e_1^{S^{i-1}}$. Since f is an idempotent it must be a sum of distinct e_i. Now $\mathfrak{E}_0 = (f^{S^k}, f^{S^{k+1}}, \cdots, f^{S^{k+r+1}})$ for every k, and if $e_1^{S^i}$ is a component of f the quantity e_1 is a component of $f^{S^{-i}}$. Hence there is no loss of generality if we take

$$f = e_1 + e_{i_2} + \cdots + e_{i_\tau} \quad (1 < i_2 < \cdots < i_\tau \leqq t). \tag{8}$$

But then every f^{S^i} is a sum of τ distinct idempotents $(e_{i_k})^{S^i}$, and no e_j occurs in two f^{S^i} for distinct values of $i = 0, \cdots, r - 1$ since such f^{S^i} are orthogonal.

The sum of all the r quantities f^{S^i} is unity, and hence $t = \tau r$. Now $f = f^{S^r}$ implies that the τ distinct quantities $e_1, e_1^{S^r}, e_1^{S^{2r}}, \cdots, e_1^{S^{(\tau-1)r}}$ are all components of f and hence f is their uniquely determined sum. Thus $\mathfrak{E}_0$ is uniquely determined. Put

$$g = (e_1 + e_1^S + \cdots + e_1^{S^{r-1}}), \tag{9}$$

so that $\mathfrak{E}_1 = (g, g^{S^r}, \cdots, g^{S^{(\tau-1)r}}) \stackrel{\cdot}{\sim} 1$ has the same unity quantity as $\mathfrak{Z}$. Now $fg = e_1$, $\mathfrak{E}_0\mathfrak{E}_1$ contains

$$f^{S^i} g^{S^{rj}} = e_1^{S^{i+rj}} \qquad (i = 0, \cdots, r-1; j = 0, \cdots, \tau - 1), \tag{10}$$

$\mathfrak{E}_0\mathfrak{E}_1 = \mathfrak{E}_0 \times \mathfrak{E}_1 = \mathfrak{E}$. Then $\mathfrak{Z} = \mathfrak{E}_0 \times \mathfrak{Z}_0$, $\mathfrak{Z}_0 = \mathfrak{E}_1 \times \mathfrak{Y}$ a semi-field. Now we have already shown that $(e_i\mathfrak{Y})^S = e_{i+1}\mathfrak{Y}$. It follows that $(g\mathfrak{Y})^{S^r} = g^{S^r}\mathfrak{Y}$, and that $\mathfrak{Z}_0 = g\mathfrak{Y} \oplus \cdots \oplus g^{S^{(\tau-1)r}}\mathfrak{Y} = \mathfrak{Z}_0^{S^r}$. Hence S^r induces an automorphism T in $\mathfrak{Z}_0$. But $\mathfrak{E}_1 = (g, g^T, \cdots, g^{T^{\tau-1}})$ is T-irreducible and so is $\mathfrak{Z}_0$.

Conversely if $t = \tau r$ we define $\mathfrak{E}_0$ as above with $f = e_1 + e_1^{S^r} + \cdots + e_1^{S^{r(\tau-1)}}$ and have our result.

The order of an automorphism of a field $\mathfrak{Y}$ divides the degree of $\mathfrak{Y}$ by Theorem 2.32. Applying this together with the result above we obtain the

COROLLARY. *The order σ of S is rq where q is the order of the automorphism T of $\mathfrak{Z}_0$. Moreover σ divides the order of the semi-field $\mathfrak{Z}$ over $\mathfrak{F}$.*

5. Cyclic semi-fields. A semi-field $\mathfrak{Z} = \mathfrak{E} \times \mathfrak{Y}$ of Theorem 6.5 is called a *cyclic semi-field* if the field $\mathfrak{Y}$ is a cyclic field over $\mathfrak{F}$. Then all automorphisms of $\mathfrak{Y}$ over $\mathfrak{F}$ are in the center of the automorphism group of $\mathfrak{Y}$ over $\mathfrak{F}$ and we may use the result just preceding Theorem 6.4. Such algebras have two interesting characterizations the first of which is given as

Theorem 9. *An algebra $\mathfrak{Z}$ of order n over $\mathfrak{F}$ is a cyclic semi-field if and only if $\mathfrak{Z}$ is a commutative semi-simple algebra with an automorphism S over $\mathfrak{F}$ of order n such that $\mathfrak{Z}$ is S-irreducible.*

For if $\mathfrak{Z} = \mathfrak{E} \times \mathfrak{Y}$ with $\mathfrak{Y}$ cyclic we define an automorphism S of $\mathfrak{Z}$ by (6) and (7) with T a generating automorphism of $\mathfrak{Y}$. Then by Theorem 6.4, S has order n and $\mathfrak{Z}$ is S-irreducible. Conversely, if $\mathfrak{Z}$ is S-irreducible we apply Theorem 6.3 to obtain $\mathfrak{Z} = \mathfrak{E} \times \mathfrak{Y}$ where $\mathfrak{Y}$ has order m over $\mathfrak{F}$, S^t induces an automorphism T of order m in $\mathfrak{Y}$. Since $\mathfrak{Z}$ is commutative and semi-simple, $\mathfrak{Y}$ is a field and must be cyclic by the corollary of Theorem 2.32.

A second useful characterization may be stated as

Theorem 10. *An algebra $\mathfrak{Z}$ of order n over $\mathfrak{F}$ is a cyclic semi-field if and only if there exists a scalar extension $\mathfrak{K}$ of $\mathfrak{F}$ and an automorphism S of order n over $\mathfrak{F}$ of $\mathfrak{Z}$ such that*

$$\mathfrak{Z}_{\mathfrak{K}} = (e, e^S, \cdots, e^{S^{n-1}}) \stackrel{\cdot}{\sim} 1. \tag{11}$$

For if $\mathfrak{Z}$ is a cyclic semi-field we have $\mathfrak{Z} = \mathfrak{E} \times \mathfrak{Y}$, $\mathfrak{E} = (e_1, e_1^S, \cdots, e_1^{S^{t-1}}) \stackrel{\cdot}{\sim} 1$, where $\mathfrak{Y}$ is a cyclic field of degree m over $\mathfrak{F}$ with generating automorphism T

induced by S^t and S is defined by (6) and (7). By Theorem 2.27 there exists a scalar extension $\mathfrak{K}$ of $\mathfrak{F}$ such that $\mathfrak{Y}_{\mathfrak{K}} = (f, f^{S^t}, \cdots, f^{S^{t(m-1)}}) \dot{\sim} 1$. Then we have (11) with $e = e_1 f$. Conversely, if (11) holds the algebra $\mathfrak{Z}$ is a commutative separable algebra with an automorphism S of order n. If $\mathfrak{Z} = \mathfrak{B} \oplus \mathfrak{C}$ such that $\mathfrak{B} = \mathfrak{B}^S$, $\mathfrak{C} \neq 0$, the unity quantity of $\mathfrak{B}$ has the property $E = E^S$. But if $E = \alpha_1 e + \alpha_2 e^S + \cdots + \alpha_n e^{S^{n-1}} = E^S = \alpha_n e + \alpha_1 e^S + \cdots + \alpha_{n-1} e^{S^{n-1}}$ for α_i in $\mathfrak{K}$ the α_i are all equal, $E = \alpha_1 \cdot 1$ which is impossible. Hence $\mathfrak{Z}$ is S-irreducible and is a cyclic semi-field by Theorem 6.9.

Since every diagonal algebra is a direct sum of total matric algebras of order one, the field $\mathfrak{K}$ of Theorem 6.10 is a splitting field of $\mathfrak{Z}$.

An automorphism S of a cyclic semi-field $\mathfrak{Z}$ of order n over $\mathfrak{F}$ will be called a *generating automorphism* of $\mathfrak{Z}$ if S has order n and $\mathfrak{Z}$ is S-irreducible. Then we have already observed that $\mathfrak{Z} = \mathfrak{E} \times \mathfrak{Y}$, $\mathfrak{E} = (e_1, e_1^S, \cdots, e_1^{S^{t-1}}) \dot{\sim} 1$, S^t induces a generating automorphism (in the sense of both the present definition and also that of Chapter AIX) of the cyclic field $\mathfrak{Y}$, and S is given by (6), (7). Moreover, there then exists a field $\mathfrak{K}$ such that $\mathfrak{Z}_{\mathfrak{K}} = (e, e^S, \cdots, e^{S^{n-1}})$. The proof above implies immediately

Theorem 11. *Let $\mathfrak{Z}$ be a cyclic semi-field with generating automorphism S. Then $z = z^S$ for z in $\mathfrak{Z}$ if and only if z is in $\mathfrak{F}$. If $\mathfrak{K}$ is any scalar extension of $\mathfrak{F}$ the algebra $\mathfrak{Z}_{\mathfrak{K}}$ is a cyclic semi-field with generating automorphism S.*

We may also easily prove further elementary properties of cyclic semi-fields. The first of these is

Theorem 12. *Let $\mathfrak{Z}$ be a cyclic semi-field with generating automorphism S. Then $\mathfrak{Z}$ has a normal basis, that is, $\mathfrak{Z} = (u, u^S, \cdots, u^{S^{n-1}})$ over $\mathfrak{F}$.*

For by the lemma on page A201 we have $\mathfrak{Y} = (v, v^T, \cdots, v^{T^{m-1}})$ where $\mathfrak{Z} = \mathfrak{E} \times \mathfrak{Y}$, $\mathfrak{E} = (e_1, e_1^S, \cdots, e_1^{S^{t-2}})$, S^t induces the generating automorphism T in the cyclic field $\mathfrak{Y}$. Then the quantity $u = e_1 v$ has the property of the theorem. For it may be easily verified that $u^{S^k} = e_1^{S^i} v^{T^j}$ for $k = jt + i$, $i = 0, 1, \cdots, t-1$; $j = 0, 1, \cdots, m-1$.

We may define the *norm* of a quantity z of $\mathfrak{Z}$ relative to $\mathfrak{F}$ to be the product

$$N_{\mathfrak{Z}|\mathfrak{F}}(z) = zz^S \cdots z^{S^{n-1}}. \tag{12}$$

Now $z = y_1 e_1 + \cdots + y_t e_t$, $z^{S^t} = y_1^T e_1 + \cdots + y_t^T e_t$, $zz^{S^t} \cdots z^{S^{t(m-1)}} = N_{\mathfrak{Y}|\mathfrak{F}}(y_1)e_1 + \cdots + N_{\mathfrak{Y}|\mathfrak{F}}(y_t)e_t = N_0$ in $\mathfrak{E}$. It follows that

$$\gamma = N_{\mathfrak{Z}|\mathfrak{F}}(z) = N_0 N_0^S \cdots N_0^{S^{t-1}} = N_{\mathfrak{Y}|\mathfrak{F}}(y_1 \cdots y_t),$$

since the norm function in the cyclic field $\mathfrak{Y}$ is multiplicative. We have now shown that the norm of any quantity of $\mathfrak{Z}$ is in $\mathfrak{F}$ and is the norm of a quantity of $\mathfrak{Y}$ and is thus independent of the particular generating automorphism S of $\mathfrak{Z}$. Conversely if $\gamma = N_{\mathfrak{Y}|\mathfrak{F}}(y)$, y in $\mathfrak{Y}$, then $N_{\mathfrak{Z}|\mathfrak{F}}(ye_1 + e_2 + \cdots + e_t) = N_{\mathfrak{Y}|\mathfrak{F}}(y_1) = \gamma$. We have proved

Theorem 13. *A quantity γ of $\mathfrak{F}$ is the norm of a quantity of $\mathfrak{Z}$ if and only if γ is the norm of a quantity of the corresponding cyclic field $\mathfrak{Y}$.*

Subfields of an algebra with a unity quantity were defined in Section 4.4. Let $\mathfrak{X}$ be a subfield of a semi-field $\mathfrak{Z} = \mathfrak{E} \times \mathfrak{Y}$ so that every quantity of $\mathfrak{X}$ is uniquely expressible in the form $x = y_1 e_1 + \cdots + y_t e_t$ for y_i in $\mathfrak{Y}$. If $f(\lambda)$ is the minimum function of x then $f(x) = f(y_1)e_1 + \cdots + f(y_t)e_t = 0$ so that $f(y_1) = \cdots = f(y_t) = 0$. Thus $\mathfrak{F}(x)$ is equivalent to a subfield of $\mathfrak{Y}$. If $\mathfrak{Z}$ is cyclic, $\mathfrak{Y}$ is a cyclic field, every quantity of $\mathfrak{Y}$ is separable over $\mathfrak{F}$, every x of $\mathfrak{X}$ is separable over $\mathfrak{F}$. Hence $\mathfrak{X}$ is separable. But then $\mathfrak{X} = \mathfrak{F}(x)$ for some x and we have proved

Theorem 14. *The subfields of a cyclic semi-field $\mathfrak{Z} = \mathfrak{E} \times \mathfrak{Y}$ are all cyclic fields equivalent to subfields of $\mathfrak{Y}$.*

The result above implies that $\mathfrak{Y}$ is a subfield of $\mathfrak{Z}$ of maximal possible degree and that every other such subfield of $\mathfrak{Z}$ is equivalent to $\mathfrak{Y}$. We shall speak of all subfields of the maximum degree as *maximal subfields* of $\mathfrak{Z}$.

6. Automorphisms of a direct product. Let $\mathfrak{A}$ be an algebra with a unity quantity and $\mathfrak{A} = \mathfrak{B} \times \mathfrak{C}$. According to our usual convention we then assume that $\mathfrak{B}$ and $\mathfrak{C}$ are subalgebras of $\mathfrak{A}$, each containing the unity quantity of $\mathfrak{A}$. If S is any given automorphism of $\mathfrak{B}$, the correspondence in $\mathfrak{A}$ generated by the partial correspondences

$$b \leftrightarrow b^S, \qquad c \leftrightarrow c$$

for every b of $\mathfrak{B}$ and c of $\mathfrak{C}$ is clearly an automorphism S_0 of $\mathfrak{A}$. We may think of S_0 as an extension in $\mathfrak{A}$ of S in $\mathfrak{B}$, and thus also of S as the automorphism in $\mathfrak{B}$ induced by S_0 of $\mathfrak{A}$. Thus every group $\mathfrak{G}$ of automorphisms of $\mathfrak{B}$ may be replaced by a group $\mathfrak{G}_0$ of automorphisms of $\mathfrak{A}$ such that

$$c^{S_0} = c, \qquad \mathfrak{B}^{S_0} = \mathfrak{B}$$

for every c of $\mathfrak{C}$ and S_0 of $\mathfrak{G}_0$.

In a similar fashion every group $\mathfrak{H}$ of automorphisms T of $\mathfrak{C}$ may be thought of as the automorphism group of $\mathfrak{C}$ induced by a group $\mathfrak{H}_0$ of automorphisms T_0 of $\mathfrak{A}$ such that $\mathfrak{C}^{T_0} = \mathfrak{C}$, $b^{T_0} = b$ for every T_0 of $\mathfrak{H}$ and b of $\mathfrak{B}$. The partial correspondences

$$b \leftrightarrow b^S, \qquad c \leftrightarrow c^T$$

then define correspondences of $\mathfrak{A}$ which are clearly the product automorphisms $S_0 T_0 = T_0 S_0$ of $\mathfrak{A}$. The set of all such automorphisms is clearly the group $\mathfrak{G}_0 \times \mathfrak{H}_0$. Hence we have essentially constructed the formal direct product $\mathfrak{G} \times \mathfrak{H}$, replacing $\mathfrak{G}$ by the equivalent $\mathfrak{G}_0$, $\mathfrak{H}$ by $\mathfrak{H}_0$ where now $\mathfrak{G}_0$ and $\mathfrak{H}_0$ are subgroups of $\mathfrak{G}_0 \times \mathfrak{H}_0$.

There is no loss of generality if we replace $\mathfrak{G}$ by $\mathfrak{G}_0$, $\mathfrak{H}$ by $\mathfrak{H}_0$ in our remaining discussions. When we do so we are merely thinking of every group of automorphisms of the direct factor $\mathfrak{B}$ of $\mathfrak{A} = \mathfrak{B} \times \mathfrak{C}$ as the group induced by a group of automorphisms of $\mathfrak{A}$ leaving all the quantities of $\mathfrak{C}$ unaltered, and similarly for $\mathfrak{C}$. We shall thus use the same notations for certain automorphisms of $\mathfrak{A} = \mathfrak{B} \times \mathfrak{C}$ and those induced in its direct factors $\mathfrak{B}$ and $\mathfrak{C}$.

7. Uniqueness of direct factorization. When we studied the direct product $\mathfrak{A} = \mathfrak{B} \times \mathfrak{C}$ with $\mathfrak{B}$ normal we were able to show that if also $\mathfrak{A} = \mathfrak{B} \times \mathfrak{C}_0$ then $\mathfrak{C} = \mathfrak{C}_0$. However if $\mathfrak{B}$ is not normal $\mathfrak{C}$ is not unique. This was observed already in the case of semi-fields $\mathfrak{Z}$ where $\mathfrak{Z} = \mathfrak{C} \times \mathfrak{Y} = \mathfrak{C} \times \mathfrak{Y}_0$ with $\mathfrak{Y} \neq \mathfrak{Y}_0$. Our further study will in fact be concerned with the possible factorizations $\mathfrak{Z} \times \mathfrak{C} = \mathfrak{Z} \times \mathfrak{W}_k$ where $\mathfrak{Z}$ and $\mathfrak{C}$ are cyclic semi-fields, and $\mathfrak{W}_k$ will also be a cyclic semi-field satisfying certain preassigned conditions and fixed uniquely by these conditions. The method of accomplishing this uniqueness will now be described.

Let $\mathfrak{A} = \mathfrak{B} \times \mathfrak{C}$ where $\mathfrak{B}$ has a group $\mathfrak{G}$ of automorphisms S over $\mathfrak{F}$ such that the only quantities of $\mathfrak{B}$ unaltered by all the automorphisms of $\mathfrak{G}$ are the quantities of $\mathfrak{F}$. Let $\mathfrak{C} = (v_1, \cdots, v_q)$ over $\mathfrak{F}$. Then every quantity of $\mathfrak{A}$ is expressible uniquely in the form $a = b_1v_1 + \cdots + b_qv_q$ with the b_i in $\mathfrak{B}$. We identify $\mathfrak{G}$ with the corresponding group of automorphisms of $\mathfrak{A}$ in the manner discussed above and have $v_i = v_i^S$ for every S of $\mathfrak{G}$. Then $a = a^S$ for every S of $\mathfrak{G}$ if and only if

$$a - a^S = \sum_{i=1}^{q} (b_i - b_i^S)v_i = 0,$$

that is, $b_i - b_i^S = 0$ for $i = 1, \cdots, q$ and every S of $\mathfrak{G}$. But then the b_i are in $\mathfrak{F}$, a is in $\mathfrak{C}$. It follows that $\mathfrak{C}$ is the uniquely determined set of all quantities of $\mathfrak{A}$ unaltered by all the automorphisms of $\mathfrak{G}$. Then if $\mathfrak{A} = \mathfrak{B} \times \mathfrak{C} = \mathfrak{B} \times \mathfrak{C}_0$ and $\mathfrak{C}_0$ also consists of quantities unaltered by the automorphisms of $\mathfrak{G}$ we have $\mathfrak{C} = \mathfrak{C}_0$.

Observe that $\mathfrak{C}$ has been shown in $\mathfrak{A}$ to be a function of the algebra-group pair $\mathfrak{B}$, $\mathfrak{G}$. Hence we shall study algebra-group pairs $\mathfrak{B}$, $\mathfrak{G}$ and $\mathfrak{C}$, $\mathfrak{H}$ and the direct product $\mathfrak{A} = \mathfrak{B} \times \mathfrak{C}$. In this direct product which has a group $\mathfrak{G} \times \mathfrak{H}$ of automorphisms, the expressibility of the given $\mathfrak{A}$ in the form $\mathfrak{B} \times \mathfrak{C}_0$ is possible only in the fashion $\mathfrak{A} = \mathfrak{B} \times \mathfrak{C}$ and we say that $\mathfrak{C}_0$ is thus uniquely determined. In fact the pair $\mathfrak{C}$, $\mathfrak{H}$ is unique in the sense of algebra-group equivalence. We shall study later other direct factorizations $\mathfrak{G}_1 \times \mathfrak{H}_1$ of $\mathfrak{G} \times \mathfrak{H}$ with the property that $\mathfrak{G}_1 \neq \mathfrak{G}$ but $\mathfrak{G}_1$ induces the same group in $\mathfrak{B}$ as does $\mathfrak{G}$. Then we shall be able to express our given $\mathfrak{A} = \mathfrak{B} \times \mathfrak{C}$ uniquely in a form $\mathfrak{A} = \mathfrak{B} \times \mathfrak{C}_1$ where $\mathfrak{B}$, $\mathfrak{G}_1$ is equivalent to $\mathfrak{B}$, $\mathfrak{G}$ but $\mathfrak{C}_1 \neq \mathfrak{C}$, $\mathfrak{C}_1$ will not in general even be equivalent to $\mathfrak{C}$. Here $\mathfrak{C}_1$ will be the unique set of all quantities of $\mathfrak{B} \times \mathfrak{C}$ unaltered by the automorphisms of $\mathfrak{G}_1$. Also $\mathfrak{C}_1$, $\mathfrak{H}_1$ will be unique in the sense that if we replace $\mathfrak{B}$, $\mathfrak{G}$ and $\mathfrak{C}$, $\mathfrak{G}$ by respectively equivalent pairs then $\mathfrak{C}_1$, $\mathfrak{H}_1$ will be replaced by an equivalent pair.

We shall not consider the general problem outlined above but shall proceed to a consideration merely of the case of pairs $\mathfrak{Z}$, S and $\mathfrak{C}$, T where $\mathfrak{Z}$ and $\mathfrak{C}$ are cyclic semi-fields with respective generating automorphisms S and T.

8. Direct products of cyclic semi-fields. Let $\mathfrak{Z}$ be a cyclic semi-field with generating automorphism S, $\mathfrak{C}$ be a cyclic semi-field of the same order n as $\mathfrak{Z}$ and with generating automorphism T. We form the direct product $\mathfrak{J} =$

$\mathfrak{Z} \times \mathfrak{C}$ which is an algebra with the direct product $[S] \times [T]$ as a group of automorphisms. By Theorem 6.11 and Section 6.7, $\mathfrak{Z}$ is the set of all quantities of $\mathfrak{J}$ unaltered by T and similarly for $\mathfrak{C}$ and S. Moreover if a is in $\mathfrak{J}$ then $a = a^S = a^T$ if and only if a is in $\mathfrak{F}$. We now prove

Theorem 15. *The set $\mathfrak{W}_k$ of all quantities of $\mathfrak{J} = \mathfrak{Z} \times \mathfrak{C}$ unaltered by ST^k is a cyclic semi-field of order n over $\mathfrak{F}$ with generating automorphism induced by the automorphism T of $\mathfrak{J}$. Moreover, ST^k induces the same generating automorphism in $\mathfrak{Z}$ as S and*

$$\mathfrak{J} = \mathfrak{Z} \times \mathfrak{C} = \mathfrak{Z} \times \mathfrak{W}_k .$$

Observe that we have essentially expressed the group $[S] \times [T]$ in the form $[ST^k] \times [T]$ which is possible since $ST = TS$, $(ST^k)^r = I$ if and only if $S^r = T^{kr} = I$, that is, n divides r. Since $z = z^T$ it follows that $z^{ST^k} = z^S$ for every z of $\mathfrak{Z}$. Our definition of $\mathfrak{W}_k$ implies that it is a commutative algebra over $\mathfrak{F}$ with the same unity quantity as $\mathfrak{J}$, $\mathfrak{Z}$, $\mathfrak{C}$. If $a = a^{ST^k}$ then $a^T = a^{ST^{k+1}} = (a^T)^{ST^k}$. Hence $\mathfrak{W}_k = \mathfrak{W}_k^T$, T induces an automorphism of $\mathfrak{W}_k$. If $\mathfrak{W}_k$ were T-reducible then there would be an idempotent $e \neq 1$ in $\mathfrak{W}_k$ such that $e = e^T$. But $e = e^{ST^k}$, $e = e^{T^{-k}}$, $e = e^S$ is in $\mathfrak{F}$, $e = 1$. Hence $\mathfrak{W}_k$ is T-irreducible. By Theorem 6.7 it is a separable algebra.

To complete our proof we let $\mathfrak{Z} = (z_1, \cdots, z_n)$ over $\mathfrak{F}$ where, by Theorem 6.12, we may take $z_{i+1} = z_1^{S^i}$ $(i = 0, \cdots, n-1)$. Every quantity of $\mathfrak{J}$ has the unique form

$$a = z_1 c_1 + \cdots + z_n c_n \tag{13}$$

with c_i in $\mathfrak{C}$ and

$$a^{ST^k} = z_1 c_n^{T^k} + z_2 c_1^{T^k} + \cdots + z_n c_{n-1}^{T^k} = a \tag{14}$$

if and only if

$$c_1 = c_n^{T^k}, \qquad c_2 = c_1^{T^k}, \cdots, c_n = c_{n-1}^{T^k} . \tag{15}$$

The last $n - 1$ of these equations determine $c_2, \cdots, c_n$ uniquely in terms of c_1. The first implies only that $c_1 = c_1^{T^{nk}}$ which is a consequence of $T^n = I$. But then if $\mathfrak{C} = (u_1, \cdots, u_n)$ with $u_{i+1} = u_1^{T^i}$ we have $c_1 = \sum_{i=1}^{n} \alpha_i u_i$ where the α_i are in $\mathfrak{F}$, and $a = a^{ST^k}$ if and only if $a = \sum_{i=1}^{n} \alpha_i w_i^{(k)}$,

$$w_i^{(k)} = z_1 u_i + z_2 u_i^{T^k} + \cdots + z_n u_i^{T^{(n-1)k}} \qquad (i = 1, \cdots, n). \tag{16}$$

The $w_i^{(k)}$ are clearly linearly independent in $\mathfrak{F}$ and form a basis of $\mathfrak{W}_k$. This proves $\mathfrak{W}_k$ of order n over $\mathfrak{F}$. Also $(w_i^{(k)})^T = w_{i+1}^{(k)}$, $w_i^{(k)T^r} = w_i$ if and only if n divides r. Hence T has order n in $\mathfrak{W}_k$. By Theorem 6.9, $\mathfrak{W}_k$ is a cyclic semi-field with generating automorphism induced by T.

To prove that $\mathfrak{J} = \mathfrak{Z} \times \mathfrak{W}_k$ we use a common splitting field $\mathfrak{K}$ of both $\mathfrak{Z}$ and $\mathfrak{C}$ and may write $\mathfrak{Z}_{\mathfrak{K}} = (e, e^S, \cdots, e^{S^{n-1}})$ over $\mathfrak{K}$, $\mathfrak{C}_{\mathfrak{K}} = (f, f^T, \cdots, f^{T^{n-1}})$ over $\mathfrak{K}$. Observe that the linear set $\mathfrak{Z}\mathfrak{W}_k$ is carried into itself by every S^iT^j since $\mathfrak{Z}^{S^iT^j} = \mathfrak{Z}^{S^i} = \mathfrak{Z}$, $\mathfrak{W}_k^{S^iT^j} = \mathfrak{W}_k^{T^{j-ki}} = \mathfrak{W}_k$. It is clear that $\mathfrak{W}_k$ contains the quantities

$$(17) \qquad g_i^{(k)} = f_i e_1 + f_{i+k} e_2 + \cdots + f_{i+(n-1)k} e_n ,$$

where we have of course written $e_{j+1} = e^{S^j}$, $f_{j+1} = f^{T^j}$. The set $(\mathfrak{J}\mathfrak{W}_k)_{\mathfrak{K}}$ contains $g_i^{(k)} e_1 = f_i e_1$ and hence also $(f_i e_1)^{S^{j-1}} = f_i e_j$. But the quantities $f_i e_j$ form a basis of $\mathfrak{J}_{\mathfrak{K}}$ so that $(\mathfrak{Z}\mathfrak{W}_k)_{\mathfrak{K}}$ has order n^2, so does $\mathfrak{Z}\mathfrak{W}_k$, $\mathfrak{Z}\mathfrak{W}_k = \mathfrak{J} = \mathfrak{Z} \times \mathfrak{W}_k$. This proves our theorem.

Note that a normal basis of $\mathfrak{W}_k$ is given by (16) where the z_j and u_i form respective normal bases of $\mathfrak{Z}$ and $\mathfrak{C}$. We shall frequently refer to this basis.

9. Cyclic systems. The relation of equivalence of algebra-group pairs is a formal equivalence relation and we may put all pairs equivalent to a given one into a class. We will be interested only in pairs $\mathfrak{Z}$, $[S]$ where $\mathfrak{Z}$ is a cyclic semi-field with generating automorphism S, and shall designate the corresponding class by

$$\mathcal{Z} = \{\mathfrak{Z}, S\},$$

and call $\mathcal{Z}$ a *cyclic system*. We shall use capital script letters for cyclic systems both in what follows and in the next chapter.

The order n of $\mathfrak{Z}$ will be called the *degree* of $\mathcal{Z}$. The degree m of a maximal subfield of $\mathfrak{Z}$ is then a uniquely determined divisor of n and we shall call m the *index* of $\mathcal{Z}$. Observe that if $\mathcal{Z} = \{\mathfrak{Z}, S\}$ and $\mathcal{C} = \{\mathfrak{C}, T\}$ are given the corresponding system $P_k(\mathcal{Z}, \mathcal{C}) = \{\mathfrak{W}_k, T\}$ of Theorem 6.15 is uniquely determined for every integer k. We shall use this notation later.

If $\mathfrak{Z} = \mathfrak{C} \times \mathfrak{Y}$ has generating automorphism S then S^t induces an automorphism in the cyclic field $\mathfrak{Y}$. Then the pair $\mathfrak{Y}$, S^t is unique in the sense of algebra-group equivalence, as was remarked in Section 6.2, and we designate the corresponding uniquely determined cyclic system by

$$\mathcal{Y} = \{\mathfrak{Y}, S^t\}.$$

If now $\mathcal{Z}_1$ is a cyclic system of any degree over $\mathfrak{F}$ determining the same $\mathcal{Y}$ as $\mathcal{Z}$ we shall call $\mathcal{Z}$ and $\mathcal{Z}_1$ *similar* and write

$$\mathcal{Z} \sim \mathcal{Z}_1 .$$

The relation of similarity among cyclic systems is a formal equivalence relation. All cyclic systems similar to a given one form a class and each class contains one cyclic system for every degree $n = mt$. Here t has all possible positive integral values and m is the common index of each cyclic system of the class. Moreover $\mathcal{Y}$ is the unique cyclic system of the class for which $n = m$. Note the marked analogy between this situation and the corresponding one for similarity of normal simple algebras. There are indeed closer relations which will be discussed in Chapter VII.

If $\mathfrak{Z}$ is a diagonal algebra the class $\mathcal{Z} = \{\mathfrak{Z}, S\}$ has index one, we have been writing $\mathfrak{Z} \sim 1$, and we now indicate the property by writing

$$\mathcal{Z} \sim 1.$$

This will be shown to be closely connected with the analogous concept for normal simple algebras. We shall also use the notation $\mathcal{J}$ for this class. Note that every diagonal algebra $\mathfrak{E}$ of the same order as $\mathfrak{Z}$ defines an element $\mathfrak{E}$, T of the class $\{\mathfrak{Z}, S\}$, where T is any generating automorphism of $\mathfrak{E}$.

In Theorem 6.8 we observed that if $\mathcal{Z}_1 = \{\mathfrak{Z}_1, S_1\}$ is similar to $\mathcal{Z} = \{\mathfrak{Z}, S\}$ such that the degree of $\mathcal{Z}_1$ is a multiple rn of the degree n of $\mathcal{Z}$, then $\mathfrak{Z}_1 = \mathfrak{E}_1 \times \mathfrak{Z}_0$, $\mathfrak{E}_1 \sim 1$ of order r, $\mathcal{Z}_0 = \{\mathfrak{Z}_0, S_1^r\} = \mathcal{Z}$. We use this result in the derivation of

Theorem 16. *Let $\mathcal{Z}$ and $\mathcal{C}$ be cyclic systems of the same degree and index. Then if $\mathcal{Z}_1 \sim \mathcal{Z}$ and $\mathcal{C}_1 \sim \mathcal{C}$, where $\mathcal{Z}_1$ and $\mathcal{C}_1$ have the same degrees, we have*

$$P_k(\mathcal{Z}_1, \mathcal{C}_1) \sim P_k(\mathcal{Z}, \mathcal{C}). \tag{18}$$

Write $\mathfrak{Z} = \mathfrak{E} \times \mathfrak{Y}$, $\mathfrak{C} = \mathfrak{E}_0 \times \mathfrak{X}$ for diagonal algebras $\mathfrak{E}$ and $\mathfrak{E}_0$ of the same order t, $\mathcal{Z} = \{\mathfrak{Z}, S\} \sim \mathcal{Y} = \{\mathfrak{Y}, S^t\}$, $\mathcal{C} = \{\mathfrak{C}, T\} \sim \mathcal{X} = \{\mathfrak{X}, T^t\}$. Then $\mathfrak{J} = \mathfrak{Z} \times \mathfrak{C} = \mathfrak{E} \times \mathfrak{E}_0 \times \mathfrak{Y} \times \mathfrak{X}$. Now S and T are respective generating automorphisms of $\mathfrak{E}$ and $\mathfrak{E}_0$ and, by Theorem 6.15,

$$\mathfrak{E} \times \mathfrak{E}_0 = \mathfrak{E} \times \mathfrak{E}_k \tag{19}$$

where $\mathfrak{E}_k \sim 1$ is the set of all quantities of $\mathfrak{E} \times \mathfrak{E}_0$ unaltered by ST^k. We may write $\mathfrak{E} = (e_1, \cdots, e_t)$, $\mathfrak{E}_k = (f_1, \cdots, f_t)$ such that $e_i = e_1^{S^{i-1}}$, $f_j = f_1^{T^{j-1}}$.

The algebra $\mathfrak{E}_k$ is a subalgebra of the algebra $\mathfrak{W}_k$ of Theorem 6.15 defined for $\mathfrak{Z} \times \mathfrak{C}$. By Theorem 6.8 $\mathfrak{W}_k = \mathfrak{E}_k \times \mathfrak{D}_k$, $\mathfrak{D}_k$ a cyclic semi-field whose generating automorphism is that induced by T^t. Now $\mathfrak{J} = \mathfrak{E}_k \times (\mathfrak{E} \times \mathfrak{Y} \times \mathfrak{X})$ so that if $u_1, \cdots, u_\mu$ are any μ quantities of $\mathfrak{Y} \times \mathfrak{X}$ linearly independent in $\mathfrak{F}$, the quantities $f_1e_1u_1, \cdots, f_1e_1u_\mu$ are also linearly independent in $\mathfrak{F}$. Then if b is in $\mathfrak{Y} \times \mathfrak{X}$ and $f_1e_1b = 0$ we have $b = 0$. Every quantity a of $e_1f_1\mathfrak{J}$ has the form e_1f_1b with b in $\mathfrak{Y} \times \mathfrak{X}$. If a is in $e_1f_1\mathfrak{D}_k$ then $a = a^{(ST^k)^t} = (e_1f_1b)^{(ST^k)^t} = e_1f_1b^{S^tT^{tk}}$ if and only if $b = b^{S^tT^{tk}}$ is in the set $\mathfrak{V}_k$ defined by Theorem 6.15 for $\mathfrak{Y} \times \mathfrak{X}$. Hence $e_1f_1\mathfrak{D}_k \leqq e_1f_1\mathfrak{V}_k$. Both sets have the same order over $\mathfrak{F}$ and $e_1f_1\mathfrak{D}_k = e_1f_1\mathfrak{V}_k$. The linear independence described above implies that if v and v' are in $\mathfrak{V}_k$ and $e_1f_1v = e_1f_1v'$ then $v = v'$. Hence if d is in $\mathfrak{D}_k$ the quantity $e_1f_1d = e_1f_1v$ for v a uniquely determined quantity of $\mathfrak{V}_k$. Similarly $\mathfrak{J} = \mathfrak{Z} \times \mathfrak{W}_k = \mathfrak{E} \times \mathfrak{E}_k \times \mathfrak{Y} \times \mathfrak{D}_k$ implies that v uniquely determines the d of $e_1f_1v = e_1f_1d$. The algebras $\mathfrak{D}_k$, $e_1f_1\mathfrak{D}_k = e_1f_1\mathfrak{V}_k$, and $\mathfrak{V}_k$ are now all equivalent under the correspondences

$$d \leftrightarrow e_1f_1d = e_1f_1v \leftrightarrow v. \tag{20}$$

The automorphism T^t induces a generating automorphism in both $\mathfrak{D}_k$ and $\mathfrak{V}_k$ such that $(e_1f_1d)^{T^t} = (e_1f_1v)^{T^t}$, $d^{T^t} \leftrightarrow v^{T^t}$, and thus $\{\mathfrak{D}_k, T^t\} = \{\mathfrak{V}_k, T^t\}$, $P_k(\mathcal{Z}, \mathcal{C}) \sim P_k(\mathcal{Y}, \mathcal{X})$. The result just proved applied to $\mathcal{Z}_1$, $\mathcal{C}_1$ states that if $\mathcal{Z}_1 \sim \mathcal{Z}$ and $\mathcal{C}_1 \sim \mathcal{C}$ then $P_k(\mathcal{Z}_1, \mathcal{C}_1) \sim P_k(\mathcal{Y}, \mathcal{X})$. Hence $P_k(\mathcal{Z}_1, \mathcal{C}_1) \sim P_k(\mathcal{Z}, \mathcal{C})$ as desired.

10. The group of cyclic systems. The set $\mathcal{S}_n$ of all cyclic systems of the same degree n is closed with respect to the operation defined in

$$\mathcal{Z}\mathcal{C} = P_{-1}(\mathcal{Z}, \mathcal{C}). \tag{21}$$

We call $P_{-1}(\mathfrak{Z}, \mathfrak{C})$ the *product* of $\mathfrak{Z}$ and $\mathfrak{C}$ and have already observed that it is a cyclic system uniquely determined by $\mathfrak{Z}$ and $\mathfrak{C}$. If $\mathfrak{Z} = \{\mathfrak{Z}, S\}$ and $\mathfrak{C} = \{\mathfrak{C}, T\}$, the set $\mathfrak{W}_{-1}$ of all quantities of $\mathfrak{Z} \times \mathfrak{C}$ unaltered by ST^{-1} is clearly the same set as that unaltered by $(ST^{-1})^{-1} = TS^{-1}$. Hence $P_{-1}(\mathfrak{Z}, \mathfrak{C}) = P_{-1}(\mathfrak{C}, \mathfrak{Z})$, that is, our operation is commutative. We shall in fact prove

Theorem 17. *The set $\mathfrak{S}_n$ of all cyclic systems of the same degree n forms an abelian group with respect to the operation defined in* (21). *The identity element of the group $\mathfrak{S}_n$ is the cyclic system*

$$\mathcal{J} = \{\mathfrak{E}, T\} \sim 1, \tag{22}$$

and

$$P_{-k}(\mathfrak{Z}, \mathfrak{C}) = \mathfrak{Z}^k\mathfrak{C}. \tag{23}$$

The order of each element of the group $\mathfrak{S}_n$ is its index.

To prove $\mathfrak{S}_n$ a group we first show that $\mathfrak{Z}(\mathfrak{C}\mathfrak{D}) = (\mathfrak{Z}\mathfrak{C})\mathfrak{D}$. Write $\mathfrak{D} = \{\mathfrak{D}, U\}$ and form

$$\mathfrak{A} = (\mathfrak{Z} \times \mathfrak{C}) \times \mathfrak{D} = (\mathfrak{Z} \times \mathfrak{W}_{-1}) \times \mathfrak{D} = \mathfrak{Z} \times (\mathfrak{W}_{-1} \times \mathfrak{D})$$
$$= \mathfrak{Z} \times (\mathfrak{W}_{-1} \times \mathfrak{V}_{-1}).$$

Here $\mathfrak{W}_{-1}$ is the set of all quantities of $\mathfrak{Z} \times \mathfrak{C}$ unaltered by ST^{-1}, the quantities of $\mathfrak{W}_{-1} \times \mathfrak{D}$ are unaltered by ST^{-1}, T induces a generating automorphism in $\mathfrak{W}_{-1}$, $\mathfrak{W}_{-1} \times \mathfrak{D} = \mathfrak{W}_{-1} \times \mathfrak{V}_{-1}$ with $\mathfrak{V}_{-1}$ the set of all quantities of $\mathfrak{W}_{-1} \times \mathfrak{D}$ unaltered by TU^{-1}. Hence $\{\mathfrak{V}_{-1}, U\} = (\mathfrak{Z}\mathfrak{C})\mathfrak{D}$. We let $\mathfrak{W}$ be the set of all quantities of $\mathfrak{A}$ unaltered by both ST^{-1} and TU^{-1}. Then clearly $\mathfrak{W} \geqq \mathfrak{V}_{-1}$. Every quantity of $\mathfrak{A}$ is uniquely expressible in the form $a = a_1u_1 + \cdots + a_nu_n$ with the u_i a basis of $\mathfrak{V}_{-1}$ over $\mathfrak{F}$ and the a_i in $\mathfrak{Z} \times \mathfrak{W}_{-1}$. Then $a = a^{ST^{-1}} = a^{TU^{-1}}$ if and only if the $a_i = a_i^{ST^{-1}} = a_i^{TU^{-1}}$. But $a_i = a_i^U$, every $a_i = a_i^T = a_i^S$ is in $\mathfrak{Z} \times \mathfrak{C}$ and hence is in $\mathfrak{F}$, a is in $\mathfrak{V}_{-1}$. Hence $\mathfrak{W} = \mathfrak{V}_{-1}$. Now $\mathfrak{Z}(\mathfrak{C}\mathfrak{D}) = \{\mathfrak{W}_0, U\}$ where we apply an argument analogous to that above to show that $\mathfrak{W}_0$ is the set of all quantities of $\mathfrak{A}$ unaltered by TU^{-1} and SU^{-1}. However $ST^{-1} = (SU^{-1})(TU^{-1})^{-1}$ and $\mathfrak{W}_0 = \mathfrak{W}$, $\{\mathfrak{V}_{-1}, U\} = \{\mathfrak{W}_0, U\}$ as desired. This proves the associative law.

We have seen that $\mathfrak{Z} \times \mathfrak{C} = \mathfrak{Z} \times \mathfrak{W}_{-k}$ where in the latter direct factorization $\mathfrak{Z}$ is considered as having generating automorphism ST^{-k}, $\mathfrak{W}_{-k}$ as having generating automorphism T. But then $\mathfrak{Z}P_{-k}(\mathfrak{Z}, \mathfrak{C}) = P_{-(k+1)}(\mathfrak{Z}, \mathfrak{C})$ follows from $(ST^{-k})T^{-1} = ST^{-(k+1)}$. An immediate induction yields (23).

We continue our proof of the group property of $\mathfrak{S}_n$ by forming $\mathfrak{Z}\mathcal{J}$ and hence $\mathfrak{Z} \times \mathfrak{E}$ where $\mathfrak{E}$ is diagonal. Write $\mathfrak{E} = (f_1, \cdots, f_n)$ with the notation $f_j = f_1^{T^{j-1}}$ for all (positive, zero, or negative) integral values of j. Then $\mathfrak{Z} \times \mathfrak{E} = \mathfrak{Z} \times \mathfrak{W}_{-1}$ where we obtain a basis $w_1, \cdots, w_n$ of $\mathfrak{W}_{-1}$ over $\mathfrak{F}$ from (16) with $k = -1$, $u_i = f_i$. We take $z_1, \cdots, z_n$ to be a normal basis of $\mathfrak{Z}$ over $\mathfrak{F}$ so that $z_i = z_1^{S^{i-1}}$ and have

$$w_1 = z_1f_1 + z_2f_n + \cdots + z_nf_2. \tag{24}$$

Moreover,

$$w_2 = w_1^T = z_2 f_1 + z_3 f_n + \cdots + z_1 f_2 = w_1^S. \tag{25}$$

By the remark made in Section 6.2 about the equivalence of algebras $\mathfrak{Z}$ and certain algebras of diagonal matrices with elements in $\mathfrak{Z}$ we see that the correspondence generated by $z_i \leftrightarrow w_i$ is an equivalence of $\mathfrak{Z}$ and $\mathfrak{W}_{-1}$ such that $z^S \leftrightarrow w^T$, $\{\mathfrak{W}_{-1}, T\} = \{\mathfrak{Z}, S\}$. Hence $\mathcal{J}$ is the identity element of $\mathfrak{S}_n$.

We have already shown that $P_{-n}(\mathcal{Z}, \mathcal{J}) = \mathcal{Z}^n\mathcal{J} = \mathcal{Z}^n$. But $\mathfrak{W}_{-n}$ is the set of all quantities of $\mathfrak{Z} \times \mathfrak{E}$ unaltered by $ST^{-n} = S$ and hence is $\mathfrak{E}$, $\{\mathfrak{W}_{-n}, T\} = \mathcal{J}$, $\mathcal{Z}^n = \mathcal{J}$. Thus $\mathcal{Z}^{-1} = \mathcal{Z}^{n-1}$ exists in $\mathfrak{S}_n$ for every $\mathcal{Z}$ of $\mathfrak{S}_n$, $\mathfrak{S}_n$ is an abelian group.

It remains to determine the order r of the arbitrary group element $\mathcal{Z}$ of $\mathfrak{S}_n$. By Theorem 6.16 if $\mathcal{Z}_1 \dot{\sim} \mathcal{Z}$ then $\mathcal{Z}_1^k \dot{\sim} \mathcal{Z}^k$. Hence $\mathcal{Z}^k = \mathcal{J} \dot{\sim} 1$ if and only if $\mathcal{Y}^k \dot{\sim} 1$. Thus there is no loss of generality if we let $\mathcal{Z} = \{\mathfrak{Z}, S\}$ with $\mathfrak{Z}$ a cyclic field. We form $\mathfrak{Z} \times \mathfrak{E} = \mathfrak{Z} \times \mathfrak{W}_{-r}$ where $\{\mathfrak{W}_{-r}, T\} = \mathcal{Z}^r \dot{\sim} 1$. Hence $\mathfrak{W}_{-r}$ is diagonal. But $\mathfrak{Z} \times \mathfrak{E}$ is a semi-field and $\mathfrak{E}$ is unique by Theorem 6.3. Thus each quantity of $\mathfrak{E} = \mathfrak{W}_{-r}$ is unaltered by ST^{-r} as well as by S and hence by T^{-r}. Since $\mathcal{Z}^n = \mathcal{J}$ we have $r \leqq n$. The idempotent f_1 of $\mathfrak{E} = (f_1, \cdots, f_n)$ has the property $f_1^{T^{-r}} = f_{1-r} = f_{n+1-r} \neq f_1$ unless $r = n$. This completes our proof.

11. Powers of cyclic systems. The cyclic systems $\mathcal{Z}^k$ $(k = 0, \cdots, m-1)$ which are the powers of a cyclic system $\mathcal{Z} = \{\mathfrak{Z}, S\}$ of index m define cyclic semi-fields $\mathfrak{Z}^{(k)}$ such that $\mathcal{Z}^k = \{\mathfrak{Z}^{(k)}, S\}$. Intuition suggests that the structure of $\mathfrak{Z}^{(k)}$ must be closely connected with that of $\mathfrak{Z}$. We shall give this connection and apply our result in the next chapter to obtain one of the most important theorems on cyclic algebras. Observe that since $\mathcal{Z} \dot{\sim} \mathcal{Z}_1$ implies that $\mathcal{Z}^k \dot{\sim} \mathcal{Z}_1^k$ we need only study the case where $\mathcal{Z} = \{\mathfrak{Z}, S\}$ with $\mathfrak{Z}$ a field.

If $\mathfrak{Z}$ is any cyclic field of degree n over $\mathfrak{F}$ and g is any divisor of n there exists a cyclic subfield $\mathfrak{Z}_g$ of degree g over $\mathfrak{F}$ of $\mathfrak{Z}$. If S is a generating automorphism of $\mathfrak{Z}$ over $\mathfrak{F}$ the automorphism

$$T: \qquad z_g \leftrightarrow z_g^S \qquad (z_g \text{ in } \mathfrak{Z}_g)$$

induced by S is a generating automorphism of $\mathfrak{Z}_g$ and may be indicated by S. Thus every cyclic system $\{\mathfrak{Z}, S\}$ of degree and index n defines the cyclic systems

$$\{\mathfrak{Z}_g, S\} \tag{26}$$

for every divisor g of n. Also if r is any integer prime to g, the automorphism S^r generates the cyclic group $[S^r]$ of $\mathfrak{Z}_g$ and hence defines the cyclic systems

$$\{\mathfrak{Z}_g, S^r\}. \tag{27}$$

We then have the result stated in

Theorem 18. *Let $\mathfrak{Z}$ be a cyclic field of degree n over $\mathfrak{F}$, k be an integer, d be the g.c.d. of n and k so that $n = dg$, $k = dq$. Then if S is a generating automorphism of $\mathfrak{Z}$ we have*

$$\{\mathfrak{Z}, S\}^k \dot{\sim} \{\mathfrak{Z}_g, S^r\}, \tag{28}$$

where r is any integer chosen so that $rq \equiv 1 \pmod{g}$.

We let $\mathfrak{E} = (f_1, \cdots, f_n)$ be diagonal, $f_i = f_1^{T^{i-1}}$, and form $\mathfrak{Z} \times \mathfrak{E} = \mathfrak{Z} \times \mathfrak{W}_{-d}$. Then $\{\mathfrak{Z}, S\}^d = \{\mathfrak{Z}, S\}^d \{\mathfrak{E}, T\} = \{\mathfrak{W}_{-d}, T\}$, where $\mathfrak{W}_{-d} = (w_1, w_1^T, \cdots, w_1^{T^{n-1}})$ over $\mathfrak{F}$ and, by (16) with $u_i = f_i$, $k = -d$, $n = dg$, we have

$$w_1 = y_1 f_1 + y_2 f_{1-d} + \cdots + y_g f_{1-(g-1)d}, \tag{29}$$

as well as

$$y_1 = z_1 + z_{1+g} + \cdots + z_{1+(d-1)g}, \qquad y_i = y_1^{S^{i-1}} \qquad (i = 1, \cdots, g). \tag{30}$$

Here we are taking $z_1, \cdots, z_n$ as a normal basis of the cyclic field $\mathfrak{Z}$ over $\mathfrak{F}$, and it is clear that $y_1, \cdots, y_g$ form a normal basis of the subfield $\mathfrak{Z}_g$ over $\mathfrak{F}$. Every idempotent of the cyclic semi-field $\mathfrak{Z} \times \mathfrak{E}$ is a sum of idempotents of $\mathfrak{E}$, and the elementary computation of the proof of Theorem 6.8 shows that the only idempotents of $\mathfrak{E}$ unaltered by ST^{-d} are sums of distinct ones of the idempotents $e_0, e_0^T, \cdots, e_0^{T^{d-1}}$, where

$$e_0 = f_1 + f_{1-d} + \cdots + f_{1-(g-1)d}. \tag{31}$$

Hence $\mathfrak{W}_{-d} = \mathfrak{E}_0 \times \mathfrak{X}$, $\mathfrak{E}_0 = (e_0, e_0^T, \cdots, e_0^{T^{d-1}})$ over $\mathfrak{F}$, $\mathfrak{X}$ is a cyclic field over $\mathfrak{F}$ with generating automorphism T^d. Now it is clear that $\{\mathfrak{X}e_0, T^d\} = \{\mathfrak{X}, T^d\}$ and also that $w_1, w_1^{T^d}, \cdots, w_1^{T^{d(g-1)}}$ are in $\mathfrak{X}e_0$. It follows that $\mathfrak{X}e_0$ is the linear set $(w_1, w_1^{T^d}, \cdots, w_1^{T^{d(g-1)}})$ over $\mathfrak{F}$. Moreover $(f_{1-d})^{T^d} = f_1$, and thus

$$\begin{aligned} w_1^{T^d} &= y_2 f_1 + y_3 f_{1-d} + \cdots + y_1 f_{1-(g-1)d} = w_1^S \\ &= y_1^S f_1 + y_1^{S^2} f_{1-d} + \cdots + y_1^{S^g} f_{1-(g-1)d}. \end{aligned} \tag{32}$$

The partial correspondence $w_1 \leftrightarrow y_1$ generates an equivalence of $\mathfrak{Z}_g$ and $\mathfrak{X}$ such that by (32) $\{\mathfrak{Z}_g, S\} = \{\mathfrak{X}, T^d\}$.

We have now proved that $\{\mathfrak{Z}, S\}^d \sim \{\mathfrak{Z}_g, S\}$. Then $\{\mathfrak{Z}, S\}^k = [\{\mathfrak{Z}, S\}^d]^q \sim \{\mathfrak{Z}_g, S\}^q$, where q is prime to g. This reduces the proof of our theorem to the case k prime to n and, to avoid introducing new notations we shall consider this case. We then wish to show that $\{\mathfrak{Z}, S\}^k = \{\mathfrak{Z}, S^r\}$ where $rk \equiv 1 \pmod{n}$.

As before form $\mathfrak{Z} \times \mathfrak{E}$ and have $\{\mathfrak{Z}, S\}^k = \{\mathfrak{W}_{-k}, T\}$. Since k is prime to n, the integers $1, 1-k, 1-2k, \cdots, 1-(n-1)k$ are all distinct modulo n. These are the subscripts on the idempotents $f_j = u_j$ of (16) and the partial correspondences $w_i \leftrightarrow z_i$ generate an equivalence of $\mathfrak{Z}$ and $\mathfrak{W}_{-k}$. Now if $rk \equiv 1 \pmod{n}$ we have $2 - rk \equiv 1 \pmod{n}$, $f_{2-rk} = f_1$. Hence

$$\begin{aligned} w_1^T &\equiv z_1 f_2 + z_2 f_{2-k} + \cdots + z_n f_{2-k(n-1)} \\ &= z_{r+1} f_1 + z_{r+2} f_{1-k} + \cdots + z_{r+n} f_{1-k(n-1)} \\ &\leftrightarrow z_{r+1} = z_1^{S^r}. \end{aligned}$$

Then $\{\mathfrak{Z}, S\}^k = \{\mathfrak{W}_{-k}, T\} = \{\mathfrak{Z}, S^r\}$ as desired.

This result completes our study of cyclic semi-fields. We now pass to the theory of cyclic algebras.

CHAPTER VII

CYCLIC ALGEBRAS AND p-ALGEBRAS

1. Generalized cyclic algebras. Let $\mathfrak{M}$ be a total matric algebra of degree t over $\mathfrak{F}$ so that

$$\mathfrak{M} = (e_{ij}\,;\, i, j = 1, \cdots, t),$$

where we have the usual laws of multiplication. Clearly if $\mathfrak{E} = (e_1, \cdots, e_t)$ is any diagonal algebra over $\mathfrak{F}$ we may imbed $\mathfrak{E}$ in $\mathfrak{M}$ and in fact take $e_i = e_{ii}$ $(i = 1, \cdots, t)$. By Section A4.9 if γ is any non-zero quantity of $\mathfrak{F}$ and

(1) $$j_0 = e_{12} + e_{23} + \cdots + e_{t-1,t} + \gamma e_{t1}$$

then

(2) $$\mathfrak{M} = \mathfrak{E} + j_0\mathfrak{E} + \cdots + j_0^{t-1}\mathfrak{E},$$

(3) $$dj_0 = j_0 d^S, \qquad j_0^t = \gamma,$$

for every d of $\mathfrak{E}$. We are assuming that the unity quantities of $\mathfrak{F}$, $\mathfrak{E}$, and $\mathfrak{M}$ all coincide and that S is the generating automorphism of $\mathfrak{E}$ determined by $e_i = e_1^{S^{i-1}}$.

Every cyclic semi-field $\mathfrak{Z}$ with generating automorphism S has the form $\mathfrak{Z} = \mathfrak{E} \times \mathfrak{Y}$ with $\mathfrak{E}$ as above, S^t inducing a generating automorphism T in the cyclic field $\mathfrak{Y}$ of degree m over $\mathfrak{F}$. As in Section 5.8 we may define a cyclic algebra $(\mathfrak{Y}, T, \gamma)$ of degree m over $\mathfrak{F}$. The order of $\mathfrak{Z}$ is $n = mt$ and we may prove

Theorem 1. *The algebra $\mathfrak{A} = \mathfrak{M} \times (\mathfrak{Y}, T, \gamma)$ contains $\mathfrak{Z}$ and has the form*

(4) $$\mathfrak{A} = \mathfrak{Z} + j\mathfrak{Z} + \cdots + j^{n-1}\mathfrak{Z}$$

such that

(5) $$zj = jz^S, \qquad j^n = \gamma \qquad (z \text{ in } \mathfrak{Z}).$$

*We call $\mathfrak{A}$ a **generalized cyclic algebra** and write*

(6) $$\mathfrak{A} = (\mathfrak{Z}, S, \gamma).$$

For $\mathfrak{B} = (\mathfrak{Y}, T, \gamma) = \mathfrak{Y} + j_1\mathfrak{Y} + \cdots + j_1^{m-1}\mathfrak{Y}$ such that $j_1^m = \gamma$, $yj_1 = j_1 y^T$ for every y of $\mathfrak{Y}$. Hence $\mathfrak{B} > \mathfrak{Y}$, $\mathfrak{M} > \mathfrak{E}$, $\mathfrak{M} \times \mathfrak{B} > \mathfrak{E} \times \mathfrak{Y} = \mathfrak{Z}$. We put

(7) $$j = e_{12} + \cdots + e_{t-1,t} + j_1 e_{t1}\,.$$

The formal computations of Section A4.9 then give

(8) $$j^t = j_1\,.$$

Every quantity of $\mathfrak{Z}$ is uniquely expressible in the form $z = y_1 e_1 + \cdots + y_t e_t$ for y_i in $\mathfrak{Y}$ and S may be chosen so that $z^S = y_t^T e_1 + y_1 e_2 + \cdots + y_{t-1} e_t$. Now if $1 \leqq i < t$ we have $y_i e_i j = y_i e_{i,i+1} = j y_i e_{i+1}$. Also $y_t e_t j = y_t j_1 e_{t1} = j_1 y_t^T e_{t1} = j(y_t^T e_1)$, so that in either case $(y_i e_i) j = j(y_i e_i)^S$. Hence $zj = jz^S$ for every z of $\mathfrak{Z}$. Moreover (8) implies that $j^n = j_1^m = \gamma$. The linear set $\mathfrak{A}_0$ which is the sum $(\mathfrak{Z}, j\mathfrak{Z}, \cdots, j^{n-1}\mathfrak{Z})$ over $\mathfrak{F}$ is a subalgebra of $\mathfrak{A}$ since we have shown that $(j^r\mathfrak{Z})(j^s\mathfrak{Z}) = j^{r+s}\mathfrak{Z}$. The quantity j_1 is in $\mathfrak{A}_0$ by (8) and evidently $\mathfrak{A}_0 \geqq \mathfrak{B}$, $\mathfrak{A}_0 \geqq \mathfrak{E}$, $\mathfrak{A}_0$ contains $je_1 = j_1 e_{t1}$. But j_1 and $j_1^{-1} = \gamma^{-1} j^{t(m-1)}$ are in $\mathfrak{A}_0$, so is e_{t1}. Thus $\mathfrak{A}_0$ contains $j - j_1 e_{t1} + \gamma e_{t1} = j_0$ and hence contains $\mathfrak{M} = \mathfrak{E} + j_0\mathfrak{E} + \cdots + j_0^{t-1}\mathfrak{E}$. It follows that $\mathfrak{A}_0 \geqq \mathfrak{M}\mathfrak{B} = \mathfrak{M} \times \mathfrak{B} = \mathfrak{A}$, $\mathfrak{A}_0 = \mathfrak{A}$ is the supplementary sum (4).

A generalized cyclic algebra is then a normal simple algebra of degree n with a commutative subalgebra $\mathfrak{Z}$ of order n. We may prove $\mathfrak{Z}$ maximal, that is,

Theorem 2. *The $\mathfrak{A}$-commutator of $\mathfrak{Z}$ in the generalized cyclic algebra $\mathfrak{A} = (\mathfrak{Z}, S, \gamma)$ is $\mathfrak{Z}$.*

For $\mathfrak{A}^{\mathfrak{Z}} \leqq \mathfrak{A}^{\mathfrak{Y}}$ of degree t over $\mathfrak{Y}$ by Theorem 4.12. Also $\mathfrak{A}^{\mathfrak{Y}} \geqq \mathfrak{M} \times \mathfrak{Y}$ which also has degree t over $\mathfrak{Y}$, $\mathfrak{A}^{\mathfrak{Y}} = \mathfrak{M} \times \mathfrak{Y}$. Now if a is in $\mathfrak{A}^{\mathfrak{Z}}$ it has the form $a = \sum_{i,j=1}^{t} a_{ij} e_{ij}$ with the a_{ij} in $\mathfrak{Y}$ and $ae_{qq} = \sum_{i=1}^{t} a_{iq} e_{iq} = e_{qq} a = \sum_{j=1}^{t} a_{qj} e_{qj}$ if and only if $a_{iq} = 0$ for $i \neq q$, a is in $\mathfrak{E} \times \mathfrak{Y}$ as desired.

As in Theorem 5.9 we now see that there exists a generalized cyclic algebra $(\mathfrak{Z}, S, \gamma)$ for every cyclic semi-field $\mathfrak{Z} = \mathfrak{E} \times \mathfrak{Y}$ with generating automorphism S and for every quantity γ of $\mathfrak{F}$. For, the algebras $\mathfrak{M}$ and $(\mathfrak{Y}, T, \gamma)$ defined above are known to exist and $\mathfrak{M} \times (\mathfrak{Y}, T, \gamma) = (\mathfrak{Z}, S, \gamma)$.

Cyclic algebras were defined in Section 5.8 as normal simple algebras of degree n over $\mathfrak{F}$ with either a cyclic maximal subfield or a cyclic splitting field of degree n over $\mathfrak{F}$. The corresponding results for generalized cyclic algebras are given in the following two theorems.

Theorem 3. *A normal simple algebra $\mathfrak{A}$ of degree n over $\mathfrak{F}$ is a generalized cyclic algebra if and only if $\mathfrak{A}$ has a cyclic splitting field whose degree m divides n.*

For if $\mathfrak{Y}_0$ is cyclic of degree m and splits $\mathfrak{A}$ it must split the division algebra factor $\mathfrak{D}$ of $\mathfrak{A}$. By the corollary of Theorem 4.22, $\mathfrak{B} = \mathfrak{M}_s \times \mathfrak{D}$ contains $\mathfrak{Y}$ equivalent to $\mathfrak{Y}_0$ as maximal subfield and hence is a cyclic algebra. The degree of $\mathfrak{B}$ is m and $\mathfrak{A} \sim \mathfrak{B}$ so that $\mathfrak{A} = \mathfrak{M}_t \times \mathfrak{B}$. The converse is immediate.

Theorem 4. *A normal simple algebra $\mathfrak{A}$ of degree n over $\mathfrak{F}$ is a generalized cyclic algebra if and only if $\mathfrak{A}$ contains a cyclic semi-field $\mathfrak{Z}$ of order n over $\mathfrak{F}$ with the same unity quantity as $\mathfrak{A}$.*

For if $\mathfrak{A}$ contains $\mathfrak{Z} = \mathfrak{E} \times \mathfrak{Y}$ the algebra $\mathfrak{A}^{\mathfrak{Y}}$ is normal simple of degree t over $\mathfrak{Y}$ and contains $\mathfrak{E}$. But by Theorem 4.2 the maximal number r of pairwise orthogonal primitive idempotents in $\mathfrak{A}^{\mathfrak{Y}}$ is at most the degree t of $\mathfrak{A}^{\mathfrak{Y}}$ and $\mathfrak{A}^{\mathfrak{Y}} = \mathfrak{M}_r \times \mathfrak{D}$, $\mathfrak{M}_r \sim 1$, $\mathfrak{D}$ a division algebra. However $\mathfrak{E} = (e_1, \cdots, e_t)$ has unity quantity $e_1 + \cdots + e_t = 1$, $r \geqq t$, $r = t$, $\mathfrak{A}^{\mathfrak{Y}} = \mathfrak{M} \times \mathfrak{Y}$ where $\mathfrak{M} \sim 1$ has degree t. Then $\mathfrak{A} = \mathfrak{M} \times \mathfrak{B}$ where $\mathfrak{B}$ is normal simple of degree m over $\mathfrak{F}$ and contains $\mathfrak{Y}$. It follows that $\mathfrak{B} = (\mathfrak{Y}, T, \gamma)$, $\mathfrak{A} = (\mathfrak{Z}, S, \gamma)$ as desired. The converse is immediate.

We observe that the generalized cyclic algebra $(\mathfrak{Z}, S, \gamma)$ is determined not only by $\mathfrak{Z}$ and γ but also in terms of S. Hence $\mathfrak{A} = (\mathfrak{Z}, S, \gamma)$ is uniquely determined, in the sense of formal equivalence, by γ and the cyclic system $\mathcal{Z} = (\mathfrak{Z}, S)$. We may therefore introduce the notation

$$\mathfrak{A} = (\mathcal{Z}, \gamma) \tag{9}$$

for such algebras. The use of this notation will simplify the statements of our theorems. Note that if $\mathcal{I}$ is the identity element of the set of all cyclic systems of degree n over $\mathfrak{F}$ then, by Theorem 7.1, $(\mathcal{I}, \gamma) \sim 1$ for every $\gamma \neq 0$ of $\mathfrak{F}$.

2. Elementary results. Our results of Section 5.9 may be readily extended to theorems on generalized cyclic algebras. We first have

Theorem 5. *An algebra $(\mathcal{Z}, 1)$ is a total matric algebra.*

For $\mathcal{Z} \dot{\sim} \mathcal{Y} = \{\mathfrak{Y}, T\}$ where $\mathfrak{Y}$ is a cyclic field. By Theorem 5.10, $(\mathcal{Y}, 1) = (\mathfrak{Y}, T, 1) \sim 1$. Hence by Theorem 7.1, $(\mathcal{Z}, 1) \sim (\mathcal{Y}, 1) \sim 1$.

We next have

Theorem 6. *The algebra $(\mathcal{Z}, \gamma) \sim 1$ if and only if γ is the norm $N_{\mathfrak{Z}|\mathfrak{F}}(z)$ of a quantity z in a cyclic semi-field $\mathfrak{Z}$ defining $\mathcal{Z}$.*

For $(\mathcal{Z}, \gamma) \sim (\mathcal{Y}, \gamma) \sim 1$ if and only if $\gamma = N_{\mathfrak{Y}|\mathfrak{F}}(b)$, b in $\mathfrak{Y}$. Our result is then an immediate consequence of Theorem 6.13.

The theorem just proved may be regarded as a consequence of Theorem 7.5 and of the special case $\gamma = 1$ of

Theorem 7. *The algebras $(\mathcal{Z}, \gamma)$ and $(\mathcal{Z}, \delta)$ are equivalent if and only if $\delta = \gamma N_{\mathfrak{Z}|\mathfrak{F}}(z)$ for z in $\mathfrak{Z}$.*

For $(\mathcal{Z}, \gamma) \sim (\mathcal{Y}, \gamma)$, $(\mathcal{Z}, \delta) \sim (\mathcal{Y}, \delta)$, where $(\mathcal{Z}, \gamma)$ and $(\mathcal{Z}, \delta)$ are normal simple of the same degree, and similarly for $(\mathcal{Y}, \gamma)$ and $(\mathcal{Y}, \delta)$. But then $(\mathcal{Z}, \gamma) \cong (\mathcal{Z}, \delta)$ if and only if $(\mathcal{Y}, \gamma) \cong (\mathcal{Y}, \delta)$. Our result is then an immediate consequence of Theorems 5.13 and 6.13. Observe that we have also shown that $(\mathcal{Z}, \gamma) \cong (\mathcal{Z}, \delta)$ if and only if $\delta = \gamma N_{\mathfrak{Y}|\mathfrak{F}}(y)$ for y in the cyclic field $\mathfrak{Y}$.

In closing this section we observe again that any generalized cyclic algebra $\mathfrak{A} = (\mathfrak{Z}, S, \gamma)$ has the form $\mathfrak{A} = \mathfrak{Z} + j\mathfrak{Z} + \cdots + j^{n-1}\mathfrak{Z}$ with $j^n = \gamma$, $zj = jz^S$ for every z of $\mathfrak{Z}$. If r is any integer prime to n then $rr_0 = 1 + nn_0$, $j = \gamma^{-n_0}(j^r)^{r_0}$. Hence $\mathfrak{A}_0 = \mathfrak{Z} + j^r\mathfrak{Z} + \cdots + j^{r(n-1)}\mathfrak{Z} \geqq \mathfrak{A}$, $\mathfrak{A}_0 = \mathfrak{A}$. But $(j^r)^n = \gamma^r$, $zj^r = j^r z^{S^r}$ and we have

Theorem 8. *Let r be prime to the order n of $\mathfrak{Z}$. Then $(\mathfrak{Z}, S, \gamma) = (\mathfrak{Z}, S^r, \gamma^r)$.*

3. Applications of the theory of cyclic systems. We have seen in Theorem 7.1 how every cyclic system $\mathcal{Z}$ defines a generalized cyclic algebra $(\mathcal{Z}, \gamma)$ and that if $\mathcal{Y}$ is the cyclic system of equal degree and index similar to $\mathcal{Z}$ then $(\mathcal{Z}, \gamma) =$

$\mathfrak{M}_t \times (\mathfrak{Y}, \gamma)$. This result has also been expressed by $(\mathfrak{Z}, \gamma) \sim (\mathfrak{Y}, \gamma)$ under our assumption about the degree and index of $\mathfrak{Y}$. If also $\mathfrak{Z}_1 \dot{\sim} \mathfrak{Y}$ then $\mathfrak{Z}_1 \dot{\sim} \mathfrak{Z}$, $(\mathfrak{Z}_1, \gamma) \sim (\mathfrak{Y}, \gamma) \sim (\mathfrak{Z}, \gamma)$. We have now shown that the consequence of Theorem 7.1 has the alternative statement

Theorem 9. *Let* $\mathfrak{Z}_1, \dot{\sim} \mathfrak{Z}_2$. *Then* $(\mathfrak{Z}_1, \gamma) \sim (\mathfrak{Z}_2, \gamma)$.

The most fundamental of our applications may now be derived. The property is given in

Theorem 10. *Let* $\mathfrak{Z}$ *and* $\mathfrak{C}$ *be cyclic systems of the same degree* n *and* $\gamma \neq 0$, $\delta \neq 0$ *be in* $\mathfrak{F}$. *Then*

$$(\mathfrak{Z}, \gamma) \times (\mathfrak{C}, \delta) = (\mathfrak{Z}, \gamma\delta^{-1}) \times (\mathfrak{Z}\mathfrak{C}, \delta). \tag{10}$$

For $\mathfrak{Z} = \{\mathfrak{Z}, S\}$, $\mathfrak{A} = (\mathfrak{Z}, \gamma) = \mathfrak{Z} + j\mathfrak{Z} + \cdots + j^{n-1}\mathfrak{Z}$ such that $j^n = \gamma$' $zj = jz^S$ for every z of $\mathfrak{Z}$. Also $\mathfrak{B} = (\mathfrak{C}, \delta) = \mathfrak{C} + d\mathfrak{C} + \cdots + d^{n-1}\mathfrak{C}$, $d^n = \delta$' $cd = dc^T$ for every c of $\mathfrak{C}$ where $\mathfrak{C} = \{\mathfrak{C}, T\}$. By Theorem 6.15 $\mathfrak{Z} \times \mathfrak{C} = \mathfrak{Z} \times \mathfrak{W}_{-1}$ where $\mathfrak{Z}\mathfrak{C} = \{\mathfrak{W}_{-1}, T\}$, $\mathfrak{W}_{-1}$ is the set of all quantities of $\mathfrak{Z} \times \mathfrak{C}$ unaltered by ST^{-1}. Hence $\mathfrak{W}_{-1}$ is the set of all quantities of $\mathfrak{Z} \times \mathfrak{C}$ commutative with jd^{-1}. Now $\mathfrak{S} = \mathfrak{A} \times \mathfrak{B} \geqq \mathfrak{A}_1 = (\mathfrak{Z}, ST^{-1}, \gamma\delta^{-1}) = \mathfrak{Z} + jd^{-1}\mathfrak{Z} + \cdots + (jd^{-1})^{n-1}\mathfrak{Z}$. Also $\mathfrak{A} \times \mathfrak{B} \geqq \mathfrak{B}_1 = (\mathfrak{W}_{-1}, T, \delta) = \mathfrak{W}_{-1} + d\mathfrak{W}_{-1} + \cdots + d^{n-1}\mathfrak{W}_{-1}$. The algebras $\mathfrak{A}_1$ and $\mathfrak{B}_1$ are both generalized cyclic algebras of degree n over $\mathfrak{F}$ and hence are normal simple, $\mathfrak{S} = \mathfrak{A}_1 \times \mathfrak{S}^{\mathfrak{A}_1}$ by Theorem 4.6. Moreover it is clear that $a_1b_1 = b_1a_1$ for every a_1 of $\mathfrak{A}_1$, b_1 of $\mathfrak{B}_1$. Hence $\mathfrak{B}_1 \leqq \mathfrak{S}^{\mathfrak{A}_1}$. However $\mathfrak{S}^{\mathfrak{A}_1}$ has order n^2 over $\mathfrak{F}$, $\mathfrak{S}^{\mathfrak{A}_1}$ has the same order as $\mathfrak{B}_1$, $\mathfrak{S}^{\mathfrak{A}_1} = \mathfrak{B}_1$, $\mathfrak{A} \times \mathfrak{B} = \mathfrak{A}_1 \times \mathfrak{B}_1$ as desired.

Observe that our proof of Theorem 7.10 depends only upon the general theory of normal simple algebras, the theory of cyclic systems, and the structural properties of cyclic algebras as used in the proof of Theorem 7.1. It does not depend at all upon Theorem 5.11 or its corollary, Theorem 5.12, but will actually be specialized shortly to give a second and independent proof of these results.

We first put $\delta = \gamma$ in Theorem 7.10 and have $(\mathfrak{Z}, \gamma\delta^{-1}) = (\mathfrak{Z}, 1) \sim 1$. Then Theorem 7.10 yields

Theorem 11. *Let* $\mathfrak{Z}$ *and* $\mathfrak{C}$ *be cyclic systems of the same degree. Then*

$$(\mathfrak{Z}, \gamma) \times (\mathfrak{C}, \gamma) \sim (\mathfrak{Z}\mathfrak{C}, \gamma). \tag{11}$$

An evident induction gives

Theorem 12. *The direct power* $(\mathfrak{Z}, \gamma)^k \sim (\mathfrak{Z}^k, \gamma)$.

One should note that by Theorems 7.1 and 5.11 we also have $(\mathfrak{Z}, \gamma)^k \sim (\mathfrak{Z}, \gamma^k)$. In our further applications we shall restrict our attention exclusively to the case where $\mathfrak{Z}$ is a cyclic system of degree and index n so that $\mathfrak{Z} = \{\mathfrak{Z}, S\}$ where $\mathfrak{Z}$ is a cyclic field of degree n over $\mathfrak{F}$ with generating automorphism S. This will not lessen the generality of our results but will simplify their statement. Observe that our algebras $(\mathfrak{Z}, \gamma)$ are now ordinary cyclic algebras. We shall first apply Theorem 6.18 to Theorem 7.8 and obtain immediately

Theorem 13. *Let q be prime to n, $qr \equiv 1 \pmod{n}$ where n is the degree and index of $\mathfrak{Z}$. Then $(\mathfrak{Z}, \gamma) = (\mathfrak{Z}^q, \gamma^r)$.*

Theorem 6.18 will now be applied to Theorem 7.12 to give as an immediate corollary a result which has important consequences for the theory of norms in cyclic fields. We state this fundamental result as

Theorem 14. *Let $\mathfrak{Z}$ be a cyclic field of degree n over $\mathfrak{F}$ with generating automorphism S, d be the g.c.d. of $n = dg$ and an integer $k = dq$, $qr \equiv 1 \pmod{g}$. Then if $\mathfrak{Y}$ is the cyclic subfield of $\mathfrak{Z}$ of degree g over $\mathfrak{F}$, $\mathfrak{Z} = \{\mathfrak{Z}, S\}$, and $\mathfrak{Y}_r = \{\mathfrak{Y}, S^r\}$, we have*

$$(\mathfrak{Z}, \gamma)^k \sim (\mathfrak{Y}_r, \gamma). \tag{12}$$

We have evidently obtained this theorem as a rather direct consequence of the specialization, Theorem 7.12, of Theorem 7.10 and our theory of cyclic systems. We now apply Theorems 7.10 and 7.13 to give a new proof of Theorem 5.11, the result of which may be given symbolically as

$$(\mathfrak{Z}, \gamma) \times (\mathfrak{Z}, \delta) \sim (\mathfrak{Z}, \gamma\delta). \tag{13}$$

By Theorem 7.9 it is sufficient to consider the case where $\mathfrak{Z}$ has equal degree and index n. Take $q = n - 1$, $r = -1$ in Theorem 7.13 to get $(\mathfrak{Z}, \delta) = (\mathfrak{Z}^{n-1}, \delta^{-1})$. Then $(\mathfrak{Z}, \gamma) \times (\mathfrak{Z}, \delta) = (\mathfrak{Z}, \gamma) \times (\mathfrak{Z}^{n-1}, \delta^{-1}) = (\mathfrak{Z}, \gamma\delta) \times (\mathfrak{Z}^n, \delta^{-1})$. But by Theorem 6.17 $\mathfrak{Z}^n = \mathfrak{I}$, and hence by Theorem 7.1 $(\mathfrak{Z}^n, \delta^{-1}) \sim 1$ as desired.

If $\mathfrak{Y}$ is a subfield of a cyclic field $\mathfrak{Z}$ then $\mathfrak{Y} \times (\mathfrak{Z}, S, \gamma)$ has a total matric algebra direct factor whose degree is that of $\mathfrak{Y}$. It is then interesting to form the direct product $(\mathfrak{Z}, S, \gamma) \times (\mathfrak{Y}, S, \delta)$. We do this and obtain

Theorem 15. *Let $\mathfrak{Z}$, $\mathfrak{Z}$ and $\mathfrak{Y}$ be as in Theorem 7.14 and define $\mathfrak{Y} = \{\mathfrak{Y}, S\}$. Then*

$$(\mathfrak{Z}, \gamma) \times (\mathfrak{Y}, \delta) \sim (\mathfrak{Z}, \gamma\delta^d). \tag{14}$$

For $(\mathfrak{Y}, \delta) \sim (\mathfrak{Z}, \delta)^d \sim (\mathfrak{Z}, \delta^d)$ by Theorem 7.14 and (13). We use (13) to obtain $(\mathfrak{Z}, \gamma) \times (\mathfrak{Y}, \delta) \sim (\mathfrak{Z}, \gamma) \times (\mathfrak{Z}, \delta^d) \sim (\mathfrak{Z}, \gamma\delta^d)$ as desired.

4. Norms and exponents. The result of Theorem 7.14 combined with (13) gives

$$(\mathfrak{Z}, \gamma^k) \sim (\mathfrak{Y}_r, \gamma). \tag{15}$$

But then we use Theorem 7.6 to obtain a result on norms in a cyclic field which seems not to have been suspected until the proof of Theorem 7.14 was given in [22].

Theorem 16. *Let $\mathfrak{Z}$ be a cyclic field of degree n over $\mathfrak{F}$, d be the g.c.d. of $n = dg$ and an integer k, $\mathfrak{Y}$ be the cyclic subfield of degree g of $\mathfrak{Z}$. Then the kth power of a quantity γ of $\mathfrak{F}$ is the norm,*

$$\gamma^k = N_{\mathfrak{Z}|\mathfrak{F}}(z), \tag{16}$$

of a quantity z *of* $\mathfrak{Z}$ *if and only if* γ *is the norm*

$$(17) \qquad \gamma = N_{\mathfrak{Y}|\mathfrak{F}}(y)$$

of a quantity y *of* $\mathfrak{Y}$.

Our proof is merely the observation that $(\mathfrak{Z}, \gamma^k)$ is a total matric algebra if and only if the similar algebra $(\mathfrak{Y}_r, \gamma)$ is a total matric algebra. That r may be not unity does not alter the norm, $N_{\mathfrak{Y}|\mathfrak{F}}(y)$, since norms are independent of the particular generating automorphisms selected in the cyclic field $\mathfrak{Y}$.

If ρ is the exponent of a cyclic algebra $(\mathfrak{Z}, \gamma)$ then $(\mathfrak{Z}, \gamma)^\rho \sim (\mathfrak{Z}, \gamma^\rho) \sim 1$ so that γ^ρ is the norm $N_{\mathfrak{Z}|\mathfrak{F}}(z)$, z in $\mathfrak{Z}$. If also $(\mathfrak{Z}, \gamma)^\sigma \sim 1$ then we have seen that ρ divides σ, $\gamma^\sigma = N_{\mathfrak{Z}|\mathfrak{F}}(z_0)$, z_0 in $\mathfrak{Z}$. This gives

Theorem 17. *The exponent of a cyclic algebra* $(\mathfrak{Z}, S, \gamma)$ *is the least positive integer* ρ *such that* γ^ρ *is the norm of a quantity of* $\mathfrak{Z}$. *Then* ρ *divides any integer* σ *such that* $\gamma^\sigma = N_{\mathfrak{Z}|\mathfrak{F}}(z_0)$, z_0 *in* $\mathfrak{Z}$.

We also apply Theorem 7.16 to obtain

Theorem 18. *Let* $\mathfrak{Z}$ *be a cyclic field of degree* n *over* $\mathfrak{F}$ *and* e *be the largest integer such that* γ *is the norm*

$$(18) \qquad \gamma = N_{\mathfrak{Z}_e|\mathfrak{F}}(z_e)$$

with z_e *in the cyclic subfield* $\mathfrak{Z}_e$ *of* $\mathfrak{Z}$ *of degree* e *over* $\mathfrak{F}$. *Then* $n = e\rho$ *where* ρ *is the exponent of* $(\mathfrak{Z}, S, \gamma)$.

For by Theorem 7.16, $\gamma = N_{\mathfrak{Z}_e|\mathfrak{F}}(z_e)$ if and only if $\gamma^\rho = N_{\mathfrak{Z}|\mathfrak{F}}(z)$ where $n = e\rho$. When ρ takes on its minimal positive integral value, e takes on its maximal value.

Thus, while $(\mathfrak{Z}, S, \gamma)$ may not be equivalent to $(\mathfrak{Z}, S_1, \gamma)$, these algebras always have the same exponent. This follows, of course, from Theorem 7.8 since the replacement of S by S_1 is equivalent to the replacement of γ by a power of γ with exponent prime to n and this clearly does not affect the integer e of our theorem.

As a corollary of the result above we have

Theorem 19. *Let* $\mathfrak{Z}$ *be cyclic of degree* n *over* $\mathfrak{F}$, $\gamma \neq 0$ *be in* $\mathfrak{F}$. *Then the cyclic algebra* $(\mathfrak{Z}, S, \gamma)$ *has exponent* n *if and only if* γ *is not the norm*

$$(19) \qquad N_{\mathfrak{Z}_p|\mathfrak{F}}(z_p)$$

for p *ranging over all the prime divisors of* n, $\mathfrak{Z}_p$ *the cyclic subfield of* $\mathfrak{Z}$ *of degree* p *over* $\mathfrak{F}$, *and* z_p *any quantity of* $\mathfrak{Z}_p$.

For by Theorem 7.18, $(\mathfrak{Z}, S, \gamma)$ has exponent n if and only if $\gamma \neq N_{\mathfrak{Z}_e|\mathfrak{F}}(z_e)$ for any divisor $e > 1$ of n. If $e > 1$ and $\gamma = N_{\mathfrak{Z}_e|\mathfrak{F}}(z_e)$ then $e = fp$ where p is a prime divisor of n, $\gamma^f = N_{\mathfrak{Z}_e|\mathfrak{F}}(z_e^f)$. By Theorem 7.16 we have $\gamma = N_{\mathfrak{Z}_p|\mathfrak{F}}(z_p)$, a contradiction. The converse is trivial.

Observe that our theorem reduces the problem of constructing cyclic algebras of exponent n to the problem of finding, for every prime divisor of n, a quantity γ_p which is not the norm $N_{\mathfrak{Z}_p|\mathfrak{F}}(z_p)$. For if such quantities γ_p have been found we choose integers a_p such that

$$a_p \equiv 0 \pmod{r_p}, \qquad a_p \equiv 1 \pmod{p^e}, \tag{20}$$

where $n = p^e r_p$, r_p not divisible by p. Then define

$$\gamma = \prod_p \gamma_p^{a_p}. \tag{21}$$

That the a_p exist is trivial since $a_p = r_p^{p-1}$ satisfies (20). Then $\gamma = \delta^p \gamma_p \neq N_{\mathfrak{Z}_p|\mathfrak{F}}(z_p)$ as desired. The algebras constructed in this way are cyclic algebras of degree and exponent n and hence are division algebras. Conversely, every cyclic (division) algebra of degree and exponent n has this form. For if $\mathfrak{A} = (\mathfrak{Z}, S, \gamma_0)$ with $\gamma_0 \neq N_{\mathfrak{Z}_p|\mathfrak{F}}(z_p)$ then we put

$$\gamma = \prod_p \gamma_0^{a_p} = \gamma_0^{\tau},$$

where $\tau = \sum a_p$. But by (20), $\tau \equiv 1 \pmod{p_i^{e_i}}$ for every prime divisor p_i of $n = p_1^{e_1} \cdots p_t^{e_t}$ and hence $\tau \equiv 1 \pmod n$, $\tau = 1 + n\sigma$, $\gamma = \gamma_0 N_{\mathfrak{Z}|\mathfrak{F}}(\gamma_0^{\sigma})$, $\mathfrak{A} = (\mathfrak{Z}, S, \gamma)$, as desired.

We shall also show in Chapter IX that all normal division algebras of degree n over any algebraic number field $\mathfrak{K}$ are cyclic algebras of exponent n so that our result will furnish a rather simple construction of all normal division algebras over $\mathfrak{K}$.

5. Algebras of prime-power degree. The important structural property of Theorem 5.18 reduces the problem of determining all normal division algebras $\mathfrak{D}$ of degree n over $\mathfrak{F}$ to the case where n is a power p^e of a prime p. For it stated that $\mathfrak{D}$ is expressible uniquely, in the sense of equivalence, in the form

$$\mathfrak{D} = \mathfrak{D}_1 \times \cdots \times \mathfrak{D}_t, \tag{22}$$

where $\mathfrak{D}_i$ is a normal division algebra of degree $p_i^{e_i}$,

$$n = p_1^{e_1} \cdots p_t^{e_t}, \tag{23}$$

for the p_i distinct primes. Then the construction of all $\mathfrak{D}_i$ will give all algebras $\mathfrak{D}$.

There are many properties of $\mathfrak{D}$ which can be determined only if we know how to combine analogous properties of its direct factors $\mathfrak{D}_i$. In particular we may ask whether the hypothesis that the $\mathfrak{D}_i$ are cyclic implies that $\mathfrak{D}$ is cyclic, and conversely. We shall answer these questions in

Theorem 20. *Let $\mathfrak{Z}$ be a cyclic field of degree n over $\mathfrak{F}$ so that* (23) *holds and*

$$\mathfrak{Z} = \mathfrak{Z}_1 \times \cdots \times \mathfrak{Z}_t, \tag{24}$$

where $\mathfrak{Z}_i$ is the uniquely determined cyclic subfield of degree $p_i^{e_i}$ of $\mathfrak{Z}$. Then for every cyclic algebra $\mathfrak{A} = (\mathfrak{Z}, S, \gamma)$ we have

(25) $$\mathfrak{A} = (\mathfrak{Z}, S, \gamma) = \mathfrak{A}_1 \times \cdots \times \mathfrak{A}_t,$$

with

(26) $$\mathfrak{A}_i = (\mathfrak{Z}_i, S_i, \gamma),$$

where S_i is the automorphism of $\mathfrak{Z}_i$ induced by $S^{np_i^{-e_i}}$. Conversely, any direct product $\mathfrak{A}_1 \times \cdots \times \mathfrak{A}_t$ of cyclic algebras $\mathfrak{A}_i = (\mathfrak{Z}_i, S_i, \gamma_i)$ of degrees $p_i^{e_i}$, where the p_i are distinct primes ($i = 1, \cdots, t$), is a cyclic algebra

(27) $$\mathfrak{A} = (\mathfrak{Z}, S, \gamma) = (\mathfrak{Z}, S_0, \gamma_0),$$

where $\mathfrak{Z}$ is given by (24), $S^{np_i^{-e_i}}$ induces S_i in $\mathfrak{Z}_i$, $\mathfrak{A}_i = (\mathfrak{Z}_i, S_i, \gamma)$, and, alternatively, $S_0 = S_1 S_2 \cdots S_t$ induces S_i in $\mathfrak{Z}_i$ and

(28) $$\gamma_0 = \prod_{i=1}^{t} \gamma_i^{np_i^{-e_i}}.$$

For proof let $n = ab$, a and b relatively prime and hence $\mathfrak{Z} = \mathfrak{X} \times \mathfrak{Y}$ where $\mathfrak{X}$ is cyclic of degree a over $\mathfrak{F}$, $\mathfrak{Y}$ is cyclic of degree b over $\mathfrak{F}$. Then if S is a given generating automorphism of $\mathfrak{Z}$, the set $\mathfrak{Y}$ is the set of all quantities of $\mathfrak{Z}$ unaltered by S^b, $\mathfrak{X}$ the set unaltered by S^a. Now $(\mathfrak{Z}, S, \gamma) = \mathfrak{Z} + j\mathfrak{Z} + \cdots + j^{n-1}\mathfrak{Z}$, $j^n = \gamma$, $zj = jz^S$ for every z of $\mathfrak{Z}$. Also $\mathfrak{A}$ contains $\mathfrak{X} + j^b\mathfrak{X} + \cdots + j^{b(a-1)}\mathfrak{X} = (\mathfrak{X}, S^b, \gamma) = \mathfrak{B}$, as well as $\mathfrak{Y} + j^a\mathfrak{Y} + \cdots + j^{a(b-1)}\mathfrak{Y} = (\mathfrak{Y}, S^a, \gamma) = \mathfrak{C}$. Both $\mathfrak{B}$ and $\mathfrak{C}$ are normal simple subalgebras of $\mathfrak{A}$ and $\mathfrak{A}^{\mathfrak{B}} \geqq \mathfrak{C}$. The degree of $\mathfrak{A}^{\mathfrak{B}}$ is that of $\mathfrak{C}$ by Theorem 4.6 and hence $\mathfrak{A} = \mathfrak{B} \times \mathfrak{C}$. We take $a = p_1^{e_1}$, $b = p_2^{e_2} \cdots p_t^{e_t}$ and have

$$(\mathfrak{Z}, S, \gamma) = (\mathfrak{Z}_1, S^{np_1^{-e_1}}, \gamma) \times (\mathfrak{Y}, S^a, \gamma).$$

An evident induction yields our result. Conversely, take $\mathfrak{A} = \mathfrak{B} \times \mathfrak{C}$ where $\mathfrak{B} = (\mathfrak{X}, S_1, \gamma_1)$, $\mathfrak{C} = (\mathfrak{Y}, S_2, \gamma_2)$. Then by the exercise on page A186, $\mathfrak{Z} = \mathfrak{X} \times \mathfrak{Y}$ is cyclic of degree ab and has a generating automorphism S_0 such that S_0 induces S_1 in $\mathfrak{X}$, S_2 in $\mathfrak{Y}$. We may formally write $S_0 = S_1 S_2$. Now $\mathfrak{B} = \mathfrak{X} + j_1\mathfrak{X} + \cdots + j_1^{a-1}\mathfrak{X}$, $j_1^a = \gamma_1$, $xj_1 = j_1x^{S_1}$ for every x of $\mathfrak{X}$. Also the algebra $\mathfrak{C} = \mathfrak{Y} + j_2\mathfrak{Y} + \cdots + j_2^{b-1}\mathfrak{Y}$ with $j_2^b = \gamma_2$, $yj_2 = j_2y^{S_2}$ for every y of $\mathfrak{Y}$. Then $j = j_1j_2$ has the property $j^n = \gamma_1^b\gamma_2^a$ and $\mathfrak{Z} = \mathfrak{X} \times \mathfrak{Y}$ is a maximal subfield of $\mathfrak{A}$ such that $zj = jz^{S_0}$. By Theorem 5.9 $\mathfrak{A} = (\mathfrak{Z}, S_0, \gamma_1^b\gamma_2^a) = \mathfrak{Z} + j\mathfrak{Z} + \cdots + j^{n-1}\mathfrak{Z}$. An evident induction yields the last part of (27).

To derive the first part of (27) we obtain

LEMMA 1. *Let a and b be relatively prime. Then there exists an integer c such that*

(29) $$ca \equiv 1 \pmod{b}, \qquad cb \equiv 1 \pmod{a}.$$

For $aa_0 + bb_0 = 1$, $c = a_0 + gb$ satisfies the first congruence of (28) for every integer g. Then $b(a_0 + gb) - 1 = gb^2 + (ba_0 - 1) \equiv 0 \pmod{a}$ has a solution g since b^2 is prime to a.

Now let $S = S_0^c$ so that $\mathfrak{A} = (\mathfrak{Z}, S, \gamma)$ with $\gamma = (\gamma_1^b\gamma_2^a)^c$. Clearly $\gamma = \gamma_1\beta_1^a =$

$\gamma_2\beta_2^b$ with β_1 and β_2 in $\mathfrak{F}$ so that $\mathfrak{B} = (\mathfrak{X}, S^b, \gamma)$, $\mathfrak{C} = (\mathfrak{Y}, S^a, \gamma)$. However S^b induces $S_1^{cb} = S_1$ in $\mathfrak{X}$, S^a induces $S_2^{ca} = S_2$ in $\mathfrak{Y}$, $\mathfrak{B} = (\mathfrak{X}, S_1, \gamma)$, $\mathfrak{C} = (\mathfrak{Y}, S_2, \gamma)$. An evident induction yields our result.

Normal division algebras of prime-power degree p^e over $\mathfrak{F}$ fall into two distinct cases depending upon whether the characteristic of $\mathfrak{F}$ is or is not p. It does not seem that many general results of a simple nature are obtainable in the latter case. In the former case, however, we shall derive some rather remarkable algebraic theorems closely connecting the algebras considered with cyclic algebras. Our results are consequences of certain properties of pure inseparable fields over $\mathfrak{F}$ and we shall proceed to a derivation of these properties. We shall use the notations and properties of Section 2.16.

6. Lemmas on pure inseparable fields. In the remainder of this chapter we shall assume that the field of reference $\mathfrak{F}$ for all our algebras and algebraic extensions is any field of prime characteristic p. We first obtain a modification and partial restatement of the result of Lemma 2.11 in

LEMMA 2. *Let γ be in $\mathfrak{F}$ and $\mathfrak{K}$ be the simple extension of $\mathfrak{F}$ which is a stem field $\mathfrak{F}(j)$ of the equation $\lambda^{p^e} = \gamma$. Then either $\mathfrak{K} = \mathfrak{F}$ or there exists a quantity δ in $\mathfrak{F}$ such that $j^{p^f} = \delta$, $\delta \neq \epsilon^p$ for any ϵ of $\mathfrak{F}$, $\gamma = \delta^{p^{e-f}}$. The integer p^f is both the exponent and degree of $\mathfrak{K}$, a pure inseparable simple extension of $\mathfrak{F}$.*

For, either $\gamma = \delta^{p^e}$ and $\mathfrak{K} = \mathfrak{F}(\delta) = \mathfrak{F}$ or there is a maximum positive integer $g < e$ such that $\gamma = \delta^{p^g}$ for δ in $\mathfrak{F}$, $\delta \neq \epsilon^p$ for any ϵ of $\mathfrak{F}$. Evidently $j^{p^e} = \delta^{p^g}$, $f = e - g > 0$ gives $(j^{p^f} - \delta)^{p^g} = 0$ which is possible in a field $\mathfrak{K}$ if and only if $j^{p^f} = \delta$. By Lemma 2.11 the equation $j^{p^f} = \delta$ is irreducible, $\mathfrak{K}$ is pure inseparable of degree p^f over $\mathfrak{F}$. The exponent of $\mathfrak{K}$ is p^f by Theorem 2.29.

We next have

LEMMA 3. *Let $\mathfrak{K}$ be a pure inseparable extension of finite degree $n > 1$ and exponent p^e over $\mathfrak{F}$. Then $n = p^\nu$, $\nu \geqq e$, and there exist quantities $\gamma_1, \cdots, \gamma_t$ in $\mathfrak{F}$ such that $\gamma_i \neq \epsilon_i^p$ for any ϵ_i of $\mathfrak{F}$,*

$$\mathfrak{K} = \mathfrak{F}(j_1, \cdots, j_t), \qquad j_i^{p^{e_i}} = \gamma_i . \tag{30}$$

The integers t, $e_1, \cdots, e_t$ may be chosen so that $\mathfrak{K} \neq \mathfrak{F}(y_1, \cdots, y_s)$ for any $s < t$ and $y_i^{p^{f_i}} = \delta_i$ in $\mathfrak{F}$,

$$e = e_1 \geqq e_2 \geqq \cdots \geqq e_t > 0. \tag{31}$$

Also there exist fields $\mathfrak{K}_i$ such that $\mathfrak{K}_i$ is a simple pure inseparable field of degree p over $\mathfrak{K}_{i-1}$,

$$\mathfrak{K} = \mathfrak{K}_\nu > \mathfrak{K}_{\nu-1} > \cdots > \mathfrak{K}_1 > \mathfrak{K}_0 = \mathfrak{F}. \tag{32}$$

For, the definition of $\mathfrak{K}$ implies that it has the form (30) for some positive integer t. We then choose the minimum such t. It is also trivial to order the e_i, and $e = e_1$ by Theorem 2.29 and also $n = p^\nu$, $\nu \geqq e$. We use the notation of Section 2.16 to write

$$\mathfrak{K} > \mathfrak{K}^{(p)} > \cdots > \mathfrak{K}^{(p^e)} = \mathfrak{F} \tag{33}$$

so that $\mathfrak{K}^{(p^i)}$ is pure inseparable of exponent p over $\mathfrak{K}^{(p^{i+1})}$. But if $\mathfrak{L}$ is pure inseparable of exponent p over $\mathfrak{F}$ we have $\mathfrak{L} = \mathfrak{F}(z_1, \cdots, z_r)$, $z_i^p = \alpha_i$ in $\mathfrak{F}$ and if we choose r to be a minimum then $\mathfrak{L} = \mathfrak{L}_r > \mathfrak{L}_{r-1} = \mathfrak{F}(z_1, \cdots, z_{r-1})$. By Lemma 7.2, $\mathfrak{L}$ has degree p over $\mathfrak{L}_{r-1}$. An evident induction gives $\mathfrak{L}_r > \mathfrak{L}_{r-1} > \cdots > \mathfrak{L}_0 = \mathfrak{F}$, $\mathfrak{L}_i$ simple pure inseparable of degree p over $\mathfrak{L}_{i-1}$. Applying this result to $\mathfrak{K}^{(p^i)}$ over $\mathfrak{K}^{(p^{i+1})}$ we obtain our theorem.

When $\nu = e$ the sequences (32) and (33) coincide and we have the immediate corollary stated in

LEMMA 4. *Let* $\mathfrak{K} = \mathfrak{F}(j)$, $j^{p^e} = \gamma$ *in* $\mathfrak{F}$, *and* $\mathfrak{K}$ *have degree* p^e *over* $\mathfrak{F}$. *Then*

$$\mathfrak{K} > \mathfrak{K}^{(p)} > \cdots > \mathfrak{K}^{(p^e)} = \mathfrak{F}, \tag{34}$$

$\mathfrak{K}^{(p^i)} = \mathfrak{F}(j^{p^i})$, *and has degree* p *over* $\mathfrak{K}^{(p^{i+1})}$, $\mathfrak{K}$ *has degree* p *over* $\mathfrak{K}^{(p)}$.

We shall also require

LEMMA 5. *A quantity* a *of a simple pure inseparable field* $\mathfrak{K} > \mathfrak{F}$ *generates* $\mathfrak{K}$ *over* $\mathfrak{F}$ *if and only if* a *is not in* $\mathfrak{K}^{(p)}$.

For we let $\mathfrak{K} = \mathfrak{F}(j)$ as in Lemma 7.4 and may write $a = A_0 + A_1 j + \cdots + A_{p-1} j^{p-1}$ with A_i in $\mathfrak{K}^{(p)}$. If $\mathfrak{K} = \mathfrak{F}(a)$ then a in $\mathfrak{K}^{(p)}$ implies that $\mathfrak{F}(a) \leqq \mathfrak{K}^{(p)} < \mathfrak{K}$ which is impossible. Conversely, if a is not in $\mathfrak{K}^{(p)}$ at least one of $A_1, \cdots, A_{p-1}$ is not zero. But $a^{p^e} = (B_0 + B_1 x + \cdots + B_{p-1} x^{p-1})^p = \gamma$ in $\mathfrak{F}$ with the $B_i = A_i^{p^{e-1}}$ in $\mathfrak{F}$, $x = j^{p^{e-1}}$. Moreover $A_i \neq 0$ implies that $B_i \neq 0$ so that $B_0 + B_1 x + \cdots + B_{p-1} x^{p-1} = b$ is not in $\mathfrak{F}$. If $\gamma = \epsilon^p$ for ϵ in $\mathfrak{F}$ then $\epsilon^p = b^p$, $\epsilon = b$, which is impossible. By Lemma 7.2 the field $\mathfrak{F}(a)$ has degree p^e over $\mathfrak{F}$, $\mathfrak{K} = \mathfrak{F}(a)$.

There are several important connections between separable and pure inseparable extensions of $\mathfrak{F}$. The first of these is

LEMMA 6. *Let* a *be in a field* $\mathfrak{L}$ *which is separable of finite degree over* $\mathfrak{F}$, e *be any positive integer. Then there exists a pure inseparable extension* $\mathfrak{K}$ *of exponent at most* p^e *over* $\mathfrak{F}$ *such that* $a = b^{p^e}$ *for* b *in the composite* $\mathfrak{Z}$ *of* $\mathfrak{K}$ *and* $\mathfrak{L}$.

Note that by Theorem 2.31

$$\mathfrak{Z} = \mathfrak{K} \times \mathfrak{L} \tag{35}$$

for any pure inseparable $\mathfrak{K}$. To prove our lemma write $\mathfrak{L} = \mathfrak{F}(\xi)$ of degree m over $\mathfrak{F}$. By Theorem 2.28, $\mathfrak{L} = \mathfrak{F}(\xi^{p^e})$, $a = \alpha_1 + \alpha_2 \xi^{p^e} + \cdots + \alpha_m \xi^{p^e(m-1)}$ with the α_i in $\mathfrak{F}$. Put $\mathfrak{K} = \mathfrak{F}(\beta_1, \cdots, \beta_m)$, $\beta_i^{p^e} = \alpha_i$ and have $a = b^{p^e}$, $b = \beta_1 + \beta_2 \xi + \cdots + \beta_m \xi^{m-1}$. The field $\mathfrak{K}$ is evidently the field desired in our theorem.

Theorems 2.28 and 2.31 have as a consequence

LEMMA 7. *Let* $\mathfrak{L}$ *be separable of degree* m *over a pure inseparable extension* $\mathfrak{K}$ *of exponent* p^e *over* $\mathfrak{F}$, *and* $\mathfrak{L}_0$ *be the maximal separable subfield over* $\mathfrak{F}$ *of* $\mathfrak{L}$. *Then*

$$\mathfrak{L} = \mathfrak{L}_0 \times \mathfrak{K}, \tag{36}$$

$\mathfrak{L}_0$ has degree m over $\mathfrak{F}$. Moreover if ξ is any quantity such that $\mathfrak{L} = \mathfrak{K}(\xi)$ we have $\mathfrak{L}_0 = \mathfrak{F}(\xi^{p^e})$.

For $\mathfrak{L} = \mathfrak{K}(\xi)$ where ξ is a root of $f(\lambda) = \lambda^m + \alpha_1\lambda^{m-1} + \cdots + \alpha_m = 0$, α_i in $\mathfrak{K}$, $f(\lambda)$ a separable polynomial. Then by Theorem 2.28, $\mathfrak{L} = \mathfrak{K}(\xi^{p^e})$, ξ^{p^e} is a root of the separable equation $f^{(p^e)}(\lambda) = \lambda^m + \beta_1\lambda^{m-1} + \cdots + \beta_m = 0$, $\beta_i = \alpha_i^{p^e}$ in $\mathfrak{F}$. The polynomial $f^{(p^e)}(\lambda)$ is irreducible in $\mathfrak{K}$ and has coefficients in $\mathfrak{F}$ and hence must be irreducible in $\mathfrak{F}$. Then $\mathfrak{L}_1 = \mathfrak{F}(\xi^{p^e})$ is separable of degree m over $\mathfrak{F}$. By Theorem 2.31 the composite of $\mathfrak{L}_1$ and $\mathfrak{K}$ is their direct product $\mathfrak{L}_1 \times \mathfrak{K}$. This subfield of $\mathfrak{L}$ has degree m over $\mathfrak{K}$ and we must have $\mathfrak{L} = \mathfrak{L}_1 \times \mathfrak{K}$. That $\mathfrak{L}_0 = \mathfrak{L}_1$ follows from Theorem 2.31.

LEMMA 8. *Let the field $\mathfrak{L}$ of Lemma 7.7 be normal over $\mathfrak{K}$ and have automorphism group $\mathfrak{G} = (S_1, \cdots, S_m)$. Then the field $\mathfrak{L}_0$ of that lemma is normal over $\mathfrak{F}$ with group $\mathfrak{G}_0 = (S_{10}, \cdots, S_{m0})$ equivalent to $\mathfrak{G}$ and in fact chosen so that S_{i0} is the correspondence induced in $\mathfrak{L}_0$ by S_0, that is, S_{i0} is*

$$x_0 \leftrightarrow x_0^{S_{i0}} = x_0^{S_i} \tag{37}$$

for every x_0 of $\mathfrak{L}_0$.

The conjugates over $\mathfrak{F}$ of each quantity of $\mathfrak{L}_0$ are separable over $\mathfrak{F}$, hence in $\mathfrak{L}_0$. Thus $\mathfrak{L}_0$ is normal over $\mathfrak{F}$. If $\xi^{S_i} = \gamma_1^{(i)} + \gamma_2^{(i)}\xi + \cdots + \gamma_m^{(i)}\xi^{m-1}$ with the $\gamma_j^{(i)}$ in $\mathfrak{K}$ then $(\xi^{S_i})^{p^e} = (\xi^{p^e})^{S_i} = \delta_1^{(i)} + \delta_2^{(i)}\xi^{p^e} + \cdots + \delta_m^{(i)}\xi^{p^e(m-1)}$ is in $\mathfrak{L}_0$. The correspondence generated by $\xi^{p^e} \leftrightarrow (\xi^{p^e})^{S_i}$ clearly induces an automorphism S_{i0} of $\mathfrak{L}_0 = \mathfrak{F}(\xi^{p^e})$ with the desired property of our lemma.

We shall close our theory of inseparable fields with

LEMMA 9. *Let $\mathfrak{L}$ be separable of degree m over $\mathfrak{K}$, where $\mathfrak{K}$ is a pure inseparable simple extension of $\mathfrak{F}$. Then there exists a quantity b in $\mathfrak{L}$ such that the norm,*

$$a = N_{\mathfrak{L}|\mathfrak{K}}(b), \tag{38}$$

generates $\mathfrak{K}$ over $\mathfrak{F}$.

The result is trivial if $\mathfrak{F}$ is a finite field since then $\mathfrak{K} = \mathfrak{F}$, $b = 1 = a$. Hence assume that $\mathfrak{F}$ is infinite.

By Lemma 7.7 we may write $\mathfrak{L} = \mathfrak{L}_0 \times \mathfrak{K}$, $\mathfrak{L}_0 = \mathfrak{F}(\xi)$, ξ a root of a separable polynomial

$$f(x) = \sum_{i=0}^{p-1} x^i[a_i(x^p)].$$

The quantities $a_i(x^p)$ are polynomials in x^p with coefficients in $\mathfrak{F}$ and at least one $a_i(x^p) \neq 0$ for $i > 0$ since otherwise $f(x)$ would be inseparable. Let r be one such value of i, write $\mathfrak{K} = \mathfrak{F}(j)$, $j^{p^e} = \gamma$ in $\mathfrak{F}$, and replace x by λj. Then $a_r(\lambda^p j^p)$ is a polynomial in j^p with coefficients polynomials of $\mathfrak{F}[\lambda]$ not all zero and since $\mathfrak{F}$ is infinite we may choose a non-zero quantity α of $\mathfrak{F}$ such that $a_r(\alpha^p j^p) \neq 0$. Then if $b = \xi - \alpha j$ we have

$$f(\xi) = f(b + \alpha j) = b^m + c_1 b^{m-1} + \cdots + f(\alpha j), \tag{39}$$

with the c_i in $\mathfrak{K}$ and the constant term

$$(40) \qquad c_m = f(\alpha j) = \sum_{i=0}^{p-1} b_i j^i, \qquad b_i = \alpha^i a_i(\alpha^p j^p).$$

Now $b_r \neq 0$, $a = N_{\mathfrak{L}|\mathfrak{K}}(b) = (-1)^m f(\alpha j)$ is not in $\mathfrak{K}^{(p)}$. Hence by Lemma 7.5 $\mathfrak{K} = \mathfrak{F}(a)$ as desired.

7. Elementary properties of *p*-algebras. An algebra $\mathfrak{A}$ over $\mathfrak{F}$ will be called a *p-algebra* if p is the characteristic of $\mathfrak{F}$ and $\mathfrak{A}$ is a normal simple algebra over $\mathfrak{F}$ whose degree is a power of p. When $\mathfrak{F}$ is a finite field of characteristic p, every p-algebra over $\mathfrak{F}$ is not only a total matric algebra but also a cyclic algebra and the following theorems on p-algebras are trivially true. Hence we need only construct our proofs for the case where $\mathfrak{F}$ is infinite. We may apply Lemma 7.6 to obtain a result which is the foundation of our theory of p-algebras.

Theorem 21. *Every p-algebra $\mathfrak{A}$ of index p^e has a pure inseparable splitting field of finite degree and exponent at most p^e over $\mathfrak{F}$.*

For $\mathfrak{A} \sim \mathfrak{D}$, a normal division algebra of degree p^e over $\mathfrak{F}$ whose splitting fields coincide with those of $\mathfrak{A}$. If $e = 1$ we apply Theorem 4.32 to obtain a field $\mathfrak{L}$ of degree r, prime to p, over $\mathfrak{F}$ such that $\mathfrak{D}_\mathfrak{L}$ has a cyclic maximal subfield $\mathfrak{Z}$. Then $\mathfrak{D}_\mathfrak{L} = (\mathfrak{Z}, S, \gamma)$, γ in $\mathfrak{L}$. Clearly $\mathfrak{L}$ is separable and by Lemma 7.6 there exists a field $\mathfrak{K}$ which is pure inseparable of exponent one or p over $\mathfrak{F}$ such that $\gamma = \beta^p$, β in the field $\mathfrak{J} = \mathfrak{L} \times \mathfrak{K}$. Then $\mathfrak{D}_\mathfrak{J} = (\mathfrak{Z}_\mathfrak{J}, S, \beta^p) \sim 1$ by Theorem 5.14. The degree of $\mathfrak{J}$ over $\mathfrak{K}$ is r and if $\mathfrak{K}$ does not split $\mathfrak{D}$ then $\mathfrak{D}_\mathfrak{K}$ is a division algebra, $(\mathfrak{D}_\mathfrak{K})_\mathfrak{J} = \mathfrak{D}_\mathfrak{J}$ must be a division algebra by the corollary of Theorem 4.20. This is contrary to proof.

We have now completed the case $e = 1$ of our theorem and shall make an induction on e. By Theorem 4.31 there exists a field $\mathfrak{L}$ of degree r, prime to p, over $\mathfrak{F}$ and a separable field $\mathfrak{N}$ of degree p^{e-1} over $\mathfrak{L}$ such that $\mathfrak{D}_\mathfrak{N} \sim \mathfrak{B} = (\mathfrak{Z}, S, \gamma)$ of degree p over $\mathfrak{N}$. Then $\mathfrak{N}$ is separable over $\mathfrak{F}$, and, by Lemma 7.6, there exists a field $\mathfrak{K}$ which is inseparable of exponent at most p over $\mathfrak{F}$ such that $\gamma = \beta^p$ with β in $\mathfrak{J} = \mathfrak{K} \times \mathfrak{N}$. But then $\mathfrak{D}_\mathfrak{J} \sim (\mathfrak{Z}_\mathfrak{J}, S, \beta^p) \sim 1$, $\mathfrak{J}$ splits $\mathfrak{D}$. If $\mathfrak{D}_\mathfrak{K}$ were a division algebra the degree of the splitting field $\mathfrak{J}$ over $\mathfrak{K}$ would necessarily be divisible by p^e whereas it is $p^{e-1}r$. Hence $\mathfrak{D}_\mathfrak{K}$ has index $p^f < p^e$. By the hypothesis of our induction there exists a pure inseparable field $\mathfrak{K}_0$ of exponent at most p^f over $\mathfrak{K}$ such that $\mathfrak{K}_0$ splits $\mathfrak{D}_\mathfrak{K}$. Then $\mathfrak{K}_0$ is pure inseparable of exponent at most $p^{f+1} \leqq p^e$ over $\mathfrak{F}$ and splits $\mathfrak{D}$. This completes our proof.

A field $\mathfrak{F}$ of characteristic p is called *perfect* if there are no pure inseparable fields of degree greater than unity over $\mathfrak{F}$, that is, if every quantity of $\mathfrak{F}$ has a pth root in $\mathfrak{F}$. Theorem 7.21 then gives

Theorem 22. *Every p-algebra over a perfect field is a total matric algebra.*

As in the special case of finite fields we see that the theory of p-algebras over perfect fields of characteristic p is a trivial theory and we shall assume henceforth that $\mathfrak{F}$ is not perfect. This includes the assumption that $\mathfrak{F}$ is infinite. We shall also require another restriction on $\mathfrak{F}$ and first derive the preliminary

LEMMA 10. *Let γ be in $\mathfrak{F}$ of characteristic p and let $\mathfrak{K} = \mathfrak{F}(\gamma^{1/p})$ split a division p-algebra $\mathfrak{D}$. Then there exists a cyclic field $\mathfrak{Z}$ of degree p over $\mathfrak{F}$ such that $\mathfrak{D} = (\mathfrak{Z}, S, \gamma)$.*

For $\mathfrak{D}$ contains a quantity j such that $j^p = \gamma$, and, by Theorem 4.17, a quantity x such that $xj = j(x + 1)$. By the argument preceding Theorem 4.18, $\mathfrak{Z} = \mathfrak{F}(x)$ is a separable subfield of prime degree p of $\mathfrak{D}$, $j^{-1}xj = x + 1$, $x \leftrightarrow x + 1$ generates a non-identical automorphism S of $\mathfrak{Z}$. The order of S is then p and $\mathfrak{Z}$ is cyclic. By Theorem 5.9 $\mathfrak{D} = \mathfrak{Z} + j\mathfrak{Z} + \cdots + j^{p-1}\mathfrak{Z} = (\mathfrak{Z}, S, \gamma)$.

We use this lemma in the proof of

Theorem 23. *Let $\mathfrak{F}$ be a field of characteristic p such that there exist no cyclic fields of degree p over $\mathfrak{F}$. Then every p-algebra over $\mathfrak{F}$ is a total matric algebra.*

For let $\mathfrak{D}$ be a division p-algebra over $\mathfrak{F}$. We apply Theorem 7.21 to obtain a pure inseparable splitting field of degree p^ν over $\mathfrak{F}$ of $\mathfrak{D}$. By Lemma 7.3 we have the sequence of fields (32) and there exists an integer τ such that $\mathfrak{K}_\tau$ splits $\mathfrak{D}$ but $\mathfrak{K}_{\tau-1}$ does not split $\mathfrak{D}$. The field $\mathfrak{K}_\tau = \mathfrak{K}_{\tau-1}(\gamma^{1/p})$, γ in $\mathfrak{K}_{\tau-1}$. By Lemma 7.10, $\mathfrak{D}_{\mathfrak{K}_{\tau-1}} \sim (\mathfrak{Z}, S, \gamma)$ over $\mathfrak{K}_{\tau-1}$, where $\mathfrak{Z}$ is cyclic of degree p over $\mathfrak{K}_{\tau-1}$. But Lemma 7.8 states that $\mathfrak{Z} = \mathfrak{Z}_0 \times \mathfrak{K}_{\tau-1}$, $\mathfrak{Z}_0$ is cyclic of degree p over $\mathfrak{F}$, contrary to our hypothesis of the non-existence of such fields.

We shall henceforth assume the existence of cyclic fields of degree p over $\mathfrak{F}$ as well as the property that $\mathfrak{F}$ is not perfect. Then there exist cyclic fields of arbitrary degree p^e over $\mathfrak{F}$. In fact Theorem A9.8 implies directly that if $\mathfrak{Z}_f$ is cyclic of degree p^f over $\mathfrak{F}$ and e is any integer greater than f there exists a cyclic field $\mathfrak{Z}_e$ of degree p^e over $\mathfrak{F}$ and with $\mathfrak{Z}_f$ as subfield. This result and Theorem 7.14 with $k = p^{e-f}$, $q = r = 1$, have as an immediate consequence

LEMMA 11. *Let $\mathfrak{A}$ be a cyclic p-algebra $(\mathfrak{Z}_f, S, \delta)$ of degree p^f over $\mathfrak{F}$. Then if $e \geqq f$ there exists a cyclic field $\mathfrak{Z}$ of degree p^e over $\mathfrak{F}$ with $\mathfrak{Z}_f$ as subfield and such that*

$$\mathfrak{A} \sim (\mathfrak{Z}, S, \delta^{p^{e-f}}). \tag{41}$$

Lemma 7.11 will be used in the proof of the principal theorem of the next section. We shall also use its immediate consequence,

LEMMA 12. *Let $\mathfrak{A}$ be a p-algebra of degree p^e over $\mathfrak{F}$ and $\mathfrak{A} \sim \mathfrak{B}$ where $\mathfrak{B}$ is cyclic of degree $p^f \leqq p^e$. Then $\mathfrak{A}$ is cyclic.*

For if $\mathfrak{B} = (\mathfrak{Z}_f, S, \delta)$ we let $\mathfrak{Z}$ be a cyclic field of degree p^e over $\mathfrak{F}$, $\mathfrak{Z} \geqq \mathfrak{Z}_f$. Put $\gamma = \delta^\nu$, $\nu = p^{e-f}$ so that, by Lemma 7.11, $(\mathfrak{Z}, S, \gamma) \sim \mathfrak{B}$. But then $(\mathfrak{Z}, S, \gamma) \sim \mathfrak{A}$ and has the same degree as $\mathfrak{A}$, $\mathfrak{A} \cong (\mathfrak{Z}, S, \gamma)$ and $\mathfrak{A}$ is cyclic.

8. p-algebras with simple, pure inseparable splitting fields. We shall now apply the properties of pure inseparable fields derived in Section 7.6. Our first and basic result will be stated as

Theorem 24. *Let $n = p^e$, γ be in $\mathfrak{F}$, $\mathfrak{A}$ be a p-algebra with $\mathfrak{K} = \mathfrak{F}(\gamma^{1/n})$ as splitting field such that $\mathfrak{K}^{(p)} = \mathfrak{F}(\gamma^{p/n})$ does not split $\mathfrak{A}$. Then there exists a cyclic field $\mathfrak{Z}$ of degree n over $\mathfrak{F}$ such that*

$$\mathfrak{A} \sim (\mathfrak{Z}, S, \gamma). \tag{42}$$

Let us assume first that n is the degree of $\mathfrak{K}$ over $\mathfrak{F}$. Our hypotheses imply $e > 0$. If $e = 1$ the index of $\mathfrak{A}$ is p and our theorem is an interpretation of Lemma 7.10. Hence assume our theorem true for p-algebras with splitting fields $\mathfrak{K}$ of our theorem and of degree $p^i < p^e$.

Write $\mathfrak{K}_0 = \mathfrak{F}(\gamma^{1/p})$ so that $\mathfrak{K}$ has degree $m = p^{e-1}$ over $\mathfrak{K}_0$. Then $\mathfrak{K}$ splits $\mathfrak{A}_{\mathfrak{K}_0}$ but $\mathfrak{K}^{(p)}$ over $\mathfrak{K}_0$ does not and hence $\mathfrak{A}_{\mathfrak{K}_0}$ is similar to $(\mathfrak{W}_0, T_0, \gamma^{1/p})$, a cyclic algebra of degree m over $\mathfrak{K}_0$. Now $\mathfrak{A}$ is similar to an algebra $\mathfrak{B}$ of degree $n = p^e$ over $\mathfrak{F}$, $\mathfrak{B}$ has a subfield equivalent to $\mathfrak{K}$, and by Theorem 4.16, $\mathfrak{B} > (\mathfrak{W}, T, y)$ which is cyclic of degree m over $\mathfrak{Y} = \mathfrak{F}(y)$, $y^p = \gamma$. Here the $\mathfrak{B}$-commutator of $\mathfrak{Y}$ is actually $(\mathfrak{W}, T, y)$. Then $\mathfrak{C} = \mathfrak{W} + j\mathfrak{W} + \cdots + j^{m-1}\mathfrak{W}$, $wj = jw^T$, $j^m = y$, $j^n = \gamma$. The field $\mathfrak{W} = \mathfrak{Z}_0 \times \mathfrak{Y}$ over $\mathfrak{F}$ by Lemma 7.8 with $\mathfrak{Z}_0$ cyclic of degree m over $\mathfrak{F}$, $z_0 \leftrightarrow z_0^T$ is a generating automorphism of $\mathfrak{Z}_0$ and we shall use T to denote it. The algebra $\mathfrak{B}^{\mathfrak{Z}_0} = \mathfrak{P}$ is now a p-algebra of degree p over $\mathfrak{Z}_0$ and contains $\mathfrak{Y}$. Hence $\mathfrak{Z}_0(\gamma^{1/p})$ splits $\mathfrak{P}$. By Theorem 4.17 there exists a quantity $\bar{x}$ in $\mathfrak{P}$ such that $\bar{x}y = y(\bar{x} + 1)$.

Since $j^{-1}\mathfrak{Z}_0 j = \mathfrak{Z}_0$ the quantity $j^{-1}bj$ is in $\mathfrak{B}^{\mathfrak{Z}_0}$ for every b of $\mathfrak{B}^{\mathfrak{Z}_0}$. Hence $j^{-1}\bar{x}j$ is in $\mathfrak{B}^{\mathfrak{Z}_0}$. But $yj = jy$, $y^{-1}(j^{-1}\bar{x}j)y = j^{-1}(y^{-1}\bar{x}y)j = j^{-1}(\bar{x} + 1)j = j^{-1}\bar{x}j + 1$, $y^{-1}(j^{-1}\bar{x}j - \bar{x})y = j^{-1}\bar{x}j - \bar{x}$. Thus $w = j^{-1}\bar{x}j - \bar{x}$ is in the $\mathfrak{P}$-commutator of $\mathfrak{Z}_0(y) = \mathfrak{W}$. This algebra is $\mathfrak{W}$ and w is in $\mathfrak{W}$. But $j^{-1}wj = w^T$ for every w of $\mathfrak{W}$, $j^{-1}\bar{x}j = \bar{x} + w$, $j^{-2}\bar{x}j^2 = \bar{x} + w + w^T$, and an evident induction yields $j^{-m}\bar{x}j^m = \bar{x} + w + w^T + \cdots + w^{T^{m-1}} = \bar{x} + T_{\mathfrak{W}|\mathfrak{Y}}(w)$. However $j^m = y$, $y^{-1}\bar{x}y = \bar{x} + 1$,

$$T_{\mathfrak{W}|\mathfrak{Y}}(w) = 1.$$

Theorem A9.4 implies the existence of a quantity β in $\mathfrak{Z}_0$ such that

$$T_{\mathfrak{Z}_0|\mathfrak{F}}(\beta) = T_{\mathfrak{Z}_0|\mathfrak{F}}(\beta^p) = 1.$$

However $1 = T_{\mathfrak{Z}_0|\mathfrak{F}}(\beta) = T_{\mathfrak{W}|\mathfrak{Y}}(\beta) = T_{\mathfrak{W}|\mathfrak{Y}}(w)$, $T_{\mathfrak{W}|\mathfrak{Y}}(\beta - w) = 0$. By Theorem A9.6 there exists a quantity w_0 in $\mathfrak{W}$ such that $\beta - w = w_0^T - w_0$. We define

$$x = \bar{x} + w_0 \tag{43}$$

and have $j^{-1}xj = \bar{x} + w + w_0^T = \bar{x} + \beta + w_0 = x + \beta$. We shall prove that $\mathfrak{B} = (\mathfrak{Z}, S, \gamma)$ with $\mathfrak{Z} = \mathfrak{F}(x)$, $x^S = x + \beta$. The algebra $\mathfrak{B}$ contains the sum $(\mathfrak{Z}, j\mathfrak{Z}, \cdots, j^{n-1}\mathfrak{Z})$, and it remains to show that $\mathfrak{Z}$ is cyclic of degree n over $\mathfrak{F}$.

The minimum function of x over $\mathfrak{Z}_0$ has the form $f(\lambda) = \lambda^r + \alpha_1\lambda^{r-1} + \cdots + \alpha_r$ with the α_i in $\mathfrak{Z}_0$ and, since x is in the normal simple algebra $\mathfrak{P}$ of degree p

over $\mathfrak{Z}_0$, we must have $r \leqq p$. Then $y^{-1}f(x)y = f(x+1) = 0$, $\phi(\lambda) = f(\lambda+1) - f(\lambda) = r\lambda^{r-1} + \cdots + \alpha_{r-1}$ must be identically zero by our definition of $f(\lambda)$. Now x is not in $\mathfrak{F}$ and hence $r > 0$, r must be p. It follows that

$$(44) \qquad f(\lambda+1) - f(\lambda) = 1 + \sum_{i=1}^{p-2} \alpha_i[(\lambda+1)^{p-i} - \lambda^{p-i}] + \alpha_{p-1} \equiv 0,$$

$\alpha_i = 0$ for $i = 1, \cdots, p-2$, $\alpha_{p-1} = -1$. Then $f(\lambda) = \lambda^p - \lambda - a$ for a in $\mathfrak{Z}_0$. Now $j^{-1}f(x)j = (x+\beta)^p - (x+\beta) - a^T = x + a + \beta^p - \beta - x - a^T = 0$, $\beta^p - \beta = a^T - a$. By Theorem A9.8 the equation $\lambda^p = \lambda + a$ is irreducible in $\mathfrak{Z}_0$ and defines a field $\mathfrak{F}(x) = \mathfrak{Z}$ which is cyclic of degree p^e over $\mathfrak{F}$ such that the equations $x^S = x + \beta$, $z^S = z^T$ for z in $\mathfrak{Z}_0$, generate a generating automorphism of $\mathfrak{Z}$. Clearly $\mathfrak{B}$ must equal $\mathfrak{Z} + j\mathfrak{Z} + \cdots + j^{n-1}\mathfrak{Z} = (\mathfrak{Z}, S, \gamma)$ as desired.

There remains the case where the degree of $\mathfrak{K}$ is less than p^e. This is possible only when $\mathfrak{K} = \mathfrak{F}(\delta^{1/p^f})$, $\delta^{p^{e-f}} = \gamma$. By our proof $\mathfrak{A} \sim (\mathfrak{Z}_f, S, \delta)$ of degree p^f. By Lemma 7.11, $\mathfrak{A} \sim (\mathfrak{Z}, S, \gamma)$ as desired.

Our result may be stated in an equivalent form which is sometimes easier to apply.

Theorem 25. *Let $\mathfrak{K} = \mathfrak{F}(\gamma^{1/n})$, $n = p^e$, be a splitting field of a p-algebra $\mathfrak{A}$. Then $\mathfrak{A} \sim (\mathfrak{Z}, S, \gamma)$ where $\mathfrak{Z}$ is a cyclic field of degree $p^f \leqq n$.*

For our hypothesis implies that some subfield $\mathfrak{K}^{(p^i)}$ of $\mathfrak{K}$ splits $\mathfrak{A}$ and $\mathfrak{K}^{(p^{i+1})}$ does not for some integer i. Clearly $\mathfrak{K}^{(p^i)} = \mathfrak{F}(\gamma^{1/m_i})$, $m_i = p^{e-i}$, and by Theorem 7.24, $\mathfrak{A} \sim (\mathfrak{Z}, S, \gamma)$ where $\mathfrak{Z}$ has degree $m_i = p^f$ over $\mathfrak{F}$, $f = e - i$. Conversely, let us assume Theorem 7.25 and that $\mathfrak{K}$ is as in Theorem 7.24. Then $\mathfrak{A} \sim (\mathfrak{Z}, S, \gamma)$ of degree $m = p^f$ over $\mathfrak{F}$, $\mathfrak{A}_{\mathfrak{Y}} \sim (\mathfrak{Z}_{\mathfrak{Y}}, S, \delta^m) \sim 1$ if $\mathfrak{Y} = \mathfrak{F}(\gamma^{1/m})$. But then $\mathfrak{Y} \leqq \mathfrak{K}$ splits $\mathfrak{A}$, our hypothesis implies that $\mathfrak{Y} = \mathfrak{K}$, $m = n$, $(\mathfrak{Z}, S, \gamma)$ has the degree desired in the conclusion of Theorem 7.24.

We shall also state the corollary

Theorem 26. *Let $\mathfrak{D}$ be a division p-algebra of degree $n = p^e$. Then $\mathfrak{D}$ is a cyclic algebra if and only if $\mathfrak{D}$ has as splitting field a simple pure inseparable extension $\mathfrak{F}(\gamma^{1/n})$ of degree p^e over $\mathfrak{F}$. In this case $\mathfrak{D} = (\mathfrak{Z}, S, \gamma)$.*

For if $\mathfrak{D}$ is cyclic we have $\mathfrak{D} = (\mathfrak{Z}, S, \gamma)$ and there exists a quantity j in $\mathfrak{D}$ whose minimum function is $\lambda^n - \gamma$. But then $\mathfrak{F}(j)$ is a maximal subfield of $\mathfrak{D}$, $\mathfrak{K} = \mathfrak{F}(\gamma^{1/p^e})$ is the desired splitting field. Conversely if $\mathfrak{K}$ is a simple pure inseparable extension whose degree is p^e over $\mathfrak{F}$, and $\mathfrak{K}$ splits $\mathfrak{D}$ of degree and index p^e over $\mathfrak{F}$, then $\mathfrak{K}^{(p)}$ has degree p^{e-1} over $\mathfrak{F}$ and cannot split $\mathfrak{D}$. By Theorem 7.24, $\mathfrak{D} \sim (\mathfrak{Z}, S, \gamma)$ of the same degree as $\mathfrak{D}$, $\mathfrak{D} = (\mathfrak{Z}, S, \gamma)$.

We finally state

Theorem 27. *A p-algebra $\mathfrak{A}$ is cyclic if and only if there exists a simple pure inseparable field $\mathfrak{K}$ whose degree over $\mathfrak{F}$ is at most the degree of $\mathfrak{A}$ and such that $\mathfrak{K}$ splits $\mathfrak{A}$.*

For if $\mathfrak{A} = (\mathfrak{Z}, S, \gamma)$ of degree p^e then $\mathfrak{K} = \mathfrak{F}(\gamma^{1/p^e})$ splits $\mathfrak{A}$ and, by Lemma 7.2, the degree of $\mathfrak{K}$ over $\mathfrak{F}$ is $p^f \leqq p^e$. Conversely, if $\mathfrak{K} = \mathfrak{F}(\gamma^{1/p^e})$ has degree $p^e \leqq p^g$ over $\mathfrak{F}$ and splits $\mathfrak{A}$ of degree $n = p^g$ we apply Theorem 7.25 to obtain $\mathfrak{A} \sim (\mathfrak{Z}, S, \gamma)$ of degree $p^f \leqq p^e \leqq n$. By Lemma 7.12 the algebra $\mathfrak{A}$ is cyclic.

Replacing the words "at most" in Theorem 7.27 by "equal to" provides another result the proof of which is trivial, since if $\mathfrak{K}$ is a simple pure inseparable field of degree a divisor of n over $\mathfrak{F}$ and splits $\mathfrak{A}$ then $\mathfrak{K}$ is contained in such a field of degree n over $\mathfrak{F}$ which clearly also splits $\mathfrak{A}$.

9. Similarity of p-algebras to direct products of cyclic p-algebras. The result of Theorem 7.25 has as consequence the generalization

Theorem 28. *Let $\gamma_1, \cdots, \gamma_t$ be in $\mathfrak{F}$ and $\mathfrak{K} = \mathfrak{F}(\gamma_1^{1/n_1}, \cdots, \gamma_t^{1/n_t})$ split a p-algebra $\mathfrak{A}$ for integers $n_i = p_i^{e_i}$. Then*

$$\mathfrak{A} \sim \mathfrak{B}_1 \times \cdots \times \mathfrak{B}_t, \tag{45}$$

where $\mathfrak{B}_i$ is a cyclic algebra $(\mathfrak{Z}_i, S_i, \gamma_i)$ of degree m_i, a divisor of n_i.

For Theorem 7.25 is our result in the case $t = 1$. We make an induction on t. If $\mathfrak{K}_0 = \mathfrak{F}(\gamma_1^{1/n_1}, \cdots, \gamma_{t-1}^{1/n_{t-1}})$ splits $\mathfrak{A}$ our induction implies (45) with $\mathfrak{B}_t$ of degree one. Hence let $\mathfrak{A}$ be not split by $\mathfrak{K}_0$ and thus $\mathfrak{A}_{\mathfrak{K}_0}$ have $\mathfrak{K}_0(\gamma_t^{1/n_t})$ as splitting field. By Theorem 7.25 $\mathfrak{A}_{\mathfrak{K}_0} \sim (\mathfrak{Z}_t^{(0)}, S_t, \gamma_t)$, where $\mathfrak{Z}_t^{(0)}$ is cyclic of degree $m_t = p_t^{f_t} \leqq p_t^{e_t}$ over $\mathfrak{K}_0$. Lemma 7.7 implies that $\mathfrak{Z}_t^{(0)} = \mathfrak{Z}_t \times \mathfrak{K}_0$, $\mathfrak{A}_{\mathfrak{K}_0} \sim (\mathfrak{Z}_t, S_t, \gamma_t)_{\mathfrak{K}_0}$. Then if $\mathfrak{B} = \mathfrak{A} \times (\mathfrak{Z}_t, S_t, \gamma_t^{-1})$ we have $\mathfrak{B}_{\mathfrak{K}_0} = \mathfrak{A}_{\mathfrak{K}_0} \times (\mathfrak{Z}_t, S_t, \gamma_t^{-1})_{\mathfrak{K}_0} \sim 1$ by the evident property that $(\mathfrak{Z}_t, S_t, \gamma_t^{-1})$ is reciprocal to $(\mathfrak{Z}_t, S_t, \gamma_t) = \mathfrak{B}_t$. By the hypotheses of our induction

$$\mathfrak{B} \sim \mathfrak{B}_1 \times \cdots \times \mathfrak{B}_{t-1}, \qquad \mathfrak{A} \times \mathfrak{B}_t^{-1} \sim \mathfrak{B}_1 \times \cdots \times \mathfrak{B}_{t-1},$$

$$\mathfrak{A} \times \mathfrak{B}_t^{-1} \times \mathfrak{B}_t \sim \mathfrak{A} \sim \mathfrak{B}_1 \times \cdots \times \mathfrak{B}_t$$

as desired.

Just as Theorems 7.25 and 7.24 are equivalent so are Theorem 7.28 and

Theorem 29. *Let $\mathfrak{A}$ be a p-algebra over $\mathfrak{F}$, $\mathfrak{K} = \mathfrak{F}(\gamma_1^{1/n_1}, \cdots, \gamma_t^{1/n_t})$ split $\mathfrak{A}$ for quantities γ_i in $\mathfrak{F}$ and integers $n_i = p^{e_i}$ such that no field $\mathfrak{K}_0 = \mathfrak{F}(\gamma_1^{1/m_1}, \cdots, \gamma_t^{1/m_t})$ splits $\mathfrak{A}$ for the $m_i = p^{f_i} \leqq n_i$ with at least one $m_i < n_i$. Then* (45) *holds with $\mathfrak{B}_i = (\mathfrak{Z}_i, S_i, \gamma_i)$ of degree n_i.*

For if (45) holds with the $\mathfrak{B}_i$ of degree m_i a divisor of n_i then clearly $\mathfrak{K}_0$ splits $\mathfrak{A}$ and, by the hypothesis of our Theorem 7.29, $m_i = n_i$ and we have the desired conclusion. Conversely, let us assume the property of Theorem 7.29. We let $\mathfrak{K} = \mathfrak{F}(\gamma_1^{1/n_1}, \cdots, \gamma_t^{1/n_t})$ split $\mathfrak{A}$ as in the hypothesis of Theorem 7.28 and may evidently choose integers m_i dividing n_i and a corresponding subfield $\bar{\mathfrak{K}}$ such that m_1 has its least possible value, m_2 has the least possible value for this m_1, and so on. Clearly $\bar{\mathfrak{K}}$ is a field $\mathfrak{K}$ of Theorem 7.29 and we have (45) for this set of m_i as desired.

As an immediate corollary of Theorems 7.21, Lemma 7.3, and Theorem 7.29 we have

Theorem 30. *Every p-algebra is similar to a direct product of cyclic p-algebras.*

However we may actually prove

Theorem 31. *Every p-algebra is similar to a cyclic p-algebra.*

For it is evidently sufficient to prove

LEMMA 13. *Let* $\mathfrak{A} = \mathfrak{B} \times \mathfrak{C}$ *where* $\mathfrak{B}$ *and* $\mathfrak{C}$ *are cyclic p-algebras. Then* $\mathfrak{A}$ *is a cyclic p-algebra.*

For $\mathfrak{B} = (\mathfrak{Z}, S, \gamma)$ of degree n, $\mathfrak{C} = (\mathfrak{W}, T, \delta)$ of degree m with γ and δ in $\mathfrak{F}$. If $\gamma = \epsilon^p$ for ϵ in $\mathfrak{F}$ then $\mathfrak{B} = \mathfrak{M} \times (\mathfrak{Z}_0, S, \epsilon)$ by Theorem 7.25, $\mathfrak{A} = \mathfrak{M} \times (\mathfrak{Z}_0, S, \epsilon) \times (\mathfrak{W}, T, \delta)$ is cyclic if and only if $(\mathfrak{Z}_0, S, \epsilon) \times (\mathfrak{W}, T, \delta)$ is cyclic by Lemma 7.12. Hence it is sufficient to consider the case $\gamma \neq \epsilon^p$ for ϵ in $\mathfrak{F}$. The algebra $\mathfrak{B}$ then contains a maximal subfield $\mathfrak{K} = \mathfrak{F}(j)$, $j^n = \gamma$. The field $\mathfrak{W}_{\mathfrak{K}} = \mathfrak{W} \times \mathfrak{K}$ is cyclic over $\mathfrak{K}$ since $\mathfrak{W}$ is separable and $\mathfrak{K}$ is pure inseparable and $\mathfrak{A}$ contains $\mathfrak{C} \times \mathfrak{K} = \mathfrak{C}_{\mathfrak{K}} = (\mathfrak{W}_{\mathfrak{K}}, T, \delta)$. By Lemma 7.9 there exists a quantity b in $\mathfrak{L} = \mathfrak{W}_{\mathfrak{K}}$ such that $N_{\mathfrak{L}|\mathfrak{K}}(b) = a$ generates $\mathfrak{K}$. Hence so does $j_0 = \delta a$, $\mathfrak{K} = \mathfrak{F}(j_0)$. Now $\mathfrak{C}_{\mathfrak{K}} = (\mathfrak{L}, T, \delta a)$ so that $\mathfrak{C}_{\mathfrak{K}}$ contains a quantity y_0 such that $y_0^m = \delta a = j_0$, $j_0^n = \gamma_0$ in $\mathfrak{F}$, γ_0 is not a pth power of a quantity of $\mathfrak{F}$. Then $y_0^{mn} = \gamma_0$ and the field $\mathfrak{F}(y_0)$ is a pure inseparable maximal subfield of $\mathfrak{A}$. Hence $\mathfrak{F}(y_0)$ splits $\mathfrak{A}$ and, by Theorem 7.27, $\mathfrak{A}$ is a cyclic algebra.

We shall close our chapter with a determination of the exponent of any p-algebra. The result is exceedingly simple and we state it as

Theorem 32. *The exponent of any p-algebra is the minimum of the exponents of all of its pure inseparable splitting fields.*

For let $\sigma = p^e$ be this minimum, $\rho = p^f$ be the exponent of $\mathfrak{A}$. By Theorem 7.28, $\mathfrak{A} \sim \mathfrak{B}_1 \times \cdots \times \mathfrak{B}_t$ where $\mathfrak{B}_i$ is a p-algebra of degree a divisor of $\sigma = p^e$ by Theorem 2.29. Then $\mathfrak{A}^\sigma \sim \mathfrak{B}_1^\sigma \times \cdots \times \mathfrak{B}_t^\sigma \sim 1$ by Theorem 5.17. Hence $\sigma \geqq \rho$. We now use Theorem 7.31 to write $\mathfrak{A} \sim (\mathfrak{Z}, S, \gamma)$, a cyclic algebra of exponent ρ. Then by Theorem 7.17, $\gamma^\rho = N_{\mathfrak{Z}|\mathfrak{F}}(a)$ with a in $\mathfrak{Z}$. By Lemma 7.6 there exists a field $\mathfrak{K}$ which is pure inseparable of exponent at most ρ such that $a = b^\rho$, b in $\mathfrak{Z} \times \mathfrak{K} = \mathfrak{J}$. Then $N_{\mathfrak{Z}|\mathfrak{F}}(a) = N_{\mathfrak{J}|\mathfrak{K}}(b^\rho) = [N_{\mathfrak{J}|\mathfrak{K}}(b)]^\rho = \gamma^\rho$, $\gamma = N_{\mathfrak{J}|\mathfrak{K}}(b)$, $\mathfrak{A}_{\mathfrak{K}} \sim (\mathfrak{Z}_{\mathfrak{K}}, S, \gamma) \sim 1$. Hence $\mathfrak{K}$ splits $\mathfrak{A}$, $\sigma \leqq \rho$ by definition of σ. But then $\sigma = \rho$.

This completes our general results on p-algebras. They may be interpreted as stating in particular that every p-algebra is cyclically representable, that is, similar to a cyclic algebra. It is not known, however, whether or not there exist non-cyclic p-algebras or indeed whether or not the property that $\mathfrak{A}$ is cyclically representable implies that $\mathfrak{A}$ is actually cyclic.

CHAPTER VIII

REPRESENTATIONS AND RIEMANN MATRICES

1. Representations of algebras. The exposition of the theory of simple algebras given in these LECTURES was based upon Theorem 1.3, a result obtained as a consequence of certain properties of the regular representations of an algebra. Some of the previous expositions used the theory of representations of algebras to a much greater extent and it seems desirable to prove the representation theorems here. Moreover we shall apply some of the properties obtained to the study of Riemann matrices.

We shall not attempt to follow the classical exposition of the representation theory rigidly here but instead make our proofs with the aid of our results on simple algebras. Moreover we shall study only the representations of an algebra $\mathfrak{A}$ over $\mathfrak{F}$ as an algebra of matrices with elements in $\mathfrak{F}$ instead of allowing these elements to range over an arbitrary ring $\mathfrak{R}$. For the more general discussion as well as additional aspects of the theory see [119], [444], [76], [77], [93].

Let $\mathfrak{A}$ be an algebra over $\mathfrak{F}$ with a unity quantity which we shall designate by 1, and let there be a correspondence

$$\Lambda: \qquad a \to a^{\Lambda} \qquad (a \text{ in } \mathfrak{A})$$

on $\mathfrak{A}$ to a second algebra $\mathfrak{A}^*$ over $\mathfrak{F}$ such that every quantity of $\mathfrak{A}^*$ is the correspondent a^{Λ} of an a of $\mathfrak{A}$, and

$$(1) \qquad (\alpha a + \beta b)^{\Lambda} = \alpha a^{\Lambda} + \beta b^{\Lambda}, \qquad (ab)^{\Lambda} = a^{\Lambda} b^{\Lambda}$$

for every a and b of $\mathfrak{A}$, α and β of $\mathfrak{F}$. Then we shall say that $\mathfrak{A}$ is *homomorphic* to $\mathfrak{A}^*$. Evidently $e = 1^{\Lambda}$ is the unity quantity of $\mathfrak{A}^*$. Moreover the assumption that $\mathfrak{A}^*$ is an algebra implies in particular that $e \neq 0$ so that Λ does not carry every quantity of $\mathfrak{A}$ into the zero quantity of $\mathfrak{A}^*$.

The correspondence Λ will be called a *representation* of $\mathfrak{A}$ by $\mathfrak{A}^*$, and $\mathfrak{A}^*$ will be called the *representation algebra* of Λ. Note that Λ is not uniquely determined by $\mathfrak{A}$ and $\mathfrak{A}^*$. To see this let us consider an elementary example.

Let $\mathfrak{A}$ be a diagonal algebra (e_1, e_2) over $\mathfrak{F}$ and $\mathfrak{A}^* = (e)$ over $\mathfrak{F}$ so that $\mathfrak{A}^* \cong \mathfrak{F}$. The correspondences

$$\Lambda_1: \qquad a = \alpha e_1 + \beta e_2 \to a^{\Lambda_1} = \alpha e,$$

$$\Lambda_2: \qquad a = \alpha e_1 + \beta e_2 \to a^{\Lambda_2} = \beta e,$$

are easily verified to be representations of $\mathfrak{A}$ by $\mathfrak{A}^*$. However $e_1^{\Lambda_1} = e$, $e_2^{\Lambda_1} = 0$, and $e_1^{\Lambda_2} = 0$, $e_2^{\Lambda_2} = e$. It is clear that the properties of representations are thus more than merely relative properties of two algebras.

In case Λ is a (1-1) correspondence we have already called Λ an *equivalence* of $\mathfrak{A}$ and $\mathfrak{A}^*$. Then we showed in Section 1.8 that any other equivalence of $\mathfrak{A}$ and $\mathfrak{A}^*$ is the product of Λ by an automorphism of $\mathfrak{A}^*$ (or of $\mathfrak{A}$ depending upon the order of the factors of this product). This is clearly a situation quite different from that occurring for arbitrary representations.

We now derive some elementary properties of representations.

LEMMA 1. *Let Λ be a representation of $\mathfrak{A}$ by $\mathfrak{A}^*$ and $\mathfrak{B}_\Lambda$ be the set of all quantities b of $\mathfrak{A}$ such that $b^\Lambda = 0$. Then $\mathfrak{B}$ is an ideal of $\mathfrak{A}$ and the correspondence*

$$[a] \rightarrow a^\Lambda \tag{2}$$

is an equivalence of $\mathfrak{A} - \mathfrak{B}_\Lambda$ and $\mathfrak{A}^*$.

For if α and β are in $\mathfrak{F}$, a is in $\mathfrak{A}$, and b, b_1, b_2 are in $\mathfrak{B}_\Lambda$ we have $(ab)^\Lambda = a^\Lambda b^\Lambda = 0$, $(ba)^\Lambda = b^\Lambda a^\Lambda = 0$, $(\alpha b_1 + \beta b_2)^\Lambda = \alpha b_1^\Lambda + \beta b_2^\Lambda = 0$, $\mathfrak{B}_\Lambda$ is an ideal of $\mathfrak{A}$. We designate the classes comprising $\mathfrak{A} - \mathfrak{B}_\Lambda$ in the usual fashion, $[a]$, where a is in $\mathfrak{A}$, and the properties defining $\mathfrak{A} - \mathfrak{B}_\Lambda$ clearly imply that (2) is a representation of $\mathfrak{A} - \mathfrak{B}_\Lambda$ by $\mathfrak{A}^*$. If $[a] \rightarrow 0$ then $a^\Lambda = 0$, a is in $\mathfrak{B}_\Lambda$, $[a] = [0]$, (2) is an equivalence.

We next obtain

LEMMA 2. *Every representation Λ of a simple algebra $\mathfrak{A}$ by $\mathfrak{A}^*$ is an equivalence of $\mathfrak{A}$ and $\mathfrak{A}^*$.*

For if $\mathfrak{A}$ is simple $\mathfrak{B}_\Lambda = \mathfrak{A}$ or $\mathfrak{B}_\Lambda = 0$. If $\mathfrak{B}_\Lambda = 0$ our result follows from Lemma 8.1. Otherwise $a^\Lambda = 0$ for every a of $\mathfrak{A}$ contrary to our definition of Λ.

2. Matric representations. Let $\mathfrak{L}$ be a linear set of order ν over $\mathfrak{F}$. We shall consider linear subsets $\mathfrak{L}_i$ over $\mathfrak{F}$ of $\mathfrak{L}$ and shall henceforth speak of $\mathfrak{L}$ as a *space* over $\mathfrak{F}$ and the $\mathfrak{L}_i$ as *subspaces* of $\mathfrak{L}$. The word space is not merely easier to enunciate than the phrase linear set but its usage is customary in the theory of representations.

A representation Λ of $\mathfrak{A}$ by $\mathfrak{A}^*$ will be called a *ν-rowed representation* of $\mathfrak{A}$ if $\mathfrak{A}^*$ is a subalgebra of a total matric algebra $\mathfrak{M}_\nu$, and the unity quantity 1^Λ of $\mathfrak{A}^*$ is the unity quantity I_ν of $\mathfrak{M}_\nu$. A total matric algebra $\mathfrak{M}_\nu$ of degree ν over $\mathfrak{F}$ is defined as an algebra equivalent to the algebra of all ν-rowed square matrices with elements in $\mathfrak{F}$, and we may think of $\mathfrak{M}_\nu$ as an abstract algebra and the corresponding ν-rowed matrices as merely notations for the quantities of $\mathfrak{M}_\nu$. We have already seen in Section 1.9 that $\mathfrak{M}_\nu$ may be regarded as an algebra of linear transformations on $\mathfrak{L}$, and we shall restrict our attention henceforth to representations of $\mathfrak{A}$ by $\mathfrak{A}^* \leqq \mathfrak{M}_\nu$. Call $\mathfrak{L}$ a *corresponding ν-dimensional representation space.*

If $u_1, \cdots, u_\nu$ is any basis of $\mathfrak{L}$ over $\mathfrak{F}$ we may write any quantity of $\mathfrak{L}$ in the form $x = \xi_1 u_1 + \cdots + \xi_\nu u_\nu$. Then x defines a vector (one-rowed matrix) $\underline{x} = (\xi_1, \cdots, \xi_\nu)$ whose form is uniquely determined by x and the given basis of $\mathfrak{L}$. If S is any linear transformation on $\mathfrak{L}$ then $\underline{x^S} = \underline{x}A$, where A is a ν-rowed square matrix which we have agreed to regard as merely a matrix notation of S. This really means of course that the correspondence

(3) $$S \leftrightarrow A$$

is an equivalence of $\mathfrak{M}_\nu$ and the algebra $\mathcal{M}_\nu$ of all ν-rowed square matrices defined by the particular basis of $\mathfrak{L}$. We also saw in Section 1.9 that if a change of basis is made the correspondence (3) is replaced by

(4) $$S \leftrightarrow BAB^{-1},$$

where B is the non-singular matrix of the change of basis. Moreover as B varies over all non-singular matrices the equivalences (4) give all equivalences of $\mathfrak{M}_\nu$ and $\mathcal{M}_\nu$.

Let now $a \to a^\Lambda = A$ be a ν-rowed representation Λ of an algebra $\mathfrak{A}$. Then the correspondences $a \to a^{\Lambda\Sigma} = BAB^{-1}$ are also representations of $\mathfrak{A}$ for every automorphism Σ of $\mathfrak{M}_\nu$, and we may think of all these representations as being merely different matric notations for the same representation $a \to S_a$ of $\mathfrak{A}$ by an algebra $\mathfrak{A}^*$ of linear transformations S_a. We shall call any two ν-rowed representations related in this way *equivalent representations* and shall seek the properties of representations invariant under equivalence. Note that in the example given in Section 8.1 the two representations are inequivalent. This result is a special case of

LEMMA 3. *Two ν-rowed representations Λ_1 and Λ_2 of an algebra $\mathfrak{A}$ are equivalent only if $\mathfrak{B}_{\Lambda_1} = \mathfrak{B}_{\Lambda_2}$. Thus if Λ_1 and Λ_2 are equivalent their representation algebras are equivalent.*

For if Λ_1 and Λ_2 are equivalent we have $a^{\Lambda_2} = ba^{\Lambda_1}b^{-1} = 0$ if and only if $a^{\Lambda_1} = 0$. We next obtain

LEMMA 4. *Any two ν-rowed representations of a simple algebra are equivalent.*

For by Lemma 8.2 if $\mathfrak{A}$ is simple the representations $a \to a^{\Lambda_1}$, $a \to a^{\Lambda_2}$ are equivalences. Then $a^{\Lambda_1} \leftrightarrow a^{\Lambda_2}$ is an equivalence of the representation algebras which are simple subalgebras of $\mathfrak{M}_\nu$ with the same unity quantity as $\mathfrak{M}_\nu$, and by Theorem 4.14 we have $a^{\Lambda_2} = ba^{\Lambda_1}b^{-1}$ as desired.

We next prove

LEMMA 5. *Let Λ be a representation of $\mathfrak{A}$ by a semi-simple algebra $\mathfrak{A}^*$. Then $\mathfrak{B}_\Lambda$ contains the radical $\mathfrak{N}$ of $\mathfrak{A}$, $\mathfrak{A} - \mathfrak{N} = \mathfrak{C}_1 \oplus \mathfrak{C}_2$ where $\mathfrak{C}_2$ consists of all classes $[b]$ for b in $\mathfrak{B}_\Lambda$, $\mathfrak{C}_1 \cong \mathfrak{A}^*$.*

For if b is in $\mathfrak{N}$ then the set $\mathfrak{N}^\Lambda$ of all b^Λ is evidently a nilpotent ideal of $\mathfrak{A}^*$. Since $\mathfrak{A}^*$ is semi-simple $\mathfrak{N}^\Lambda = 0$, $\mathfrak{N} \leqq \mathfrak{B}_\Lambda$. Now $\mathfrak{B}_\Lambda - \mathfrak{N}$ is an ideal of $\mathfrak{A} - \mathfrak{N}$ and thus must be a direct sum of simple components of $\mathfrak{A} - \mathfrak{N}$, $\mathfrak{A} - \mathfrak{N} =. (\mathfrak{B}_\Lambda - \mathfrak{N}) \oplus \mathfrak{C}_1$. Also $(\mathfrak{A} - \mathfrak{N}) - (\mathfrak{B}_\Lambda - \mathfrak{N}) \cong \mathfrak{C}_1 \cong \mathfrak{A} - \mathfrak{B}_\Lambda \cong \mathfrak{A}^*$ as desired

COROLLARY. *Let Λ be a representation of an algebra $\mathfrak{A}$ by a simple algebra $\mathfrak{A}^*$. Then $\mathfrak{A}^*$ is equivalent to one of the simple components of $\mathfrak{A} - \mathfrak{N}$, where $\mathfrak{N}$ is the radical of $\mathfrak{A}$.*

3. Reducibility of representations. A ν-rowed representation Λ of an algebra $\mathfrak{A}$ is called *reducible* if a basis of the corresponding representation space $\mathfrak{L}$ may be chosen so that Λ is given by $a \to A_a$ where for every a we have

$$(5) \qquad A_a = \begin{pmatrix} a_1 & 0 \\ a_3 & a_2 \end{pmatrix}.$$

Here a_1 is a μ-rowed square matrix, $0 < \mu < \nu$, a_2 is a $(\nu - \mu)$-rowed square matrix, the zero is a matrix of μ rows and $\nu - \mu$ columns of zero elements, a_3 has $\nu - \mu$ rows and μ columns. In the contrary case we call Λ an *irreducible* representation. We shall obtain a criterion for the occurrence of reducibility in terms of the subspaces of $\mathfrak{L}$.

Let $\mathfrak{H}$ be any set of linear transformations S on $\mathfrak{L}$. Then we call a subspace $\mathfrak{L}_1$ of $\mathfrak{L}$ a subspace *invariant* under $\mathfrak{H}$ if x_1^S is in $\mathfrak{L}_1$ for every x_1 of $\mathfrak{L}_1$ and S of $\mathfrak{H}$. Write $\mathfrak{L} = \mathfrak{L}_1 + \mathfrak{L}_2$ with $\mathfrak{L}_1 = (u_1, \cdots, u_\mu)$, $\mathfrak{L}_2 = (u_{\mu+1}, \cdots, u_\nu)$ by the exercise of Section A2.11. Thus every x of $\mathfrak{L}$ is uniquely expressible in the form $x = x_1 + x_2$ with $x_1 = \xi_1 u_1 + \cdots + \xi_\mu u_\mu$ in $\mathfrak{L}_1$, $x_2 = \xi_{\mu+1} u_{\mu+1} + \cdots + \xi_\mu u_\mu$ in $\mathfrak{L}_2$. Write $X_1 = (\xi_1, \cdots, \xi_\mu)$, $X_2 = (\xi_{\mu+1}, \cdots, \xi_\nu)$ so that $\underline{x} = (X_1, X_2)$. Then $\underline{x^S} = \underline{x}A$ where

$$(6) \qquad A = \begin{pmatrix} a_1 & a_4 \\ a_3 & a_2 \end{pmatrix}, \qquad \underline{x}A = (X_1 a_1 + X_2 a_3, X_1 a_4 + X_2 a_2).$$

The quantity x is in $\mathfrak{L}_1$ if and only if $X_2 = 0$, $\underline{x^S} = (X_1 a_1, X_1 a_4)$, x^S is in $\mathfrak{L}_1$ if and only if $X_1 a_4 = 0$, that is, $a_4 = 0$ since X_1 is an arbitrary vector. We have

LEMMA 6. *Let $\mathfrak{H}$ be any set of linear transformations on $\mathfrak{L}$. Then there exists a subspace of $\mathfrak{L}$ of order μ over $\mathfrak{F}$ which is invariant under $\mathfrak{H}$ if and only if $\mathfrak{L}$ has a basis such that the matrices corresponding to the transformations in $\mathfrak{H}$ all have the form* (5).

The criterion above states that a ν-rowed representation $a \to S_a$ of $\mathfrak{A}$ by an algebra $\mathfrak{A}^*$ of linear transformations S_a on $\mathfrak{L}$ is reducible if and only if $\mathfrak{L}$ has a proper subspace invariant under $\mathfrak{A}^*$. Note that the interchange of the roles of $\mathfrak{L}_1$ and $\mathfrak{L}_2$ corresponds to the choice $\mathfrak{L}_2 = (u_1, \cdots, u_{\nu-\mu})$, $\mathfrak{L}_1 = (u_{\nu-\mu+1}, \cdots, u_\nu)$, and gives

$$(7) \qquad A_a = \begin{pmatrix} a_1 & a_4 \\ 0 & a_2 \end{pmatrix}.$$

Hence Λ is reducible if and only if a basis of $\mathfrak{L}$ may be chosen so that every A_a takes the form (5), or so that every A_a takes the form (7).

4. Enveloping algebras. Let $\mathfrak{H}$ be any set of linear transformations over $\mathfrak{F}$ on a space $\mathfrak{L}$ and $\mathfrak{A}^*$ be the set of all finite sums of products $\lambda S_1 \cdots S_r$ with λ in $\mathfrak{F}$ and the S_i either in $\mathfrak{H}$ or equal to the identity transformation I_ν. Then $\mathfrak{A}^*$ is clearly an algebra over $\mathfrak{F}$ of linear transformations with I_ν as unity quantity. We shall call $\mathfrak{A}^*$ the *enveloping algebra* of $\mathfrak{H}$.

If L_1 is a subspace of $\mathfrak{L}$ invariant under $\mathfrak{A}^*$ then $\mathfrak{L}_1$ is clearly invariant under any subset $\mathfrak{H}$ of $\mathfrak{A}^*$. Conversely let $\mathfrak{L}_1$ be invariant under $\mathfrak{H}$ and $\mathfrak{A}^*$ be the enveloping algebra of $\mathfrak{H}$. Then if x is in $\mathfrak{L}_1$ we have $x^{\lambda S_1 \cdots S_r} = \lambda(x^{S_1})^{S_2 \cdots S_r}$ is in $\mathfrak{L}_1$ since x^{S_1} is in $\mathfrak{L}_1$, so is $(x^{S_1})^{S_2}$, and so on. We have proved

Theorem 1. *The subspaces invariant under the enveloping algebra of any set $\mathfrak{H}$ of linear transformations are the same as those invariant under $\mathfrak{H}$.*

In the consideration of Riemann matrices which we shall make later in the present chapter we shall discuss sets $\mathfrak{H}$ of ν-rowed square matrices A with elements in $\mathfrak{F}$ and the question of the existence of a non-singular matrix B such that every A of $\mathfrak{H}$ has the property

$$BAB^{-1} = \begin{pmatrix} a_1 & 0 \\ a_3 & a_2 \end{pmatrix}$$

as in (5). By Theorem 8.1 such a matrix B exists if and only if it exists for every A of the enveloping algebra $\mathfrak{A}^*$ of $\mathfrak{H}$. The equivalence $A \to A$ is a representation of $\mathfrak{A}^*$ by itself and we are clearly proposing the question of the reducibility of this representation.

We shall now introduce an addition to our terminology. Let $\mathfrak{A}^*$ be an algebra of linear transformations or of corresponding ν-rowed square matrices. The representation $A \to A$ of $\mathfrak{A}^*$ by $\mathfrak{A}^*$ is then reducible if and only if a non-singular matrix B exists as above. We wish a term describing this property of the algebra $\mathfrak{A}^*$ and cannot use reducible since when this term is applied to algebras it has already been given another meaning. We shall thus use the terminology customary in the theory of Riemann matrices and call $\mathfrak{A}^*$ *impure* if the representation is reducible, otherwise *pure*. If $\mathfrak{A}^*$ is the enveloping algebra of a subset $\mathfrak{H}$ we shall also call $\mathfrak{H}$ *pure* and *impure* in the respective cases.

5. Reduction to irreducible components. Let

$$\Lambda: \qquad a \to a^{\Lambda} = A_a$$

be a reducible ν-rowed representation of an algebra $\mathfrak{A}$ so that we may assume that A_a has the form (5). Then if b is also in $\mathfrak{A}$ we have $ab = c$, $\alpha a + \beta b = d$ such that

$$A_b = \begin{pmatrix} b_1 & 0 \\ b_3 & b_2 \end{pmatrix}, \qquad A_c = \begin{pmatrix} c_1 & 0 \\ c_3 & c_2 \end{pmatrix}, \qquad A_d = \begin{pmatrix} d_1 & 0 \\ d_3 & d_2 \end{pmatrix}. \tag{8}$$

But $A_c = A_aA_b$, $A_d = \alpha A_a + \beta A_b$, and direct matrix multiplication gives

$$c_1 = a_1b_1, \quad c_2 = a_2b_2, \quad d_1 = \alpha a_1 + \beta b_1, \quad d_2 = \alpha a_2 + \beta b_2. \tag{9}$$

Moreover the matrix A_a for $a = 1$ is the ν-rowed identity matrix and thus $a = 1$ in (5) implies that $a_1 = I_\mu$, $a_2 = I_{\nu-\mu}$. It follows that

$$\Lambda_1: \qquad a \to a_1$$

and

$$\Lambda_2: \qquad a \to a_2$$

are μ-rowed and $(\nu - \mu)$-rowed representations respectively of $\mathfrak{A}$. We shall call Λ_1 and Λ_2 *components* of Λ. It is trivial to show that by replacing Λ by a properly chosen equivalent representation we replace Λ_1 and Λ_2 respectively by any desired equivalent representations. Thus if Λ_1 is reducible we may write

$$a_1 = \begin{pmatrix} a_{11} & 0 \\ a_{13} & a_{12} \end{pmatrix}. \tag{10}$$

After a finite number of such steps we arrive at a representation

$$a \to a^* = \begin{pmatrix} a_{11}^* & 0 & \cdots & 0 \\ a_{21}^* & a_{22}^* & \cdots & 0 \\ \cdot & \cdot & \cdots & \cdot \\ a_{\tau 1}^* & a_{\tau 2}^* & \cdots & a_{\tau\tau}^* \end{pmatrix} \tag{11}$$

equivalent to Λ and such that the representations

$$\Lambda_i: \qquad a \to a_{ii}^* \qquad (i = 1, \cdots, \tau),$$

are irreducible representations of $\mathfrak{A}$. We call these Λ_i *irreducible components* of Λ. Note that we have found subspaces $\mathfrak{L}_i^{(0)}$ of $\mathfrak{L}$ such that $\mathfrak{L}_i^{(0)}$ is an invariant subspace of $\mathfrak{L}$ under the transformations S_a with matrices A_a, there exists no invariant subspace $\mathfrak{L}^{(0)}$ such that $\mathfrak{L}_{i-1}^{(0)} < \mathfrak{L}^{(0)} < \mathfrak{L}_i^{(0)}$, and

$$0 = \mathfrak{L}_0^{(0)} < \mathfrak{L}_1^{(0)} < \cdots < \mathfrak{L}_\tau^{(0)} = \mathfrak{L}. \tag{12}$$

If μ_i is the order of $\mathfrak{L}_i^{(0)}$ then

$$0 < \mu_1 < \mu_2 < \cdots < \mu_\tau = \nu. \tag{13}$$

Put $\nu_i = \mu_i - \mu_{i-1} > 0$ and write

$$\mathfrak{L}_i^{(0)} = \mathfrak{L}_1 + \cdots + \mathfrak{L}_i \qquad (i = 1, \cdots, \tau), \tag{14}$$

where ν_i is the order of $\mathfrak{L}_i$. Then Λ_i is a ν_i-rowed representation of $\mathfrak{A}$ by the algebra $\mathfrak{A}_i$ of all the a_{ii}^*, that is, by linear transformations on $\mathfrak{L}_i$ with matrices a_{ii}^*.

6. Decomposable representations. Let Λ be a ν-rowed representation $a \to S_a$ of $\mathfrak{A}$ and suppose that $\mathfrak{L} = \mathfrak{L}_1 + \mathfrak{L}_2$ where both $\mathfrak{L}_1$ and $\mathfrak{L}_2$ are invariant under the S_a. Combining (5) and (7) we have

$$a \to A_a = \begin{pmatrix} a_1 & 0 \\ 0 & a_2 \end{pmatrix},$$

and shall call Λ a *decomposable representation* of $\mathfrak{A}$ by $\mathfrak{A}^*$, $\mathfrak{A}^*$ a *decomposable algebra* with components the algebras $\mathfrak{A}_1$ and $\mathfrak{A}_2$ of respective matrices a_1 and a_2. Clearly $\mathfrak{L}$ may be expressed as the supplementary sum $\mathfrak{L} = \mathfrak{L}_1 + \cdots + \mathfrak{L}_\tau$ where the $\mathfrak{L}_i$ are subspaces invariant under $\mathfrak{A}^*$ and no $\mathfrak{L}_i$ is expressible as a supplementary sum of two proper subspaces both invariant under $\mathfrak{A}^*$. Correspondingly we may take Λ to be

$$a \to A_a = \text{diag}\,\{a_1, \cdots, a_\tau\}, \tag{15}$$

where the component representations

$$\Lambda_i : \qquad a \to a_i$$

are indecomposable ν_i-rowed representations of $\mathfrak{A}$ (by indecomposable algebras $\mathfrak{A}_i$ of matrices a_i, the indecomposable components of A_a). If it is possible to choose the $\mathfrak{L}_i$ so that the Λ_i are irreducible we shall call Λ a *fully decomposable* representation of $\mathfrak{A}$. Then the $\mathfrak{A}_i$ are pure algebras. We shall obtain a necessary and sufficient condition that a representation be fully decomposable and shall prove that for such a representation the indecomposable components are unique in the sense of equivalence.

7. Irreducible representations. We shall first derive the so-called Schur Lemma which we state as

LEMMA 7. *Let $a \to a_1$ and $a \to a_2$ be ν_1- and ν_2-rowed irreducible representations, respectively, of $\mathfrak{A}$ and let there be a matrix b such that*

$$ba_1 = a_2 b. \tag{16}$$

Then either $b = 0$ or $\nu_1 = \nu_2$, b is non-singular, and the representations are equivalent.

For let $b \neq 0$ so that its rank is $\mu > 0$. Clearly (16) implies that b has ν_2 rows and ν_1 columns, $\nu_2 \geqq \mu$, $\nu_1 \geqq \mu$. By Theorem A3.10 there exists a non-singular ν_1-rowed square matrix d and a non-singular ν_2-rowed square matrix c such that

$$cbd = \begin{pmatrix} I_\mu & 0 \\ 0 & 0 \end{pmatrix}. \tag{17}$$

Then $cbd\, d^{-1}a_1 d = ca_2c^{-1}cbd$ and we put

$$d^{-1} a_1 d = \begin{pmatrix} a_{11} & a_{14} \\ a_{13} & a_{12} \end{pmatrix}, \qquad ca_2 c^{-1} = \begin{pmatrix} a_{21} & a_{24} \\ a_{23} & a_{22} \end{pmatrix} \tag{18}$$

so that

$$\begin{pmatrix} a_{11} & a_{14} \\ 0 & 0 \end{pmatrix} = \begin{pmatrix} a_{21} & 0 \\ a_{23} & 0 \end{pmatrix}, \tag{19}$$

that is, $a_{14} = 0$, $a_{23} = 0$. The irreducibility of $a \to a_1$ implies the irreducibility of $a \to d^{-1}a_1 d$, which is not possible when $a_{14} = 0$ unless $\mu = \nu_1$. Similarly $\mu = \nu_2$, b is non-singular, $a_2 = ba_1b^{-1}$, and the representations are equivalent.

As an immediate corollary we have

LEMMA 8. *Let Λ be an irreducible ν-rowed representation of $\mathfrak{A}$ by $\mathfrak{A}^*$. Then the $\mathfrak{M}_\nu$-commutator of $\mathfrak{A}^*$ is a division algebra.*

We use this result in the proof of

Theorem 2. *A ν-rowed representation Λ of $\mathfrak{A}$ by $\mathfrak{A}^*$ is irreducible if and only if $\mathfrak{A}^*$ is simple and the $\mathfrak{M}_\nu$-commutator of $\mathfrak{A}^*$ is a division algebra. In this case, $\nu = nmt$ where n and m are the degree and index of $\mathfrak{A}^*$ over its centrum $\mathfrak{K}$ and t is the degree of $\mathfrak{K}$. Also, $\mathfrak{A}^* \sim \mathfrak{D}$, a normal division algebra over $\mathfrak{K}$, and $\mathfrak{M}_\nu^{\mathfrak{A}^*} = \mathfrak{D}^{-1}$.*

For let Λ be irreducible and $\mathfrak{N}$ be the radical of $\mathfrak{A}^*$. We let $\mathfrak{L}_0$ be the subspace of $\mathfrak{L}$ consisting of all finite linear combinations x_0 of quantities x^N for N in $\mathfrak{N}$ and x in $\mathfrak{L}$. Then x^{NS} is in $\mathfrak{L}_0$ for every S of $\mathfrak{A}^*$, since NS is in $\mathfrak{N}$, x_0^S is in $\mathfrak{L}_0$ for every x_0 of $\mathfrak{L}_0$ and S of $\mathfrak{A}^*$, $\mathfrak{L}_0$ is an invariant subspace of $\mathfrak{L}$. Since Λ is irreducible $\mathfrak{L}_0 = \mathfrak{L}$ or $\mathfrak{L}_0 = 0$. In the former case every x of $\mathfrak{L}$ has the form $x = \sum_{i=1}^r x_i^{N_i}$ with N_i in $\mathfrak{N}$, x_i in $\mathfrak{L}$. But then

$$x_i = \sum_{j=1}^{s} x_{ij}^{N_{ij}}, \qquad x = \sum_{k=1}^{t} x_k^{N_{1k}N_{2k}}.$$

Repeating this process we may write $x = \sum_{k=1}^q x_k^{N_{1k}N_{2k}\cdots N_{\alpha k}} = 0$ if $\mathfrak{N}^{\cdot\alpha} = 0$. Such an α exists since $\mathfrak{N}$ is nilpotent, it is not possible that every x of $\mathfrak{L}$ is zero, $\mathfrak{L}_0 = 0$. Thus $x^N = 0$ for every N of $\mathfrak{N}$, that is, $\mathfrak{N}$ consists only of the zero transformation on $\mathfrak{L}$, $\mathfrak{N} = 0$. We have proved that $\mathfrak{A}^*$ is semi-simple. If $\mathfrak{A}^*$ were not simple the centrum of $\mathfrak{A}^*$ would contain an idempotent $E \neq I_\nu$, and $\mathfrak{A}^* = E\mathfrak{A}^*E \oplus \mathfrak{C}_E$. As on page A88 we may choose a basis of $\mathfrak{L}$ so that

$$E = \begin{pmatrix} I_\mu & 0 \\ 0 & 0 \end{pmatrix}, \qquad A_a = \begin{pmatrix} a_1 & 0 \\ 0 & a_2 \end{pmatrix} \tag{20}$$

with a_1 in $E\mathfrak{A}^*E$, a_2 in $\mathfrak{C}_E$, Λ is decomposable. Hence $\mathfrak{A}^*$ is simple. The $\mathfrak{M}_\nu$-commutator of $\mathfrak{A}^*$ has already been shown in Lemma 8.8 to be a division algebra.

Conversely let $\mathfrak{A}^*$ be simple so that $\mathfrak{A}^*$ is normal simple of degree n over its centrum $\mathfrak{K}$. Let $\mathfrak{K}$ have degree t over $\mathfrak{F}$ and m be the index of $\mathfrak{A}^*$ over $\mathfrak{K}$, $n = mq$, $\mathfrak{A}^* = \mathfrak{M}_q \times \mathfrak{D}$. Then $\mathfrak{D}$ is a normal division algebra of degree m over $\mathfrak{K}$. By Theorem 4.11 the $\mathfrak{M}_\nu$-commutator of $\mathfrak{K}$ is a total matric algebra $\mathfrak{M}_\rho$ over $\mathfrak{K}$,

$$\nu = \rho t. \tag{21}$$

Also $\mathfrak{M}_\nu^{\mathfrak{A}^*} \leqq \mathfrak{M}_\rho$, $\mathfrak{M}_\rho = \mathfrak{A}^* \times \mathfrak{M}_\rho^{\mathfrak{A}^*}$ and $\mathfrak{M}_\rho^{\mathfrak{A}^*} = \mathfrak{M}_\nu^{\mathfrak{A}^*}$. Now the total matric algebra $\mathfrak{M}_\rho = \mathfrak{M}_q \times \mathfrak{D} \times \mathfrak{M}_\rho^{\mathfrak{A}^*}$ so that $\mathfrak{D} \times \mathfrak{M}_\rho^{\mathfrak{A}^*}$ is a total matric algebra and by Theorem 4.8

$$\mathfrak{M}_\nu^{\mathfrak{A}^*} = \mathfrak{M}_s \times \mathfrak{D}^{-1}, \tag{22}$$

for a total matric algebra $\mathfrak{M}_s$. We then have

$$\nu = tnms, \tag{23}$$

and $\mathfrak{M}_\nu^{\mathfrak{A}^*}$ is a division algebra if and only if $s = 1$, $\mathfrak{M}_\nu^{\mathfrak{A}^*} = \mathfrak{D}^{-1}$. If Λ were reducible we could assume that it had the form (11). However it is trivial to verify that the correspondences $A_a \to \overset{*}{a}_{ii}$ are irreducible ν_i-rowed representations of the simple algebra $\mathfrak{A}^*$ by $\mathfrak{A}_i$, hence are equivalences by Lemma 8.2. Our proof then implies that the $\mathfrak{M}_{\nu_i}$-commutator of $\mathfrak{A}_i$ is a division algebra, $\nu_i = tnms_i = tnm$.

But $\nu = \nu_1 + \cdots + \nu_\tau = tnm\tau = tnms$, $s = \tau$. Hence if $s = 1$ the representation Λ is irreducible. This proves our theorem.

8. Fully decomposable representations. If $\mathfrak{A}$ is a simple algebra of degree n and index m over its centrum $\mathfrak{K}$ and t is the degree of $\mathfrak{K}$ over $\mathfrak{F}$ the integer

$$(24) \qquad \delta = \delta(\mathfrak{A}) = nmt$$

is an integral invariant of $\mathfrak{A}$ over $\mathfrak{F}$, and $\delta(\mathfrak{A}) = \delta(\mathfrak{A}^*)$ for every representation algebra $\mathfrak{A}^*$. We may then prove

Theorem 3. *Every ν-rowed representation Λ of a simple algebra $\mathfrak{A}$ by $\mathfrak{A}^*$ is a fully decomposable equivalence of $\mathfrak{A}$ and $\mathfrak{A}^*$. Then $\nu = s\delta(\mathfrak{A})$ where s is the number of irreducible components of Λ. These components are all equivalent and the $\mathfrak{M}_\nu$-commutator of $\mathfrak{A}^*$ is a normal simple algebra of degree ms over the centrum of $\mathfrak{A}^*$ and is similar to $\mathfrak{A}^{*-1}$.*

For Lemma 8.2 states that Λ is an equivalence of $\mathfrak{A}$ and $\mathfrak{A}^*$, $\mathfrak{A}^*$ is simple. By the proof of Theorem 8.2 we have $\nu = s\delta(\mathfrak{A})$, Λ is irreducible if and only if $s = 1$. Moreover the $\mathfrak{M}_\nu$-commutator $\mathfrak{M}_\nu^{\mathfrak{A}^*}$ of $\mathfrak{A}^*$ is $\mathfrak{M}_s \times \mathfrak{D}^{-1}$, where $\mathfrak{D}$ is a normal division algebra over the centrum $\mathfrak{K}$ of $\mathfrak{A}^*$ and $\mathfrak{D}$ is similar to $\mathfrak{A}^*$. Hence the degree of $\mathfrak{M}_\nu^{\mathfrak{A}^*}$ is ms as desired. If $s > 1$ the subalgebra $\mathfrak{M}_s$ of the $\mathfrak{M}_\nu$-commutator of $\mathfrak{A}^*$ contains an idempotent $E \neq I_\nu$, every A of $\mathfrak{A}^*$ has the form $A = EAE + (I_\nu - E)A(I_\nu - E)$, and the proof of Theorem 8.2 shows that Λ is decomposable. The indecomposable components Λ_i of Λ are ν_i-rowed equivalences of $\mathfrak{A}$ and $\mathfrak{A}_i$ and must be simple, $\nu_i = s_i\delta(\mathfrak{A})$. But our proof shows that Λ_i is decomposable if $s_i > 1$. Hence $\nu_i = \delta(\mathfrak{A})$ and if we write Λ as in (15) we have $\nu = \nu_1 + \cdots + \nu_\tau = \tau\delta(\mathfrak{A})$. Hence $\tau = s$. Moreover the Λ_i are irreducible by Theorem 8.2, Λ is fully decomposable. That the Λ_i are equivalent $\delta(\mathfrak{A})$-rowed representations of $\mathfrak{A}$ was derived as Lemma 8.4.

The structure of fully decomposable representations is now determined in

Theorem 4. *A ν-rowed representation Λ of $\mathfrak{A}$ by $\mathfrak{A}^*$ is fully decomposable if and only if $\mathfrak{A}^*$ is semi-simple.*

For if Λ is fully decomposable it may be taken to have the form (15) such that the corresponding representations Λ_i of $\mathfrak{A}$ by $\mathfrak{A}_i$ are irreducible. If B is in the radical $\mathfrak{N}^*$ of $\mathfrak{A}^*$ then AB is in $\mathfrak{N}^*$ for every A of $\mathfrak{A}^*$, $AB = \text{diag}\{a_1b_1, \cdots, a_tb_t\}$, a_ib_i is nilpotent, b_i is properly nilpotent in $\mathfrak{A}_i$, b_i is in the radical of $\mathfrak{A}_i$. The $\mathfrak{A}_i$ are simple algebras and hence $b_i = 0$, $B = 0$, $\mathfrak{N}^* = 0$, $\mathfrak{A}^*$ is semi-simple. Conversely if $\mathfrak{A}^*$ is a semi-simple algebra our proof of Theorem 8.2 shows that if $\mathfrak{A}^*$ is not simple Λ may be decomposed into representations Λ_i of $\mathfrak{A}$ by the simple components of $\mathfrak{A}^*$. These representations are fully decomposable by Theorem 8.3 and hence so is Λ.

We next obtain

Theorem 5. *Let $\mathfrak{A}$ be the semi-simple direct sum*

$$(25) \qquad \mathfrak{A}_1 \oplus \cdots \oplus \mathfrak{A}_r$$

of simple algebras $\mathfrak{A}_i$. *Then every* ν*-rowed equivalence* Λ *of* $\mathfrak{A}$ *and* $\mathfrak{A}^*$ *is equivalent to*

$$(26) \qquad a \leftrightarrow A_a = \text{diag}\,\{a_1^*, \cdots, a_r^*\}$$

such that a is in $\mathfrak{A}_i$ *if and only if* $a_j^* = 0$ *for* $j \neq i$. *The correspondences*

$$(27) \qquad a_i \leftrightarrow a_i^* \qquad (a_i \text{ in } \mathfrak{A}_i),$$

are ν_i*-rowed equivalences of* $\mathfrak{A}_i$ *and the corresponding algebras* $\mathfrak{A}_i^*$ *of* ν_i*-rowed square matrices* a_i^*. *The* $\mathfrak{M}_\nu$*-commutator of* $\mathfrak{A}^*$ *is the direct sum* $\mathfrak{C}_1 \oplus \cdots \oplus \mathfrak{C}_r$ *where* $\mathfrak{C}_i$ *is the* $\mathfrak{M}_{\nu_i}$*-commutator of* $\mathfrak{A}_i^*$, *and two such representations are equivalent if and only if the integers* ν_i *are the same.*

For our proof of Theorem 8.4 implies (26) and that the correspondences (27) are the desired equivalences. Now let

$$(28) \qquad a \to B_a = \text{diag}\,\{a_{01}^*, \cdots, a_{0r}^*\}$$

be a ν-rowed equivalence Λ_0 of $\mathfrak{A}$ by $\mathfrak{A}_0^*$. If the integers ν_i are the same we apply Lemma 8.4 to obtain an equivalence replacing each a_{0i}^* by a_i^*, and Λ and Λ_0 are equivalent. Conversely if Λ and Λ_0 are equivalent there exists a non-singular matrix C such that $CA_a = B_aC$. We may then partition C so that $C = (C_{ij})$, $C_{ij}a_j^* = a_{0i}^*C_{ij}$. There is clearly no loss of generality if we assume that

$$(29) \qquad a_i^* = \text{diag}\,\{b_i^*, \cdots, b_i^*\}, \qquad a_{0i}^* = \text{diag}\,\{b_{0i}^*, \cdots, b_{0i}^*\}$$

where $a_i \to b_i^*$, $a_i \to b_{0i}^*$ are irreducible representations of $\mathfrak{A}_i$. Now if D is any matrix $(D_{\alpha\beta})$ such that $Da_j^* = a_{0i}^*D$ we have $D_{\alpha\beta}b_j^* = b_{0i}^*D_{\alpha\beta}$ and by Lemma 8.7 either $i = j$ or $D_{\alpha\beta} = 0$, $D = 0$. It follows that $C = \text{diag}\,\{C_{11}, \cdots, C_{rr}\}$, the C_{ii} are necessarily non-singular, the integers ν_i are the same. If Λ and Λ_0 are the same, our equation $CA_a = B_aC$ becomes $CA_a = A_aC$, C is in $\mathfrak{C} = \mathfrak{M}_\nu^{\mathfrak{A}^*}$, C_{ii} is in $\mathfrak{C}_i = \mathfrak{M}_{\nu_i}^{\mathfrak{A}_i^*}$, $\mathfrak{C} = \mathfrak{C}_1 \oplus \cdots \oplus \mathfrak{C}_r$ as desired.

Lemma 8.5 implies that every fully decomposable ν-rowed representation $a \to a^*$ of an algebra $\mathfrak{A}$ with radical $\mathfrak{N}$ defines an equivalence $[a] \to a^*$ where $[a]$ ranges over the classes of a semi-simple component of $\mathfrak{A} - \mathfrak{N}$. Thus Theorem 8.5 gives a criterion for the equivalence of any two fully decomposable representations of any algebra.

9. Irreducible components of arbitrary matric representations. The results of Lemma 8.5 and Theorem 8.2 enable us to obtain some results on the irreducible components of any ν-rowed representation of an algebra $\mathfrak{A}$. We let $\mathfrak{A}$ be any algebra with a unity quantity, $\mathfrak{N}$ be the radical of $\mathfrak{A}$, and write

$$(30) \qquad \mathfrak{A} - \mathfrak{N} = \mathfrak{A}_1 \oplus \cdots \oplus \mathfrak{A}_r,$$

for simple algebras $\mathfrak{A}_i$. By Theorem 2.20 there is a (1-1) correspondence between the ideals $\mathfrak{B}_0$ of $\mathfrak{A} - \mathfrak{N}$ and the ideals $\mathfrak{B}$ of $\mathfrak{A}$ containing $\mathfrak{N}$, such that $\mathfrak{B} - \mathfrak{N} = \mathfrak{B}_0$. Moreover $\mathfrak{A} - \mathfrak{B} \cong (\mathfrak{A} - \mathfrak{N}) - (\mathfrak{B} - \mathfrak{N})$. By Theorem 2.21

the only ideals of $\mathfrak{A} - \mathfrak{N}$ are direct sums of the $\mathfrak{A}_i$, and $(\mathfrak{A} - \mathfrak{N}) - \mathfrak{B}_0$ is simple if and only if $(\mathfrak{A} - \mathfrak{N}) - \mathfrak{B}_0 \cong \mathfrak{A}_i$ for some i. If Λ is an irreducible representation of $\mathfrak{A}$ by $\mathfrak{A}^*$ the ideal $\mathfrak{B}_\Lambda$ of $\mathfrak{A}$ was shown in Lemma 8.5 to contain $\mathfrak{N}$, $\mathfrak{A} - \mathfrak{B}_\Lambda$ is simple by Theorem 8.2, $\mathfrak{A} - \mathfrak{B}_\Lambda \cong \mathfrak{A}_i$. This gives

Theorem 6. *Let Λ be an irreducible ν-rowed representation $a \to a^*$ of an algebra $\mathfrak{A}$ by $\mathfrak{A}^*$ and define the $\mathfrak{A}_i$ as in* (30). *Then there is a value of i such that $a^* = 0$ for every a such that the component in $\mathfrak{A}_i$ of the corresponding class $[a]$ of $\mathfrak{A} - \mathfrak{N}$ is zero, and*

$$[a] \to a^* \qquad ([a] \text{ in } \mathfrak{A}_i), \tag{31}$$

is an equivalence of $\mathfrak{A}_i$ and $\mathfrak{A}^*$. Moreover $\nu = \delta(\mathfrak{A}_i)$.

As an immediate corollary of this result, Lemma 8.3, and Theorem 4.14 we have

Theorem 7. *Two irreducible ν-rowed representations Λ_1 and Λ_2 of an algebra $\mathfrak{A}$ are equivalent if and only if the corresponding ideals $\mathfrak{B}_{\Lambda_1}$ and $\mathfrak{B}_{\Lambda_2}$ of Lemma* 8.1 *are the same, that is,* (31) *holds with the same i for Λ_1 and Λ_2.*

We next derive

Theorem 8. *The irreducible components of any representation Λ of $\mathfrak{A}$ by $\mathfrak{A}^*$ are uniquely determined in the sense of equivalence.*

For if 1 is the unity quantity of $\mathfrak{A}$ we may write $1 = e_1 + \cdots + e_r$ for pairwise orthogonal idempotents e_i, and we see that $[e_i]$ is the unity quantity of $\mathfrak{A}_i$ of (30). If Λ_j is any irreducible component $a \to a^{\Lambda_j}$ of Λ we apply Theorem 8.6 to see that $e_k^{\Lambda_j} = 0$ for $k \neq i$, $e_i^{\Lambda_j} = I_{\delta_i}$, where $\delta_i = \delta(\mathfrak{A}_i)$. Thus in the representation Λ the quantity $e_i^\Lambda = e_i^*$ may be taken to be a ν-rowed matrix all of whose elements above and on the diagonal are zero except that $\rho_i = \delta_i s_i$ of these elements are ones, where s_i is the number of irreducible components Λ_j of Λ for which $e_i^{\Lambda_j} \neq 0$. All such components are unique in the sense of equivalence, by the last two theorems, and it remains only to see that the integers s_i are invariants. But then the characteristic function of e_i^* is $\lambda^{\nu-\rho_i}(\lambda - 1)^{\rho_i}$. The characteristic functions of e_i^* and any $Be_i^*B^{-1}$ are the same and we have proved that s_i is an invariant of Λ, the desired result.

Lemma 8.4 and Theorem 8.6 imply that if in two ν-rowed representations Λ and Λ_0 the invariants s_i are the same we may take the irreducible components the same. However this need not imply the equivalence of Λ and Λ_0 since, for example, Λ might be an equivalence of the form (11) of an algebra $\mathfrak{A}$ with a non-zero radical and a ν-rowed algebra $\mathfrak{A}^*$, Λ_0 the fully decomposable representation of $\mathfrak{A}$ derived by substituting zero matrices for the a_{ji} with $j \neq i$. We have, however,

Theorem 9. *Let Λ be a ν-rowed equivalence of $\mathfrak{A}$ and $\mathfrak{A}^*$. Then every irreducible representation of $\mathfrak{A}$ is equivalent to a component of Λ.*

For we are assuming that $\mathfrak{A}$ has a unity quantity, $\mathfrak{A} - \mathfrak{N} \neq 0$. By Theorems 8.6, 8.7, and the proof of Theorem 8.8 it is sufficient to show that the integer $s_i > 0$ for every $\mathfrak{A}_i$ of (30). If $s_i = 0$ the characteristic function of e_i^* is λ^ν, e_i^* is nilpotent. This is impossible since Λ is an equivalence of $\mathfrak{A}$ and $\mathfrak{A}^*$, $e_i^* \neq 0$ is idempotent.

The regular representation $a \to S_a$ of an algebra $\mathfrak{A}$ of order h over $\mathfrak{F}$ by the algebra $\mathfrak{A}^*$ of all right multiplications

$$S_a: \qquad x \to x^{S_a} = xa \qquad (x,\ a \text{ in } \mathfrak{A})$$

is an h-rowed equivalence of $\mathfrak{A}$ and $\mathfrak{A}^*$. Theorem 8.9 gives the properties of its irreducible component representations. However we may notice in addition that the space $\mathfrak{L}$ is $\mathfrak{A}$ itself, the invariant subspaces of $\mathfrak{A}$ are its right ideals. We shall not use these properties here but refer the reader to the references given in Section 8.1. However let us observe a consequence of Theorem 8.9 and the fact that the regular representation is an equivalence. We state the result as

Theorem 10. *Let $a \to a_i^*$ be r inequivalent irreducible δ_i-rowed representations of an algebra $\mathfrak{A}$ where r is determined as in* (30), *$\delta_i = \delta(\mathfrak{A}_i)$ as in* (24). *Then the correspondence*

$$(32) \qquad a \to \text{diag}\,\{a_1^*, \cdots, a_r^*\}$$

is a representation of $\mathfrak{A}$ uniquely determined by $\mathfrak{A}$ in the sense of equivalence, and we shall call (32) ***the reduced representation of*** $\mathfrak{A}$. *It is an equivalence if and only if $\mathfrak{A}$ is semi-simple.*

10. Scalar extensions. If Λ is a ν-rowed representation $a \to a^*$ of an algebra $\mathfrak{A} = (u_1, \cdots, u_n)$ over $\mathfrak{F}$ by $\mathfrak{A}^*$ and $\mathfrak{L}$ is any scalar extension of $\mathfrak{F}$ the correspondence

$$\xi_1 u_1 + \cdots + \xi_n u_n \to \xi_1 u_1^* + \cdots + \xi_n u_n^* \qquad (\xi_i \text{ in } \mathfrak{L})$$

is clearly a representation $\Lambda_{\mathfrak{L}}$ of $\mathfrak{A}_{\mathfrak{L}}$ by $\mathfrak{A}_{\mathfrak{L}}^*$ such that the ideal of Lemma 8.1 is $\mathfrak{B}_{\Lambda_{\mathfrak{L}}} = (\mathfrak{B}_\Lambda)_{\mathfrak{L}}$. We then make the

DEFINITION. *A representation Λ is called absolutely irreducible if $\Lambda_{\mathfrak{L}}$ is irreducible for every scalar extension $\mathfrak{L}$ of $\mathfrak{F}$.*

We then have

Theorem 11. *An irreducible ν-rowed representation Λ of $\mathfrak{A}$ by $\mathfrak{A}^*$ is absolutely irreducible if and only if $\mathfrak{A}^*$ is a total matric algebra, that is, the invariant δ of* (24) *has the value given by $m = t = 1$.*

For if $m = t = 1$ we have $\nu = n$, $\mathfrak{A}^* = \mathfrak{M}_\nu$, $\mathfrak{A}_{\mathfrak{L}}^* = (\mathfrak{M}_\nu)_{\mathfrak{L}} = \mathfrak{M}$, the $\mathfrak{M}$-commutator of $\mathfrak{A}_{\mathfrak{L}}^*$ is $\mathfrak{L}$, and Theorem 8.2 implies that $\Lambda_{\mathfrak{L}}$ is irreducible. Conversely, if Λ is absolutely irreducible we must have $mt = 1$. For otherwise the $\mathfrak{M}_\nu$-commutator of $\mathfrak{A}^*$ is $\mathfrak{D}^{-1}$ over $\mathfrak{K}$, the $(\mathfrak{M}_\nu)_{\mathfrak{L}}$-commutator of $\mathfrak{A}_{\mathfrak{L}}^*$ contains $\mathfrak{D}_{\mathfrak{L}}^{-1}$ which is not a division algebra by Theorem 1.9 for a properly chosen scalar extension field $\mathfrak{L}$ over $\mathfrak{F}$, $\Lambda_\mathfrak{L}$ is reducible.

As an immediate corollary of Theorem 8.4 and the definition of separability we have

Theorem 12. *A ν-rowed representation Λ of $\mathfrak{A}$ by $\mathfrak{A}^*$ has the property that $\Lambda_\mathfrak{L}$ is fully decomposable for every scalar extension $\mathfrak{L}$ over $\mathfrak{F}$ if and only if $\mathfrak{A}^*$ is separable over $\mathfrak{F}$.*

11. The characteristic and minimum functions. If Λ is any ν-rowed representation $a \to a^*$ of $\mathfrak{A}$ by $\mathfrak{A}^*$ we may assume that a^* has the form (11). The characteristic function,

$$f_\Lambda(\lambda, a) = |\lambda I_\nu - a^*| = \prod_{i=1}^{\tau} |\lambda I_{\nu_i} - a_{ii}^*|, \tag{33}$$

of a is then called the Λ-*characteristic function* of a. Observe that $f_\Lambda(\lambda, a)$ is a product of characteristic functions of quantities a_{ii}^* of simple algebras $\mathfrak{A}_{ii}^*$. Similarly we call the minimum function of a^* the Λ-*minimum function* of a.

The determinant of a^* is called the Λ-*norm* $N_\Lambda(a)$ and we evidently have

$$N_\Lambda(ab) = N_\Lambda(a)N_\Lambda(b), \qquad N_\Lambda(\alpha a) = \alpha^\nu N_\Lambda(a). \tag{34}$$

Moreover,

$$N_\Lambda(a) = N_{\Lambda_1}(a) \cdots N_{\Lambda_\tau}(a), \tag{35}$$

where $\Lambda_1, \cdots, \Lambda_\tau$ are the irreducible components of Λ. Similarly the sum of the diagonal elements of a^* will be called the Λ-*trace* $T_\Lambda(a)$ and we have

$$T_\Lambda(\alpha a + \beta b) = \alpha T_\Lambda(a) + \beta T_\Lambda(b) \tag{36}$$

for every α and β of $\mathfrak{F}$, a and b of $\mathfrak{A}$, and

$$T_\Lambda(a) = T_{\Lambda_1}(a) + \cdots + T_{\Lambda_\tau}(a). \tag{37}$$

Evidently $T_\Lambda(a)$ is the negative of the coefficient of $\lambda^{\nu-1}$ in $f_\Lambda(\lambda, a)$, $N_\Lambda(a)$ is the product of its constant term by $(-1)^\nu$.

We now make the

DEFINITION. *Let $\mathfrak{A}$ be an algebra over $\mathfrak{F}$, $\mathfrak{L}$ be an algebraically closed scalar extension of $\mathfrak{F}$, Λ the reduced representation of $\mathfrak{A}_\mathfrak{L}$ over $\mathfrak{L}$. Then $T_\Lambda(a)$ and $N_\Lambda(a)$ will be called the reduced trace and norm, respectively, of a for every quantity a of $\mathfrak{A}$.*

If $\mathfrak{A}$ is a separable algebra with splitting field $\mathfrak{L}_1$ then every quantity a_1 of $\mathfrak{A}_{\mathfrak{L}_1}$ may be regarded as a matrix which is a direct sum of matrices each component matrix of which is in an absolutely irreducible representation of $\mathfrak{A}_{\mathfrak{L}_1}$, $a_1 \to a_1$ is the reduced representation of $\mathfrak{A}_{\mathfrak{L}_1}$. Then the trace and determinant of the matrices $a_1 = a$ in $\mathfrak{A}$ are clearly the reduced trace and norm respectively of a.

We now consider a simple separable algebra $\mathfrak{A}$ over an infinite field $\mathfrak{F}$, so that the centrum $\mathfrak{K}$ of $\mathfrak{A}$ is a separable field $\mathfrak{K} = \mathfrak{F}(u)$ of degree t over $\mathfrak{F}$. In order to show that the reduced trace and norm functions defined above are independent of the field $\mathfrak{L}$ used in the definition and in fact have values in $\mathfrak{F}$, we shall discuss the minimum function of $\mathfrak{A}$ over $\mathfrak{F}$.

The algebra $\mathfrak{A}$ is normal simple of degree n over $\mathfrak{K}$,

$$(38) \qquad \mathfrak{A} = (u_1, \cdots, u_\nu) \text{ over } \mathfrak{K}, \qquad \mathfrak{A} = (u_1, \cdots, u_{\nu t}) \text{ over } \mathfrak{F},$$

where $\nu = n^2$ and we may normalize our basis so that $u_1 = 1$, $u_{\nu+1}$ is the generating quantity u of $\mathfrak{K}$ over $\mathfrak{F}$. If $\xi_1, \cdots, \xi_{\nu t}$ are independent indeterminates over $\mathfrak{K}$, and $\mathfrak{L}_0 = \mathfrak{F}(\xi_1, \cdots, \xi_{\nu t})$ the algebra $\mathfrak{A}_{\mathfrak{L}_0}$ contains the general quantity $y = \sum_{i=1}^{\nu t} \xi_i u_i$ of $\mathfrak{A}$ over $\mathfrak{F}$, $\mathfrak{A}_{\mathfrak{L}_0}$ is normal simple of degree n over $\mathfrak{L} = \mathfrak{K}(\xi_1, \cdots, \xi_{\nu t})$. The general quantity of $\mathfrak{A}$ over $\mathfrak{K}$ is $x = \sum_{i=1}^{\nu} \xi_i u_i$. By Theorem A10.11 the quantity x generates a subfield $\mathfrak{L}(x)$ of $\mathfrak{A}_{\mathfrak{L}_0}$ such that $\mathfrak{L}(x)$ is separable of degree n over $\mathfrak{L}$, degree nt over $\mathfrak{L}_0$. Also $\mathfrak{L}(x) = \mathfrak{L}_0(x, u)$ is separable over $\mathfrak{L}_0$ so that $\mathfrak{L}(x) = \mathfrak{L}_0(v)$ for a quantity $v = x + \alpha u$ with α in $\mathfrak{L}_0$.

If the minimum function of the quantity y of $\mathfrak{A}_{\mathfrak{L}_0}$ is $\phi(\lambda) = \phi(\lambda; \xi_1, \cdots, \xi_{\nu t})$ then $\phi(v; \xi_1, \cdots, \xi_\nu, \alpha, 0, \cdots, 0) = 0$. But v generates a field of degree nt over $\mathfrak{L}_0$ and $\psi(\lambda) = \phi(\lambda; \xi_1, \cdots, \xi_\nu, \alpha, 0, \cdots, 0)$ has coefficients in $\mathfrak{L}_0$ and degree in λ the same as that of $\phi(\lambda)$, $\phi(\lambda)$ has degree at least nt in λ.

To prove the converse we recall that the minimum function of y in any scalar extension $\mathfrak{A}_{\mathfrak{W}}$ of $\mathfrak{A}_{\mathfrak{L}_0}$ is also $\phi(\lambda)$, and use this result in the case where $\mathfrak{W}$ is a splitting field over $\mathfrak{L}_0$ of $\mathfrak{A}_{\mathfrak{L}_0}$. By Corollary II of Theorem 3.22 every quantity of $\mathfrak{A}_{\mathfrak{W}}$ may be regarded as a direct sum of t n-rowed square matrices. The characteristic function of y in this matric representation of $\mathfrak{A}_{\mathfrak{W}}$ has degree nt, the minimum function $\phi(\lambda)$ has degree at most nt, $\phi(\lambda)$ has degree nt. In Section 1.13 we called this invariant the degree of $\mathfrak{A}$ over $\mathfrak{F}$. We now complete this result by proving

Theorem 13. *Let $\mathfrak{A}$ be a simple algebra of order n^2t over $\mathfrak{F}$, where the centrum $\mathfrak{K}$ of $\mathfrak{A}$ is separable of degree t over $\mathfrak{F}$. Then the minimum function $\phi(\lambda)$ of $\mathfrak{A}$ over $\mathfrak{F}$ is irreducible of degree nt in λ and the characteristic function of $\mathfrak{A}$ is the nth power of $\phi(\lambda)$.*

For we have seen that $\mathfrak{L}_0(v)$ has degree nt over $\mathfrak{L}_0$, $\psi(\lambda)$ has degree nt, $\psi(v) = 0$, so that $\psi(\lambda)$ is irreducible in $\mathfrak{L}_0$. Then $\phi(\lambda)$ must be irreducible in $\mathfrak{L}_0$. The degree of the characteristic function of $\mathfrak{A}$ is n^2t and its irreducible factors divide $\phi(\lambda)$ and must coincide with $\phi(\lambda)$.

It is actually possible to show that if $\mathfrak{A}$ is normal simple over $\mathfrak{F}$ so that $t = 1$ the Galois group of $\phi(\lambda; \xi_1, \cdots, \xi_{\nu t})$ over $\mathfrak{L}_0$ is the symmetric group. For proof see [3], [4], [11], and page 51 of [119]. Moreover it is shown in those expositions that if $\mathfrak{F}$ is a *Hilbert irreducibility field* there exist quantities a in $\mathfrak{A}$ such that the minimum function of a has (its maximum) degree n and the symmetric Galois group. We are now able to obtain matrix interpretations of the principal trace and principal norm, as defined in Section 1.14, for algebras of the type considered in the last theorem.

Theorem 14. *Let a be a quantity of a simple separable algebra $\mathfrak{A}$ over $\mathfrak{F}$, such that $\mathfrak{A}$ has degree n over its centrum $\mathfrak{K}$, $\mathfrak{K}$ has degree t over $\mathfrak{F}$. Then the principal norm and trace of a are respectively its reduced norm and trace. Moreover if $\mathfrak{W}$ is a splitting field of $\mathfrak{A}$ over $\mathfrak{F}$ so that $\mathfrak{A}_{\mathfrak{W}}$ may be regarded as an algebra of nt-rowed square matrices over $\mathfrak{W}$, the principal function of a is the characteristic function of the corresponding matrix.*

For, the general quantity y of $\mathfrak{A}$, may be regarded as being also the general quantity of $\mathfrak{A}_{\mathfrak{W}}$, and hence is an nt-rowed square matrix whose characteristic function $f(\lambda)$ is a monic polynomial of degree nt. But $f(\lambda)$ is divisible by the minimum function $\phi(\lambda)$ of y which has degree nt, $f(\lambda) = \phi(\lambda)$. This proves the last part of the theorem, which implies the first part by the comment following the definition above of reduced norm and trace.

Note that we have proved that the reduced trace and reduced norm of each quantity of a separable simple algebra $\mathfrak{A}$ over $\mathfrak{F}$ are themselves quantities of $\mathfrak{F}$, the result mentioned after their definition. Also the matrix interpretation of the principal trace and norm furnishes immediate proofs of the linearity property of the principal trace and the multiplicative property of the principal norm.

12. The discriminant matrix. Let $\mathfrak{A}$ be an algebra of order h over a field $\mathfrak{F}$ and designate the reduced trace for $\mathfrak{A}$ by $T(a)$. If $u_1, \cdots, u_n$ are a basis of $\mathfrak{A}$ over $\mathfrak{F}$ the matrix

$$M_U = (T_{ij}), \qquad T_{ij} = T(u_i u_j) \qquad (i, j = 1, \cdots, h), \tag{39}$$

is called the *discriminant matrix* of $\mathfrak{A}$ with respect to the basis $U = (u_1, \cdots, u_h)$. We shall call its determinant D_U the *discriminant of $\mathfrak{A}$ with respect to U.*

If B is the matrix (b_{ki}) of a non-singular linear transformation

$$v_i = \sum_{k=1}^{h} u_k b_{ki}, \tag{40}$$

replacing a basis U of $\mathfrak{A}$ by a basis $V = (v_1, \cdots, v_h) = UB$, then

$$v_i v_j = \sum_{k,l=1}^{h} b_{ki}(u_k u_l) b_{lj},$$

and by (36) we have

$$T(v_i v_j) = \sum b_{ki} T_{kl} b_{lj}, \qquad T_{kl} = T(u_k u_l). \tag{41}$$

But then

$$M_V = B' M_U B, \qquad D_V = |B|^2 D_U, \tag{42}$$

where B' is the transpose of B. Clearly $D_V = 0$ if and only if $D_U = 0$. The vanishing of the discriminant of $\mathfrak{A}$ is then a property unaltered by a change of basis. It is also independent of the field of reference since $T(u_i u_j)$ is not a function of $\mathfrak{F}$ but merely of the multiplication constants of $\mathfrak{A}$. Hence D_U

is the same for every $\mathfrak{A}_{\mathfrak{L}}$, $\mathfrak{L}$ a scalar extension of $\mathfrak{F}$. We now prove the following criterion.

Theorem 15. *The discriminants of an algebra $\mathfrak{A}$ are not zero if and only if $\mathfrak{A}$ is separable.*

For let the discriminants be not zero and $\mathfrak{A}$ be inseparable. By the argument above we may choose $\mathfrak{F}$ so that $\mathfrak{A}$ has a non-zero radical $\mathfrak{N}$. We may write $\mathfrak{A} = \mathfrak{N} + \mathfrak{C}$, $\mathfrak{N} = (u_1, \cdots, u_f)$, $\mathfrak{C} = (u_{f+1}, \cdots, u_h)$ and have $u_i u_j$ in $\mathfrak{N}$ for every $i \leqq f$ or $j \leqq f$. The trace of a nilpotent matrix is zero and thus

$$M_U = \begin{pmatrix} 0 & 0 \\ 0 & M_0 \end{pmatrix}, \qquad M_0 = (T_{ij}) \quad (i, j = f + 1, \cdots, h). \tag{43}$$

Then $D_U = 0$ contrary to hypothesis. Conversely let $\mathfrak{A}$ be semi-simple so that $\mathfrak{A} = \mathfrak{A}_1 \oplus \cdots \oplus \mathfrak{A}_r$ for simple algebras $\mathfrak{A}_i$ of degree ν_i. The reduced representation Λ of $\mathfrak{A}$ is an equivalence of $\mathfrak{A}$ and the algebra $\mathfrak{A}^*$ of all matrices

$$\text{diag } \{a_1^*, \cdots, a_r^*\},$$

where $a_i \rightarrow a_i^*$ is an irreducible ν_i-rowed equivalence of $\mathfrak{A}_i$ and the corresponding algebra $\mathfrak{A}_i^*$ of the a_i^*. We may take a basis of $\mathfrak{A}$ to be $(u_1^{(1)}, \cdots, u_{h_1}^{(1)}, \cdots, u_1^{(r)}, \cdots, u_{h_r}^{(r)})$, where $U_k = (u_1^{(k)}, \cdots, u_{h_k}^{(k)})$ is a basis of $\mathfrak{A}_i$. Then $u_i^{(k)} u_j^{(l)} = 0$ for $k \neq l$, and

$$M_U = \text{diag } \{M_{U_1}, \cdots, M_{U_r}\}. \tag{44}$$

It follows that the discriminant (for a properly chosen basis) of any semi-simple algebra is the product of the discriminants of its simple components. If $\mathfrak{A}$ is separable we may choose $\mathfrak{F}$ so that each $\mathfrak{A}_i$ is a ν_i-rowed total matric algebra. This reduces the proof of our theorem to the case where $\mathfrak{A}$ is a ν-rowed total matric algebra. By Theorem A4.15 we have $\mathfrak{A} = \mathfrak{E} + \mathfrak{E}y + \cdots + \mathfrak{E}y^{\nu-1}$ where $\mathfrak{E}$ is the diagonal algebra $(e_1, \cdots, e_\nu)$, $ye_{i+1} = e_i y$, $y^\nu = I_\nu$. Hence $T(e_i y^j) = 0$ if $n > j > 0$, $T(e_i) = 1$. Then we compute

$$M_U = \begin{bmatrix} I_\nu & 0 & \cdots & 0 & 0 \\ 0 & 0 & \cdots & 0 & B_2 \\ 0 & 0 & \cdots & B_3 & 0 \\ \cdot & \cdot & \cdots & \cdot & \cdot \\ 0 & B_\nu & \cdots & 0 & 0 \end{bmatrix}, \; |B_i| = \pm 1, \; |M_U| = \pm 1. \tag{45}$$

This completes our proof.

The result just obtained has an important consequence for the theory of the arithmetics of algebras. We shall make this application in Section 9.2.

13. Generalized Riemann matrices. Let $\mathcal{C}$ be an algebraically closed non-modular field $\mathcal{C} = \mathcal{R}(i)$ where $i^2 = -1$ and $\mathcal{R}$ is ordered closed according to the definition of Section A5.10. It was proved in [57], [58] that our hypothesis of the existence of $\mathfrak{R}$ is always fulfilled. For simplicity of terminology we shall speak both here and in Section 10.12 of the quantities of $\mathfrak{C}$ as *complex numbers* $a + bi$ where a and b in $\mathfrak{R}$ are called *real numbers*. We also call $a + bi$ *imaginary* if $b \neq 0$, *pure imaginary* if $a = 0$, $b \neq 0$. The field $\mathfrak{C}$ is a quadratic (cyclic) field over $\mathfrak{R}$ with generating automorphism $a + bi \leftrightarrow \overline{a + bi} = a - bi$, and we call $\overline{a + bi}$ the *conjugate* of $a + bi$. The theory of symmetric and Hermitian, skew and skew-Hermitian matrices of Chapter AV is now valid and we shall assume the results of that theory. We write $\bar{A}'$ for the usual conjugate transpose of any matrix A.

We next assume that $\mathfrak{F} \leqq \mathfrak{C}$, $\mathfrak{S} \leqq \mathfrak{R}$ such that $\mathfrak{F} = \mathfrak{S}$ or $\mathfrak{F} = \mathfrak{S}(\theta)$, $\theta^2 = \rho < 0$ in $\mathfrak{S}$. Then in the latter case $\mathfrak{F}$ is a quadratic field over $\mathfrak{S}$, $\bar{\theta} = -\theta$, the conjugate operation of $\mathfrak{C}$ induces a generating automorphism of $\mathfrak{F}$ over $\mathfrak{S}$. In the former case $\bar{a} = a$ for every a of $\mathfrak{F}$ so that $a \leftrightarrow \bar{a}$ is the identity automorphism of $\mathfrak{F}$ over $\mathfrak{S}$ and is again a generating automorphism of $\mathfrak{F}$ over $\mathfrak{S}$.

A ν-rowed square matrix Ω with elements in $\mathfrak{C}$ is called a *generalized Riemann matrix* (abbreviated henceforth as GR-matrix) over $\mathfrak{F}$ if there exists a matrix C with elements in $\mathfrak{F}$ such that

$$\bar{C}' = \epsilon C, \qquad \epsilon = \pm 1, \qquad \Gamma = \Omega C \tag{46}$$

is a positive definite Hermitian matrix. The matrix C is called a *principal matrix* of Ω and is not unique in general. Clearly both C and Ω are non-singular. We call ϵ the *type number* of Ω. The study of such matrices was introduced by Weyl [440] as a generalization of that of the older Riemann matrices. A somewhat more general definition is given in [43] but the most interesting and fruitful definition appears to be that given above. See also [43], [441], [41], [14], [38] for sources and results which we shall not derive here.

A matrix Ω_0 is said to be *isomorphic* (in $\mathfrak{F}$) to Ω if there exists a non-singular matrix B with elements in $\mathfrak{F}$ such that $B\Omega B^{-1} = \Omega_0$. Then $\Gamma_0 = B\Gamma\bar{B}' = B\Omega B^{-1}BC\bar{B}'$ is positive definite, $C_0 = BC\bar{B}' = \epsilon\bar{C}_0'$ and Ω_0 is a GR-matrix with the same type number as Ω.

We call a GR-matrix Ω_0 *impure* if it is isomorphic to a matrix of the form

$$\begin{pmatrix} \Omega_1 & 0 \\ \Omega_3 & \Omega_2 \end{pmatrix}, \tag{47}$$

where Ω_1 is a μ-rowed square matrix for $0 < \mu < \nu$. Otherwise we shall call Ω_0 *pure*. If Ω_0 is impure and is isomorphic to (47) with $\Omega_3 = 0$ we shall call Ω_0 *decomposable*. We may then prove

Theorem 16. *Every impure GR-matrix Ω_0 isomorphic to* (47) *is decomposable and is in fact isomorphic to*

$$\Omega = \begin{pmatrix} \Omega_1 & 0 \\ 0 & \Omega_2 \end{pmatrix}. \tag{48}$$

Moreover the matrices Ω_1 and Ω_2 are GR-matrices with the same type number as Ω_0.

For we may let Ω_0 be given by (47) with principal matrix C_0 and write

$$C_0 = \begin{pmatrix} C_1 & C_4 \\ C_3 & C_2 \end{pmatrix}, \qquad \Gamma_0 = \begin{pmatrix} \Gamma_1 & \Gamma_4 \\ \Gamma_3 & \Gamma_2 \end{pmatrix} = \Omega_0 C_0, \tag{49}$$

such that

$$\bar{C}_1' = \epsilon C_1, \qquad \bar{C}_2' = \epsilon C_2, \qquad \bar{\Gamma}_1' = \Gamma_1, \qquad \bar{\Gamma}_2' = \Gamma_2, \tag{50}$$

while

$$C_4 = \epsilon \bar{C}_3', \qquad \Gamma_4 = \bar{\Gamma}_3'. \tag{51}$$

Then direct computation of $\Gamma_0 = \Omega_0 C_0$ gives

$$\Gamma_1 = \Omega_1 C_1 = \epsilon C_1 \bar{\Omega}_1', \qquad \Gamma_2 = \Omega_3 C_4 + \Omega_2 C_2, \tag{52}$$

$$\Gamma_3 = \Omega_3 C_1 + \Omega_2 C_3, \qquad \Gamma_4 = \Omega_1 C_4. \tag{53}$$

But then (51) implies that

$$\Omega_3 C_1 + \Omega_2 C_3 = \epsilon C_3 \bar{\Omega}_1'. \tag{54}$$

Since Γ_0 is positive definite we may apply Theorem A5.16 to see that Γ_1 and Γ_2 are positive definite, Ω_1 is a μ-rowed GR-matrix with ϵ as type number. Hence C_1 is non-singular. Also $\bar{\Gamma}_1' = \bar{C}_1'\bar{\Omega}_1' = \epsilon C_1 \bar{\Omega}_1' = \Omega_1 C$, $\bar{\Omega}_1' = \epsilon C_1^{-1}\Omega_1 C_1$. Substituting in the result of solving (54) for Ω we have

$$G = C_3 C_1^{-1}, \qquad \Omega_3 = G\Omega_1 - \Omega_2 G. \tag{55}$$

Then we define

$$B = \begin{pmatrix} I_\mu & 0 \\ -G & I_{\nu-\mu} \end{pmatrix}, \qquad B^{-1} = \begin{pmatrix} I_\mu & 0 \\ G & I_{\nu-\mu} \end{pmatrix}, \tag{56}$$

and (55) implies

$$B\Omega_0 B^{-1} = \begin{pmatrix} \Omega_1 & 0 \\ -G\Omega_1 + \Omega_3 & \Omega_2 \end{pmatrix} \begin{pmatrix} I_\mu & 0 \\ G & I_{\nu-\mu} \end{pmatrix} = \begin{pmatrix} \Omega_1 & 0 \\ 0 & \Omega_2 \end{pmatrix}, \tag{57}$$

as desired. Applying (52) with $\Omega_3 = 0$ we have $\Gamma_2 = \Omega_2 C_2$, Ω_2 is a GR-matrix with C_2 as principal matrix, ϵ as type number. This proves our theorem.

Let us now examine the result just obtained from the point of view of representation theory. A GR-matrix Ω_0 has only a finite number of elements and these span a linear set $(\xi_1, \cdots, \xi_\sigma)$ over $\mathfrak{F}$. Thus

$$\Omega_0 = A_1\xi_1 + \cdots + A_\sigma \xi_\sigma, \tag{58}$$

where the A_j are ν-rowed square matrices with elements in $\mathfrak{F}$. The enveloping algebra of the A_i is an algebra $\mathfrak{A}_{\Omega_0}$ over $\mathfrak{F}$. Now $B\Omega_0 B^{-1}$ has the form (47) if and only if $\mathfrak{A}_{\Omega_0}$ is impure. Our result in Theorem 8.16 then states that $\mathfrak{A}_{\Omega_0}$

is fully decomposable. Moreover in view of Theorems 8.4, 8.5, and 8.16, we see that Ω_0 is isomorphic to

$$\Omega = \text{diag } \{\Omega_{01}, \cdots, \Omega_{0\tau}\}, \tag{59}$$

where

$$\Omega_{0i} = \text{diag } \{\Omega_i, \cdots, \Omega_i\} \qquad (i = 1, \cdots, \tau), \tag{60}$$

for pure GR-matrices Ω_i with the same type number ϵ as Ω_0, such that Ω_i and Ω_j are not isomorphic for $i \neq j$.

The principal problem concerning a GR-matrix Ω is that of determining the multiplication algebra $\mathfrak{A}$ of Ω, that is, the algebra of all matrices A with elements in $\mathfrak{F}$ such that $A\Omega = \Omega A$. But by (58) with $\Omega = \Omega_0$ we have $A\Omega_0 = \Omega_0 A = \sum (AA_i - A_iA)\xi_i = 0$ if and only if $AA_i = A_iA$. Thus $\mathfrak{A}$ is actually the $\mathfrak{M}_\nu$-commutator of $\mathfrak{A}_\Omega$. By Theorem 8.5, $\mathfrak{A} = \mathfrak{A}_1 \oplus \cdots \oplus \mathfrak{A}_\tau$ is the semi-simple direct sum of simple algebras $\mathfrak{A}_i$ such that $\mathfrak{A}_i$ is the $\mathfrak{M}_{\nu_i}$-commutator of the algebra $\mathfrak{A}_{\Omega_i}$ of ν_i-rowed matrices defined by Ω_{0i}. Moreover $\mathfrak{A}_i = \mathfrak{M}^{(i)} \times \mathfrak{D}^{(i)}$, where $\mathfrak{D}^{(i)}$ is equivalent to the division algebra which is the $\mathfrak{M}_{\mu_i}$-commutator of the pure μ_i-rowed matric algebra $\mathfrak{A}_{\Omega_i}$ defined by the pure GR-matrix Ω_i. Here $\mathfrak{M}^{(i)}$ is a total matric algebra whose degree is the number of components Ω_i in (60).

The statement above reduces the problem of determining the multiplication algebra $\mathfrak{A}$ of any GR-matrix Ω to the case where Ω is pure, $\mathfrak{A}$ is a division algebra $\mathfrak{D}$. However we have seen that it is generally simpler to be able to consider any algebra $\mathfrak{A} \sim \mathfrak{D}$. This may clearly be accomplished by considering instead of Ω and $\mathfrak{D}$ the multiplication algebra $\mathfrak{M} \times \mathfrak{D}$ of the GR-matrix diag $\{\Omega, \cdots, \Omega\}$. We shall study this problem further in Chapter X.

CHAPTER IX

RATIONAL DIVISION ALGEBRAS

1. Algebras over an algebraic number field. An algebraic number field is a field whose quantities are ordinary complex number roots of equations with rational coefficients. The extensive theory of algebraic numbers is a theory of such fields of finite degree over the field $\mathfrak{R}$ of all rational numbers. Any field $\mathfrak{F}$ of finite degree over $\mathfrak{R}$ is equivalent to an algebraic number field, and the results of the theory of algebraic numbers are then applicable.

If $\mathfrak{A} = (u_1, \cdots, u_n)$ is an algebra over $\mathfrak{F}$ and $\mathfrak{F}$ is a field of degree r over $\mathfrak{R}$, then $\mathfrak{F} = (w_1, \cdots, w_r)$ over $\mathfrak{R}$, $\mathfrak{A}$ is the algebra $(u_1w_1, \cdots, u_nw_r)$ over $\mathfrak{R}$. It is thus natural to call all such algebras $\mathfrak{A}$ *rational algebras*. When $\mathfrak{A}$ is a division algebra over $\mathfrak{F}$ it is also a division algebra over $\mathfrak{R}$ and conversely, and we shall speak of all such algebras as *rational division algebras*.

We see now that the problem of determining all division algebras over a field of finite degree over $\mathfrak{R}$ is that of determining all rational division algebras; that this latter problem is the problem of determining all normal division algebras $\mathfrak{D}$ over $\mathfrak{F}$, where the centrum $\mathfrak{F}$ of $\mathfrak{D}$ is an algebraic number field of finite degree over $\mathfrak{R}$. The solution of this problem is one of the principal achievements of a decade in which the theory of linear algebras has been particularly rich in interesting results. We shall give an exposition of the solution here. The treatment is necessarily partly algebraic and partly arithmetic, and we shall make it as self-contained as seems possible at the time of writing. Moreover we shall indicate precisely and explicitly what arithmetic results are presupposed.

2. Integral domains of an algebra. Let $\mathfrak{F}$ be a field which is the quotient field of an integral domain which we shall henceforth designate by $J_{\mathfrak{F}}$. This notation is not to be confused with that of the scalar extension $\mathfrak{A}_{\mathfrak{K}}$ of an algebra $\mathfrak{A}$ since we do not define the symbol J itself. Let us call the quantities of $J_{\mathfrak{F}}$ the *integers* of $\mathfrak{F}$.

We shall assume that $\mathfrak{F}$ is algebraic of finite degree over the quotient field of a u.f. integral domain as defined in Section A2.8. This hypothesis implies the

Generalized Gauss Lemma. *Let $f(x)$ be a monic polynomial of $J_{\mathfrak{F}}[x]$ and $f(x) = g(x) \cdot h(x)$ for monic polynomials $g(x)$ and $h(x)$ of $\mathfrak{F}[x]$. Then $g(x)$ and $h(x)$ are in $J_{\mathfrak{F}}[x]$.*

This result was proved as Theorem A2.16 in the case where $J_{\mathfrak{F}}$ is itself a unique factorization integral domain. The proof given on page 147 of [142] may be readily shown valid for our present $J_{\mathfrak{F}}$ if one uses Theorem A2.16.

The case of major interest for our work is that where $\mathfrak{F}$ is an algebraic extension of a p-adic field $\mathfrak{R}_p$ as defined in Chapter AXII. For this case $J_{\mathfrak{F}}$ was shown in Chapter AXII actually to be a unique factorization integral domain so that the additional verification will be unnecessary for the purposes of our present exposition.

Let $\mathfrak{A}$ be an algebra over $\mathfrak{F}$ with a unity quantity. A quantity a of $\mathfrak{A}$ will be called an *integral quantity* (or *integer*) of $\mathfrak{A}$ if it is a root of some monic polynomial $t(x)$ of $J_{\mathfrak{F}}[x]$. The minimum function $g(x)$ of a divides $t(x)$ and, by the Gauss Lemma, is also a monic polynomial of $J_{\mathfrak{F}}[x]$. But then the characteristic and principal polynomials of a, defined in Section 1.13 and 8.11, are all monic polynomials of $J_{\mathfrak{F}}[x]$. For they are products of $g(x)$ by monic divisors in $\mathfrak{F}(x)$ of $g(x)$.

Every quantity of $\mathfrak{F}$ is the quotient of two integers of $\mathfrak{F}$, and it is evident that if $\alpha_1, \cdots, \alpha_r$ are any quantities of $\mathfrak{F}$, finite in number, there exists an integer $\alpha \neq 0$ in $\mathfrak{F}$ such that $\alpha\alpha_1, \cdots, \alpha\alpha_r$ are in $J_{\mathfrak{F}}$. Let then $\mathfrak{A} = (u_1, \cdots, u_n)$ over $\mathfrak{F}$, $u_iu_j = \sum_{k=1}^{n} \gamma_k^{(ij)} u_k$, with $\gamma_k^{(ij)}$ in $\mathfrak{F}$. Then we choose α in $J_{\mathfrak{F}}$ so that all of the quantities $\alpha\gamma_k^{(ij)}$ are in $J_{\mathfrak{F}}$, put $v_i = \alpha u_i$, and see that $\mathfrak{A} = (v_1, \cdots, v_n)$ over $\mathfrak{F}$, $v_iv_j = \sum_{k=1}^{n} \alpha\gamma_k^{(ij)} v_k$. The regular representation of $\mathfrak{A}$ is then an algebra of n-rowed square matrices such that the v_i are represented as matrices with elements in $J_{\mathfrak{F}}$. But then the characteristic functions of these matrices are monic polynomials of $J_{\mathfrak{F}}[x]$, and the v_i are integers of $\mathfrak{A}$. We have proved

LEMMA 1. *Every algebra $\mathfrak{A}$ over $\mathfrak{F}$ has a basis of integral quantities.*

A set $\mathfrak{J}$ of quantities of $\mathfrak{A}$ is called a *$J_{\mathfrak{F}}$-array* if $\mathfrak{J}$ is a ring of integral quantities containing $J_{\mathfrak{F}}$ as well as a basis over $\mathfrak{F}$ of $\mathfrak{A}$. We shall not use the usual terminology of *$J_{\mathfrak{F}}$-order* here, as our customary usages of the term order (of a linear set and of a p-adic number) might cause confusion.

A $J_{\mathfrak{F}}$-array is called *maximal* if it is not properly contained in any other $J_{\mathfrak{F}}$-array of $\mathfrak{A}$. The theory of integral sets of rational algebras of [142], [119] is concerned with the study of such maximal arrays.

Let $\mathfrak{J}$ be any $J_{\mathfrak{F}}$-array of $\mathfrak{A}$ and $\mathfrak{B} \leqq \mathfrak{A}$. Then we call $\mathfrak{B}$ a *right $\mathfrak{J}$-ideal of* $\mathfrak{A}$ if

$$b_1a_1 + b_2a_2$$

is in $\mathfrak{B}$ for every b_1 and b_2 of $\mathfrak{B}$, a_1 and a_2 of $\mathfrak{J}$, $\mathfrak{B}$ contains a quantity β of $\mathfrak{F}$, and there exists a $\beta_0 \neq 0$ in $\mathfrak{F}$ such that $\beta_0 b$ is in $\mathfrak{J}$ for every b of $\mathfrak{B}$. *Left $\mathfrak{J}$-ideals* and (two-sided) *$\mathfrak{J}$-ideals* are defined similarly. Then $\mathfrak{B}$ is an *ideal* of $\mathfrak{J}$ if $\mathfrak{B} \leqq \mathfrak{J}$.

If $\mathfrak{B}$ is a $\mathfrak{J}$-ideal, then $\mathfrak{B}$ contains βa for the fixed defining β of $\mathfrak{B}$ and every a of $\mathfrak{J}$. Then $\mathfrak{B}$ contains $\alpha\beta a$ for any integer α of $J_{\mathfrak{F}}$, and hence $\mathfrak{B}$ contains a basis $u_1, \cdots, u_n$ of $\mathfrak{A}$ over $\mathfrak{F}$ with the u_i integral. If $b = \sum_{i=1}^{n} \alpha_i u_i$ is in $\mathfrak{B}$, then $bu_j = \sum_{i=1}^{n} \alpha_i u_i u_j$, and

$$T(\beta_0 b u_j) = \beta_0 \sum_{i=1}^{n} \alpha_i T(u_i u_j). \tag{1}$$

By Theorem 8.15 the matrix whose element in the ith row and jth column is $T(u_iu_j)$ has non-zero determinant D if $\mathfrak{A}$ is semi-simple. Here $T(u_iu_j)$ is the reduced trace, and since $\beta_0 bu_j$ and $u_i u_j$ are integral the traces $T(\beta_0 bu_j)$ and $T(u_i u_j)$ are in $J_{\mathfrak{F}}$. We then solve (1) for the α_i and determine them as the quotients of integers of $\mathfrak{F}$ by the fixed integer $\beta_0^n D$ of $\mathfrak{F}$. By the result* of pages 87, 101 of [414] we have

Theorem 1. *Every $\mathfrak{J}$-ideal $\mathfrak{B}$ of a semi-simple algebra $\mathfrak{A}$ over $\mathfrak{F}$ has the form*

$$\mathfrak{B} = (w_1, \cdots, w_n) \text{ over } J_{\mathfrak{F}}. \tag{2}$$

In particular every $J_{\mathfrak{F}}$-array $\mathfrak{J}$ has this form.

3. The p-adic fields $\mathfrak{R}_p$. The present theory of rational division algebras is dependent upon properties of p-adic number fields. Certain of these properties were derived in Chapters AXI and AXII, and we shall base our exposition upon the results of these chapters. To fix our notation we shall recall some of the concepts and properties in the notation we now adopt, and shall derive a number of additional results.

Let p be any prime rational integer, so that every non-zero rational number a has the form

$$a = p^{\lambda} b,$$

where b is a rational number whose numerator and denominator are integers prime to p, λ is an integer. Let ρ be a positive real number less than unity and define

$$V_p(a) = \rho^{\lambda}, \qquad V_p(0) = 0, \tag{3}$$

for every rational non-zero a. Then V_p defines a non-archimedean valuation of $\mathfrak{R}$. Conversely every non-archimedean valuation of $\mathfrak{R}$ is obtainable in this way from some prime p. Moreover the derived field $\mathfrak{R}_p$ of $\mathfrak{R}$ with respect to V_p is a field which is complete with respect to an extension in $\mathfrak{R}_p$ of V_p, and which is independent of ρ in the sense of analytic equivalence.

Let λ be a rational integer, a_i range independently over the set of integers $0, 1, \cdots, p-1$ such that $a_0 \neq 0$, and define

$$\alpha_k = p^{\lambda}(a_0 + a_1 p + \cdots + a_{k-1} p^{k-1}). \tag{4}$$

Then the sequence $\{\alpha_k\}$ is regular with respect to V_p and defines a quantity α of $\mathfrak{R}_p$. This α may be called the *limit*,

$$\lim_{k \to \infty} \alpha_k,$$

and thus regarded as the *sum* $\alpha = p^{\lambda}(a_0 + a_1 p + \cdots)$ of an infinite series *convergent with respect to* V_p. Conversely every quantity of $\mathfrak{R}_p$ has such an

* This states that if $\mathfrak{M}$ is a linear set of finite order over $J_{\mathfrak{F}}$ and $\mathfrak{B}$ is a linear subset of $\mathfrak{M}$, then $\mathfrak{B}$ has a finite basis over $J_{\mathfrak{F}}$. This is an arithmetic result not properly belonging in the present exposition. The author hopes that it will be given in English in a future text on ideals, algebraic numbers, and the arithmetic of algebras.

infinite series representation. Let us write $\lambda = \lambda(\alpha)$ for the exponent of p in this representation of α, and call $\lambda(\alpha)$ the *order* of α.

The set of quantities α of $\mathfrak{R}_p$ with $\lambda(\alpha) \geqq 0$ forms an integral domain $J_{\mathfrak{R}_p}$ whose units are those quantities u of $\mathfrak{R}_p$ with $\lambda(u) = 0$. We call the quantities of $J_{\mathfrak{R}_p}$ the *integers* of $\mathfrak{R}_p$. Every ideal of $J_{\mathfrak{R}_p}$ is a principal ideal and is a power of the unique prime principal ideal (p). The residue class ring

$$(5) \qquad H_{\mathfrak{R}_p} = J_{\mathfrak{R}_p} - (p) = GF(p)$$

is the finite field whose elements are the residue classes $[\alpha]$ of the quantities α of $J_{\mathfrak{R}_p}$ modulo (p), and we see that these classes are thus $[0], [1], \cdots, [p-1]$. Hence we see that every unit α of $J_{\mathfrak{R}_p}$ has the property $\alpha \equiv \alpha_k \pmod{(p)^k}$ with α_k given by (4) for the $a_i = 0, 1, \cdots, p-1$.

As a consequence of this discussion we may prove in particular that $\mathfrak{R}_p$ contains no primitive pth root of unity. For let ζ be in $\mathfrak{R}_p$ so that $\zeta^p = 1$ implies that ζ is a unit of $J_{\mathfrak{R}_p}$, $\zeta^p \equiv 1 \pmod{(p)}$, $(\zeta - 1)^p \equiv 0 \pmod{(p)}$, $\zeta \equiv 1 \pmod{(p)}$. Thus in the α_k of (4) for $\zeta = \alpha$ we have $\lambda = 0$, $a_0 = 1$. Let r be the first positive integer for which $a_r \neq 0$ so that $\zeta \equiv 1 + a_r p^r + a_{r+1}p^{r+1} \pmod{(p)^{r+2}}$, $\zeta^p \equiv 1 + p(a_r p^r) \pmod{(p)^{r+2}}$ and, since $\zeta^p = 1$, we have $a_r p^{r+1} \equiv 0 \pmod{(p)^{r+2}}$, $a_r \equiv 0 \pmod{(p)}$, a contradiction. This proves that every $a_r = 0$, $\zeta = 1$ as desired.

4. Arithmetic theory of division algebras over $\mathfrak{R}_p$. We shall proceed to a consideration of the integral quantities of a division algebra $\mathfrak{D}$ of order N over $\mathfrak{R}_p$, and shall call such algebras *p-adic division algebras*. Every A of $\mathfrak{D}$ generates a field $\mathfrak{K} = \mathfrak{R}_p(A)$ of finite degree (a divisor of N) over $\mathfrak{R}_p$, and we have seen that A is integral if and only if its minimum function $g(x)$ is a monic polynomial of $J_{\mathfrak{R}_p}[x]$. By Section A12.10 this is true if and only if $N_{\mathfrak{K}|\mathfrak{R}_p}(A)$ is integral. But a power of $N_{\mathfrak{K}|\mathfrak{R}_p}(A)$ is the reduced norm $N(A)$ defined in Section 8.11, and hence A is an integer of $\mathfrak{D}$ if and only if $N(A)$ is an integer of $\mathfrak{R}_p$. We now prove

Theorem 2. *The set $J_{\mathfrak{D}}$ of all integers of $\mathfrak{D}$ is a maximal $J_{\mathfrak{R}_p}$-array of $\mathfrak{D}$.*

For $J_{\mathfrak{D}}$ clearly contains $J_{\mathfrak{R}_p}$ and, by Lemma 9.1, a basis of $\mathfrak{D}$ over $\mathfrak{R}_p$. If A and B are in $J_{\mathfrak{D}}$ then $N(A)$ and $N(B)$ are in $J_{\mathfrak{R}_p}$, $N(AB) = N(A)N(B)$ is in $J_{\mathfrak{R}_p}$, AB is in $J_{\mathfrak{D}}$. If A and B are any non-zero quantities of $J_{\mathfrak{D}}$, then $N(A) = p^{\lambda_1}u_1$, $N(B) = p^{\lambda_2}u_2$ for non-negative integers λ_1 and λ_2 and units u_1 and u_2 of $J_{\mathfrak{R}_p}$. There is no loss of generality if we assume that $\lambda_1 \geqq \lambda_2$. Then $N(AB^{-1}) = p^{\lambda_1 - \lambda_2}u_1u_2^{-1}$ is in $J_{\mathfrak{R}_p}$, AB^{-1} is in $J_{\mathfrak{D}}$. The minimum function $\gamma(\lambda)$ of AB^{-1} is an irreducible monic polynomial of $J_{\mathfrak{R}_p}[\lambda]$, and $AB^{-1} + 1$ is a root of the polynomial $\gamma(\lambda - 1)$ and is integral. But then $(AB^{-1} + 1)B = A + B$ is in $J_{\mathfrak{D}}$. This proves that $J_{\mathfrak{D}}$ is a ring, and we have now verified that $J_{\mathfrak{D}}$ is a $J_{\mathfrak{R}_p}$-array. It contains every integer of $\mathfrak{D}$ and is thus maximal.

The ring $J_{\mathfrak{D}}$ is an integral domain and U is a unit of $J_{\mathfrak{D}}$ if and only if $N(U)$ is a unit of $J_{\mathfrak{R}_p}$. For, $N(U) = p^{\lambda}u$, $N(U^{-1}) = p^{-\lambda}u^{-1}$ is integral if and only if $\lambda = 0$. If A is in $J_{\mathfrak{D}}$ but is not a unit, then $N(A) = p^{\lambda}u$, where u is a unit of $J_{\mathfrak{R}_p}$, and $\lambda > 0$, and we let $A = P$ be a quantity such that the order $\lambda = F_0 > 0$ is the least possible. Then $N(P) = p^{F_0}u_0$ and we shall prove

LEMMA 2. *Every quantity A of $\mathfrak{D}$ is uniquely expressible in the forms*

$$A = P^L U = U_0 P^L, \tag{6}$$

for units U and U_0 of $J_{\mathfrak{D}}$ and a rational integer $L = L(A)$ which we shall call the ***order*** *of A.*

For $N(A) = p^\lambda u$, $\lambda = F_0 L + R$ where $0 \leqq R < F_0$. Then $N(P^{-L}A) = N(AP^{-L}) = p^R u u_0^{-L} = p^R u_1$ for a unit u_1 of $J_{\mathfrak{R}_p}$. Our definition of F_0 implies that $R = 0$, $P^{-L}A = U$ and $AP^{-L} = U_0$ are units of $J_{\mathfrak{D}}$. Clearly $L = L(A)$ is a uniquely determined rational integral function of A, U and U_0 are unique.

Every quantity P_0 of $J_{\mathfrak{D}}$ for which $L(P_0) = 1$ is called a *prime quantity* of $\mathfrak{D}$. The quantity P described above is such a prime quantity, any two prime quantities P and P_0 are related by $P_0 = PU = U_0 P$ for units U and U_0 of $J_{\mathfrak{D}}$, and any prime quantity of $J_{\mathfrak{D}}$ may be used as the P of (6).

As immediate consequences of Lemma 9.2 we have

LEMMA 3. *A quantity A of $\mathfrak{D}$ is integral if and only if $L(A) \geqq 0$. The units of $J_{\mathfrak{D}}$ are those quantities A with $L(A) = 0$.*

LEMMA 4. *The function $L(A)$ has the properties $L(AB) = L(A) + L(B)$, $L(A + B) \geqq M$, where M is the minimum of $L(A)$ and $L(B)$.*

We next obtain

Theorem 3. *The quantities A of $\mathfrak{D}$ with $L(A) > 0$ form an ideal $\mathfrak{p}_{\mathfrak{D}}$ of $J_{\mathfrak{D}}$.*

For if $L(A) > 0$ and $L(B) > 0$ then Lemma 9.4 implies that $L(A + B) > 0$, $A + B$ is in $\mathfrak{p}_{\mathfrak{D}}$. If A is in $\mathfrak{p}_{\mathfrak{D}}$ and B is in $J_{\mathfrak{D}}$, then $L(B) \geqq 0$, $L(A) > 0$, $L(AB) = L(BA) > 0$. Now $\mathfrak{p}_{\mathfrak{D}} \leqq J_{\mathfrak{D}}$, p is in $\mathfrak{p}_{\mathfrak{D}}$ since $\lambda(p) = L(p)F_0$ is the degree of the minimum function of $\mathfrak{D}$ over $\mathfrak{R}_p$ and so $L(p)$ is positive. This satisfies our definition of Section 9.2.

We next prove

Theorem 4. *The $J_{\mathfrak{D}}$-ideals $\mathfrak{Q}$ are all powers $(\mathfrak{p}_{\mathfrak{D}})^t$ of $\mathfrak{p}_{\mathfrak{D}}$, and $\mathfrak{p}_{\mathfrak{D}} = (P)$, $\mathfrak{Q} = (P^t)$. Hence $\mathfrak{p}_{\mathfrak{D}}$ is the only divisorless ideal of $J_{\mathfrak{D}}$.*

For A is in $\mathfrak{p}_{\mathfrak{D}}$ if and only if $L(A) > 0$, $A = PB$ where $B = P^{L-1}U$ is in $J_{\mathfrak{D}}$. Hence $\mathfrak{p}_{\mathfrak{D}} \leqq (P)$, P and (P) are in $\mathfrak{p}_{\mathfrak{D}}$, $\mathfrak{p}_{\mathfrak{D}} = (P)$. Our definition of $J_{\mathfrak{D}}$-ideals implies the existence of a quantity b of $J_{\mathfrak{R}_p}$ such that $B = bA$ is integral for every A of $\mathfrak{Q}$. Now $A = b^{-1}B$, $L(A) = L(B) - L(b) \geqq -L(b)$. Hence there exists a rational integer t such that $L(A) \geqq t$ for every A of $\mathfrak{Q}$, and $L(A_0) = t$ for some A_0 of $\mathfrak{Q}$. But then $A_0 = P^t U$ is in $\mathfrak{Q}$, so is $P^t = A_0 U^{-1}$. Hence every A of $\mathfrak{Q}$ has the form $A = P^t B$ for B in $J_{\mathfrak{D}}$, $\mathfrak{Q} \leqq (P^t)$, $\mathfrak{Q} = (P^t)$, and the theorem is proved. Note that the integer t may be negative.

The $J_{\mathfrak{D}}$-ideal $(p) = \mathfrak{p}_{\mathfrak{D}}^E$ where the positive integer $E = L(p)$. If a and b are integers of $\mathfrak{R}_p$ such that $a - b$ is in (p), then $a - b$ is in $\mathfrak{p}_{\mathfrak{D}} \geqq (p)$. By

Theorem 9.1 the set $J_{\mathfrak{D}} = (w_1, \cdots, w_N)$ over $J_{\mathfrak{R}_p}$, and thus every integer of $\mathfrak{D}$ has the form $A = a_1w_1 + \cdots + a_Nw_N$ with the a_i in $J_{\mathfrak{R}_p}$. Consequently

$$A \equiv x_1w_1 + \cdots + x_Nw_N \pmod{\mathfrak{p}_{\mathfrak{D}}},$$

where the x_i range over $0, 1, \cdots, p - 1$ independently and the notation $A \equiv B \pmod{\mathfrak{p}_{\mathfrak{D}}}$ has the usual meaning that $A - B$ is in $\mathfrak{p}_{\mathfrak{D}}$. It follows that the residue-class ring $H_{\mathfrak{D}} = J_{\mathfrak{D}} - \mathfrak{p}_{\mathfrak{D}}$ is an algebra of some finite order F over $H_{\mathfrak{R}_p}$, where $H_{\mathfrak{R}_p} = J_{\mathfrak{R}_p} - (p)$. Now AB is in $\mathfrak{p}_{\mathfrak{D}}$ for A and B in $J_{\mathfrak{D}}$ if and only if $L(AB) = L(A) + L(B) > 0$, one of $L(A)$ and $L(B)$ is positive, one of A and B is in $\mathfrak{p}_{\mathfrak{D}}$. Hence $H_{\mathfrak{D}}$ has no zero divisors and is a finite division algebra. From Theorem 9.4 we have already seen that $\mathfrak{p}_{\mathfrak{D}}$ is a divisorless ideal and, using the definitions of page A301, is also a *prime* ideal. Applying Theorem 4.28 and the fact that $H_{\mathfrak{D}}$ has p^F elements we have

Theorem 5. *The residue-class ring $J_{\mathfrak{D}} - \mathfrak{p}_{\mathfrak{D}}$ is the finite field*

$$H_{\mathfrak{D}} = GF(p^F),$$

of degree F over its prime subfield $H_{\mathfrak{R}_p}$.

The quantities of $H_{\mathfrak{D}}$ are a fixed set of $p^F = q_{\mathfrak{D}}$ residue classes $[A]$ for A in $J_{\mathfrak{D}}$. Then every integer of $\mathfrak{D}$ is congruent modulo $\mathfrak{p}_{\mathfrak{D}}$ to one and only one of the $q_{\mathfrak{D}}$ quantities

$$x_1w_1 + \cdots + x_Fw_F \qquad (0 \leqq x_i < p), \tag{7}$$

where we arrange the w_i so that $H_{\mathfrak{D}}$ has $[w_1], \cdots, [w_F]$ as a basis over $H_{\mathfrak{R}_p}$. We may then obtain

Theorem 6. *Let $A_0, \cdots, A_{k-1}$ range independently over the $q_{\mathfrak{D}}$ quantities* (7) *Then the $q_{\mathfrak{D}}^k$ quantities*

$$A^{(k)} = A_0 + A_1P + \cdots + A_{k-1}P^{k-1} \tag{8}$$

range over integers of $\mathfrak{D}$ such that every integer A of $\mathfrak{D}$ is congruent modulo $\mathfrak{p}_{\mathfrak{D}}^k$ to one of (8).

The result is trivial for $k = 1$ and we assume it true for residue-classes modulo $\mathfrak{p}_{\mathfrak{D}}^{k-1}$. Then if A is in $J_{\mathfrak{D}}$ we have $A - (A_0 + A_1P + \cdots + A_{k-2}P^{k-2}) = B_0$ in $\mathfrak{p}_{\mathfrak{D}}^{k-1}$, $B_0 = BP^{k-1}$ for an integer B of $J_{\mathfrak{D}}$. Then $B \equiv A_{k-1} \pmod{\mathfrak{p}_{\mathfrak{D}}}$ where A_{k-1} is one of the quantities (7), $B = A_{k-1} + B_1P$ for B_1 in $J_{\mathfrak{D}}$, $B_0 = A_{k-1}P^{k-1} + B_1P^k$, $A = (A_0 + A_1P + \cdots + A_{k-1}P^{k-1}) + B_1P^k$ as desired.

The algebra $\mathfrak{D}$ may now be seen to be a ring complete with respect to

$$V(0) = 0, \qquad V(A) = \rho^{L(A)},$$

where ρ is a positive real number less than unity, $L(A)$ is defined as in Lemma 9.2. Moreover $A = P^LU$ where U is the limit of an infinite series defined by the sequence $\{A^{(k)}\}$ of (8) for U. We shall not use this property except for the case where $\mathfrak{D}$ is commutative, and for this case we already saw in Section A12.11 that $\mathfrak{D}$ is complete with respect to V.

DEFINITION. *The exponent E of $(p) = \mathfrak{p}_{\mathfrak{D}}^{E}$ will be called the ramification order of $\mathfrak{D}$ over $\mathfrak{R}_p$, and the exponent F of $q_{\mathfrak{D}} = p^F$ will be called the residue-class degree of $\mathfrak{D}$ over $\mathfrak{R}_p$.*

The two integral invariants of $\mathfrak{D}$ just defined are exceedingly important and we prove a first property of them in

Theorem 7. *The order of $\mathfrak{D}$ over $\mathfrak{R}_p$ is the product of its residue-class degree and ramification order. Moreover, the residue-class degree F is $F_0 m$ where F_0 is the order of the quantity $N(P)$ of $J_{\mathfrak{R}_p}$, and m is the degree of $\mathfrak{D}$ over its centrum.*

For $J_{\mathfrak{D}} = (w_1, \cdots, w_N)$ over $J_{\mathfrak{R}_p}$ and Theorem 9.6 states that the number of residue classes of integers of $J_{\mathfrak{D}}$ modulo $\mathfrak{p}_{\mathfrak{D}}^{E} = (p)$ is p^{EF}. If A is in $J_{\mathfrak{D}}$ then $A = a_1 w_1 + \cdots + a_N w_N$ for a_i in $J_{\mathfrak{R}_p}$, every $a_i \equiv x_i \pmod{\mathfrak{p}_{\mathfrak{D}}^{E}}$, where x_i is one of the integers $0, 1, \cdots, p-1$. Thus the p^N quantities

$$(9) \qquad X = x_1 w_1 + \cdots + x_N w_N \qquad (0 \leqq x_i < p),$$

for rational integral x_i, are a set of quantities one of which is congruent modulo $\mathfrak{p}_{\mathfrak{D}}^{E}$ to any given A of $J_{\mathfrak{D}}$. If X and $Y = y_1 w_1 + \cdots + y_N w_N$ are two distinct quantities (9) such that $X \equiv Y \pmod{\mathfrak{p}_{\mathfrak{D}}^{E}}$ then $X - Y = Z = z_1 w_1 + \cdots + z_N w_N \equiv 0 \pmod{\mathfrak{p}_{\mathfrak{D}}^{E}}$ for $0 \leqq |z_i| < p$ and at least one $z_i \neq 0$, $z_1 w_1 + \cdots + z_N w_N = Bp$ for B in $J_{\mathfrak{D}}$, $B = b_1 w_1 + \cdots + b_N w_N$ for b_i in $J_{\mathfrak{R}_p}$. Then $\sum_{i=1}^{N} (z_i - pb_i) w_i = 0$, and the linear independence of the w_i in $\mathfrak{R}_p$ implies that $z_i = pb_i$, $z_i \equiv 0 \pmod{(p)}$ which is false. Thus the p^N quantities (9) are incongruent modulo $\mathfrak{p}_{\mathfrak{D}}^{E}$, and we have shown that $p^N = p^{EF}$, $N = EF$ as desired. By Theorem 8.13 the minimum function of $\mathfrak{D}$ over $\mathfrak{R}_p$ has degree mt where t is the degree of the centrum of $\mathfrak{D}$ and $N = m^2 t$. Then $N(p) = p^{mt}$, $p = P^E U$, $N(p) = N(P^E U) = p^{EF_0} u$, $EF_0 = mt$, $EF_0 m = m^2 t = N = EF$, $F_0 m = F$ as desired. This completes our proof.

We next let $\mathfrak{D}_1$ be a subalgebra over $\mathfrak{R}_p$ of $\mathfrak{D}$, so that our definitions imply that $\mathfrak{D}_1$ is a division algebra over $\mathfrak{R}_p$ with the same unity quantity as $\mathfrak{D}$. By Theorem 1.6 the order N_1 of $\mathfrak{D}_1$ over $\mathfrak{R}_p$ divides the order N of $\mathfrak{D}$ over $\mathfrak{R}_p$, and we have

$$(10) \qquad N = N_1 N_2 .$$

We may call N_2 the *order* of $\mathfrak{D}$ over $\mathfrak{D}_1$ since we saw in fact that $\mathfrak{D} = \mathfrak{D}_1 + \mathfrak{D}_1 u_2 + \cdots + \mathfrak{D}_1 u_{N_2}$. In fact we may prove

LEMMA 5. *Let $\mathfrak{D}_1$ be a subalgebra over $\mathfrak{R}_p$ of $\mathfrak{D}$. Then the order, ramification order, and residue-class degree of $\mathfrak{D}_1$ over $\mathfrak{R}_p$ are respective divisors of the corresponding integers for $\mathfrak{D}$.*

For $(p) = \mathfrak{p}_{\mathfrak{D}}^{E} = \mathfrak{p}_{\mathfrak{D}_1}^{E_1}$, $\mathfrak{p}_{\mathfrak{D}_1} = \mathfrak{p}_{\mathfrak{D}}^{E_2}$, $(p) = \mathfrak{p}_{\mathfrak{D}}^{E_1 E_2}$, and $E = E_1 E_2$ as desired. If A_1 and B_1 are in $J_{\mathfrak{D}_1}$, and $A_1 \equiv B_1 \pmod{\mathfrak{p}_{\mathfrak{D}_1}}$, then $\mathfrak{p}_{\mathfrak{D}_1} = \mathfrak{p}_{\mathfrak{D}}^{E_2}$ implies that $A_1 \equiv B_1 \pmod{\mathfrak{p}_{\mathfrak{D}}}$. Conversely $A_1 \equiv B_1 \pmod{\mathfrak{p}_{\mathfrak{D}}}$ implies that $A_1 - B_1$ is not prime to $\mathfrak{p}_{\mathfrak{D}_1}$, $A_1 \equiv B_1 \pmod{\mathfrak{p}_{\mathfrak{D}_1}}$. Thus the quantities of $J_{\mathfrak{D}_1}$ defining distinct elements of $H_{\mathfrak{D}_1}$ define distinct elements of $H_{\mathfrak{D}}$, and it is clear that

$$H_{\mathfrak{D}} \geqq H_{\mathfrak{D}_1} \tag{11}$$

in the sense of equivalence. Now $H_{\mathfrak{D}} = GF(p^F)$, $H_{\mathfrak{D}_1} = GF(p^{F_1})$, and F and F_1 are the respective degrees of these fields over $H_{\mathfrak{R}_p}$. But then F_1 divides $F = F_1F_2$. Now $N_1 = E_1F_1$ divides $N = EF$ (a second proof of this result).

Observe that $\mathfrak{p}_{\mathfrak{D}_1} = \mathfrak{p}_{\mathfrak{D}}^{E_2}$, so that we are justified in calling E_2 the *ramification order* of $\mathfrak{D}$ over $\mathfrak{D}_1$. Similarly $q_{\mathfrak{D}_1} = p^{F_1}$, $q_{\mathfrak{D}} = p^F = (q_{\mathfrak{D}_1})^{F_2}$, and we shall call F_2 the *residue-class degree* of $\mathfrak{D}$ over $\mathfrak{D}_1$. We then have immediately

LEMMA 6. *The order of $\mathfrak{D}$ over $\mathfrak{D}_1$ is the product of the ramification order of $\mathfrak{D}$ over $\mathfrak{D}_1$ by the residue-class degree of $\mathfrak{D}$ over $\mathfrak{D}_1$.*

5. The Hensel Lemma. Let $\mathfrak{F}$ be any field of degree ν over $\mathfrak{R}_p$. Then we shall call $\mathfrak{F}$ a p-adic field. Such fields are instances of division algebras over $\mathfrak{R}_p$ and the results just derived may all be interpreted for this case. We then designate the set of all integers of $\mathfrak{F}$ by $J_{\mathfrak{F}}$, its unique prime ideal by $\mathfrak{p}_{\mathfrak{F}}$, the residue-class field $J_{\mathfrak{F}} - \mathfrak{p}_{\mathfrak{F}}$ by $H_{\mathfrak{F}} = GF(q_{\mathfrak{F}})$, $q_{\mathfrak{F}} = p^{\phi}$. Here ϕ is the residue class degree over $\mathfrak{R}_p$ of $\mathfrak{F}$, and $(p) = \mathfrak{p}_{\mathfrak{F}}^{\epsilon}$, ϵ is the ramification order of $\mathfrak{F}$ over $\mathfrak{R}_p$. We shall use these notations in what follows.

In Section A12.11 we showed that the norm $N_{\mathfrak{F}|\mathfrak{R}_p}(\mathfrak{p}_{\mathfrak{F}}) = (p)^{\phi_0}$ such that $\phi_0\epsilon = \nu$. But then $\phi = \phi_0$,

$$N_{\mathfrak{F}|\mathfrak{R}_p}(\mathfrak{p}_{\mathfrak{F}}) = (p)^{\phi}, \qquad N_{\mathfrak{F}|\mathfrak{R}_p}(\pi) = p^{\phi}u,$$

where π is a prime quantity of $J_{\mathfrak{F}}$, u is a unit of $J_{\mathfrak{R}_p}$.

We also have an additional result for $\mathfrak{F}$. Let $f(x)$ be any polynomial of $J_{\mathfrak{F}}[x]$ and replace the coefficients of $f(x)$ by corresponding residue classes modulo $\mathfrak{p}_{\mathfrak{F}}$. The resulting polynomial of $H_{\mathfrak{F}}[x]$ will be denoted* by $\bar{f}(x)$. The proof of the lemma of Section A12.10 is now valid, the only changes required being the replacement of p in the moduli of the congruences by $\mathfrak{p}_{\mathfrak{F}} = (\pi)$, and p in the left members of the congruences by π. We leave the details of this verification to the reader and state this so-called *Hensel Lemma* as

LEMMA 7. *Let $f(x)$, $g_0(x)$, $h_0(x)$ be polynomials of $J_{\mathfrak{F}}[x]$ such that*

$$\bar{f}(x) = \bar{g}_0(x)\bar{h}_0(x),$$

where $\bar{g}_0(x)$ and $\bar{h}_0(x)$ are relatively prime polynomials of $H_{\mathfrak{F}}[x]$. Then there exist polynomials $g(x)$, $h(x)$ in $J_{\mathfrak{F}}[x]$ such that

$$\bar{g}(x) = \bar{g}_0(x), \qquad \bar{h}(x) = \bar{h}_0(x), \qquad f(x) = g(x)h(x).$$

Observe that the lemma of Section A12.10 is the case $\mathfrak{F} = \mathfrak{R}_p$ of Lemma 9.7. We shall use Lemma 9.7 later to obtain important results on the structure of p-adic fields.

6. Division algebras over any p-adic field. We let $\mathfrak{D}$ be a division algebra of order n over a p-adic field $\mathfrak{F}$ of degree ν over $\mathfrak{R}_p$, so that $\mathfrak{D}$ is a division

* Note, however, that we will use $[a]$ to designate the class in $J_{\mathfrak{F}} - p_{\mathfrak{F}}$ defined by a in $J_{\mathfrak{F}}$.

algebra of order

$$N = \nu n \tag{12}$$

over $\mathfrak{R}_p$. If a quantity a of $\mathfrak{D}$ is integral over $\mathfrak{R}_p$ it is clearly integral over $\mathfrak{F}$. Conversely, every a of $\mathfrak{D}$ generates a subfield $\mathfrak{K} = \mathfrak{F}(a)$ of $\mathfrak{D}$ and if a is integral over $\mathfrak{F}$ then $b = N_{\mathfrak{K}|\mathfrak{F}}(a)$ is an integer of $\mathfrak{F}$. Then $N_{\mathfrak{F}|\mathfrak{R}_p}(b)$ is an integer of $\mathfrak{R}_p$, $N_{\mathfrak{K}|\mathfrak{R}_p}(a) = N_{\mathfrak{F}|\mathfrak{R}_p}(b)$ so that a is an integer of $\mathfrak{D}$ over $\mathfrak{R}_p$. We have proved that the set $J_{\mathfrak{D}}$ of all integers of $\mathfrak{D}$ over $\mathfrak{F}$ is the set of all integers of $\mathfrak{D}$ over $\mathfrak{R}_p$. Moreover, then $J_{\mathfrak{D}}$ is a maximal $J_{\mathfrak{F}}$-array, $\mathfrak{p}_{\mathfrak{D}} = (P)$ is the only prime $J_{\mathfrak{D}}$-ideal of $\mathfrak{D}$, every $J_{\mathfrak{D}}$-ideal of $\mathfrak{D}$ is a power $\mathfrak{p}_{\mathfrak{D}}^L = (P^L)$.

The quantities of $J_{\mathfrak{D}}$ all have the form (6), and if π is a prime quantity of $\mathfrak{F}$ then $(\pi) = \mathfrak{p}_{\mathfrak{F}}$,

$$(\pi) = \mathfrak{p}_{\mathfrak{D}}^e, \tag{13}$$

where $e = L(\pi)$ has already been called the ramification order of $\mathfrak{D}$ over $\mathfrak{F}$. Moreover $(p) = \mathfrak{p}_{\mathfrak{F}}^{\epsilon} = \mathfrak{p}_{\mathfrak{D}}^E$,

$$E = \epsilon e. \tag{14}$$

We also showed in Lemma 9.5 that $q_{\mathfrak{D}} = (q_{\mathfrak{F}})^f = p^{\phi f}$ where

$$H_{\mathfrak{D}} = GF(q_{\mathfrak{F}}^f) \geqq H_{\mathfrak{F}}, \tag{15}$$

f is the residue-class degree of $\mathfrak{D}$ over $\mathfrak{F}$ and is the degree of $H_{\mathfrak{D}}$ over $H_{\mathfrak{F}}$,

$$F = \phi f, \qquad n = ef. \tag{16}$$

Thus we obtain anew the result of Lemma 9.6 for $\mathfrak{D}_1 = \mathfrak{F}$ and state it as

Theorem 8. *The order of $\mathfrak{D}$ over $\mathfrak{F}$ is the product of the ramification order of $\mathfrak{D}$ over $\mathfrak{F}$ and the residue-class degree of $\mathfrak{D}$ over $\mathfrak{F}$.*

We next apply Lemma 9.6, and the reader may verify trivially that we obtain

Theorem 9. *Let $\mathfrak{D}_1$ be a subalgebra over $\mathfrak{F}$ of a division algebra $\mathfrak{D}$ over $\mathfrak{F}$. Then the order, ramification order, and residue-class degree of $\mathfrak{D}$ over $\mathfrak{F}$ are the products of the corresponding integers for $\mathfrak{D}$ over $\mathfrak{D}_1$ by those defined for $\mathfrak{D}_1$ over $\mathfrak{F}$.*

We now make the

DEFINITION. *A division algebra $\mathfrak{D}$ of order ν is said to be unramified or ramified over $\mathfrak{F}$ according as its ramification order e over $\mathfrak{F}$ is or is not unity. If $e = \nu$ is the order of $\mathfrak{D}$ over $\mathfrak{F}$ we shall say that $\mathfrak{D}$ is completely ramified over $\mathfrak{F}$.*

It is then evident that if $\mathfrak{D}$ is unramified (completely ramified), so is every subalgebra over $\mathfrak{F}$ of $\mathfrak{D}$. Applying this result to the case of fields we obtain a result which we shall state as

Theorem 10. *Every subfield $\mathfrak{K}_1$ over $\mathfrak{F}$ of an unramified (completely ramified) extension $\mathfrak{K}$ of $\mathfrak{F}$ is unramified (completely ramified) over $\mathfrak{F}$ and $\mathfrak{K}$ is unramified (completely ramified) over $\mathfrak{K}_1$. Conversely, if $\mathfrak{K}$ is unramified (completely rami-*

fied) over $\mathfrak{K}_1$ and $\mathfrak{K}_1$ is unramified (completely ramified) over $\mathfrak{F}$, then $\mathfrak{K}$ is unramified (completely ramified) over $\mathfrak{F}$.

We shall now begin a deeper study of the structure of algebraic extensions of finite degree over a p-adic field.

7. Structure of fields of finite degree over a p-adic field. Let $\mathfrak{K}$ be a field of degree n over any p-adic field $\mathfrak{F}$. If $\mathfrak{K}$ is unramified over $\mathfrak{F}$, the prime ideal $\mathfrak{p}_{\mathfrak{F}} = (\pi)$ remains a prime ideal when we extend $\mathfrak{F}$ to $\mathfrak{K}$ so that $\mathfrak{p}_{\mathfrak{K}} = (\pi)$. But when $\mathfrak{K}$ is completely ramified we have $(\pi) = \mathfrak{p}_{\mathfrak{K}}^n$, $H_{\mathfrak{K}} = J_{\mathfrak{K}} - \mathfrak{p}_{\mathfrak{K}} = H_{\mathfrak{F}}$. We now derive

Theorem 11. *Let $\mathfrak{K}$ be a field of degree n over a p-adic field $\mathfrak{F}$ so that n is the product ef of the ramification order e and residue-class degree f of $\mathfrak{K}$ over $\mathfrak{F}$. Write $\tau = q_{\mathfrak{F}}^f - 1$. Then there exists a primitive τth root of unity ζ in $\mathfrak{K}$ such that the corresponding residue class $[\zeta]$ generates $H_{\mathfrak{K}}$, $\mathfrak{K}$ is completely ramified of degree e over $\mathfrak{W} = \mathfrak{F}(\zeta)$, $\mathfrak{W}$ is unramified of degree f over $\mathfrak{F}$, and*

$$(17) \qquad \mathfrak{K} = \mathfrak{W}(P),$$

for every prime quantity P of $J_{\mathfrak{K}}$.

For proof we apply Theorem A7.25 to obtain a generating quantity $\xi = [\zeta_0]$ of $H_{\mathfrak{K}}$ where ζ_0 is in $J_{\mathfrak{K}}$, ξ is a primitive τth root of unity, $H_{\mathfrak{K}} = H_{\mathfrak{F}}(\xi) = H_{\mathfrak{R}_p}(\xi)$. The polynomial $x^\tau - [1]$ of $H_{\mathfrak{K}}[x]$ has ξ as a root and hence the form $(x - \xi)\bar{g}_0(x)$ for a $g_0(x)$ of $J_{\mathfrak{K}}[x]$. Now $x^\tau - [1]$ is separable since p does not divide τ, $\bar{g}_0(x)$ is prime to $x - \xi$, and we apply Lemma 9.7 for polynomials of $J_{\mathfrak{K}}[x]$ and obtain

$$(18) \qquad x^\tau - 1 = (x - \zeta)g(x), \quad [\zeta] = \xi, \quad \bar{g}(x) = \bar{g}_0(x),$$

for $g(x)$ in $J_{\mathfrak{K}}[x]$. But then $\zeta^\tau = 1$. Now $\zeta^\sigma = 1$ implies that $\xi^\sigma = [\zeta^\sigma] = [1]$' which is not true unless τ divides σ. Hence ζ is a primitive τth root of unity as desired.

The polynomial $x^\tau - [1] = \Gamma_0(x)\Delta_0(x)$ in $H_{\mathfrak{F}}[x]$, where $\Gamma_0(x)$ is the minimum function of degree f of ξ over $H_{\mathfrak{F}}$ and $\Delta_0(\xi) \neq 0$ since $x^\tau - [1]$ is separable. Apply Lemma 9.7 for polynomials of $J_{\mathfrak{F}}[x]$ to obtain the factorization $x^\tau - 1 = \Gamma(x)\Delta(x)$ for monic polynomials $\Gamma(x)$ and $\Delta(x)$ in $J_{\mathfrak{F}}[x]$ such that $\Gamma(x)$ has degree f, $\bar{\Gamma}(x) = \Gamma_0(x)$, $\bar{\Delta}(x) = \Delta_0(x)$. The roots $1, \zeta, \cdots, \zeta^{\tau-1}$ of $\Gamma(x)\cdot\Delta(x) = 0$ are τ quantities incongruent modulo $\mathfrak{p}_{\mathfrak{K}}$. Thus $\bar{\Delta}(\xi) \neq 0$ implies that $\Delta(\zeta) \neq 0$, $\Gamma(\zeta) = 0$. The minimum function $\gamma(x)$ of ζ over $\mathfrak{F}$ is a monic divisor in $J_{\mathfrak{F}}[x]$ of $\Gamma(x)$ by Theorem A2.16, and clearly $\bar{\gamma}(\xi) = 0$. If $\gamma(x)$ were a proper divisor of $\Gamma(x)$ the degree of the monic polynomial $\bar{\gamma}(x)$ would be less than that of $\Gamma_0(x)$, a contradiction of the fact that $\Gamma_0(x)$ is the minimum function of ξ. Hence $\gamma(x) = \Gamma(x)$, $\mathfrak{W} = \mathfrak{F}(\zeta)$ has degree f over $\mathfrak{F}$.

By Theorem 9.9 the residue-class degree f_0 of $\mathfrak{W}$ over $\mathfrak{F}$ divides f. Now $\mathfrak{p}_{\mathfrak{K}}$ divides $\mathfrak{p}_{\mathfrak{W}}$, and $A \equiv B \pmod{\mathfrak{p}_{\mathfrak{W}}}$ implies that $A \equiv B \pmod{\mathfrak{p}_{\mathfrak{K}}}$. The $q_{\mathfrak{F}}^f$ quantities $0, 1, \zeta, \cdots, \zeta^{\tau-1}$ of $\mathfrak{W}$ are incongruent modulo $\mathfrak{p}_{\mathfrak{K}}$ and hence also modulo $\mathfrak{p}_{\mathfrak{W}}$. It follows that $q_{\mathfrak{F}}^{f_0} \geqq q_{\mathfrak{F}}^f$, $f_0 \geqq f$, f_0 divides f, $f_0 = f$. Thus $\mathfrak{W}$ is unramified over $\mathfrak{F}$.

We now apply Theorem 9.6 to see that every integer A of $\mathfrak{K}$ is congruent modulo $\mathfrak{p}_{\mathfrak{K}}^k$ to $A_0 + A_1P + \cdots + A_{k-1}P^{k-1}$, where the A_i range independently over any desired representatives of the residue classes of $J_{\mathfrak{K}}$ modulo $\mathfrak{p}_{\mathfrak{K}}$. It follows that we may take the A_i to range over the quantities $0, 1, \zeta, \cdots, \zeta^{\tau-1}$ of $\mathfrak{W}$. Let $\mathfrak{T}$ be the set of p^{eF} quantities B obtained in this way for $k = e$ so that every A of $J_{\mathfrak{K}}$ has the property $A \equiv B \pmod{\mathfrak{p}_{\mathfrak{K}}^e}$, $B \equiv 0 \pmod{\mathfrak{p}_{\mathfrak{K}}^e}$ if and only if

$$B = 0, \qquad B = \sum_{i=0,\cdots,\tau-1}^{j=0,\cdots,e-1} a_{ij}\zeta^i P^j \quad .$$

for $a_{ij} = 0$ or 1. If $B_0 + B_1\pi + \cdots + B_{r-1}\pi^{r-1} \equiv 0 \pmod{\mathfrak{p}_{\mathfrak{K}}^{er}}$, where π is a prime quantity of $J_{\mathfrak{F}}$, $\mathfrak{p}_{\mathfrak{F}} = (\pi)$, and the B_i are in $\mathfrak{T}$, then $B_0 \equiv 0 \pmod{\mathfrak{p}_{\mathfrak{K}}^e}$, $B_0 = 0$, $(B_1 + B_2\pi + \cdots + B_{r-1}\pi^{r-2})\,\pi \equiv 0 \pmod{\mathfrak{p}_{\mathfrak{K}}^{er}}$ if and only if $B_1 + B_2\pi + \cdots + B_{r-1}\pi^{r-2} \equiv 0 \pmod{\mathfrak{p}_{\mathfrak{K}}^{e(r-1)}}$. Then a similar proof yields $B_1 = 0$, and ultimately $B_0 = B_1 = \cdots = B_{r-1} = 0$. It follows that every integer A of $J_{\mathfrak{K}}$ defines a unique sequence $\{B_i\}$ of quantities in $\mathfrak{T}$ such that

$$A \equiv B_0 + B_1\pi + \cdots + B_{r-1}\pi^{r-1} \equiv \beta_{r0} + \beta_{r1}P + \cdots + \beta_{r,e-1}P^{e-1} \pmod{\mathfrak{p}_{\mathfrak{K}}^{er}},$$

for every r. Here we have used the form above for the B_i as polynomials in ζ and P of degree at most $e - 1$ in P and the property that π is in $\mathfrak{W}$ to see that the β_{ri} are in $\mathfrak{W}$. But then there exists a sequence of quantities $\beta_i^{(k)}$ in $\mathfrak{W}$ such that $A \equiv \beta_0^{(k)} + \beta_1^{(k)}P + \cdots + \beta_{e-1}^{(k)}P^{e-1} \pmod{\mathfrak{p}_{\mathfrak{K}}^k}$, where we take $\beta_i^{(k)} = \beta_{ri}$ if $e(r-1) < k \leqq er$. Clearly the $\beta_i^{(k)}$ converge on quantities β_i of the complete field $\mathfrak{W}$, $A = \beta_0 + \beta_1P + \cdots + \beta_{e-1}P^{e-1}$. Since every quantity of $\mathfrak{K}$ is the product of an integral power of π by an integer A of $\mathfrak{K}$, we have shown that $\mathfrak{K} = \mathfrak{W}(P)$. The degree of $\mathfrak{K}$ over $\mathfrak{W}$ is the quotient e of its degree ef over $\mathfrak{F}$ by the degree f of $\mathfrak{W}$ over $\mathfrak{F}$. By Theorem 9.9 the residue-class degree of $\mathfrak{K}$ over $\mathfrak{W}$ is unity, $\mathfrak{K}$ is completely ramified over $\mathfrak{W}$. This completes the proof of our theorem.

As an immediate corollary we have

Theorem 12. *All unramified fields of the same degree over $\mathfrak{F}$ are equivalent over $\mathfrak{F}$.*

Hence we may speak of any unramified field $\mathfrak{W}$ of degree f over $\mathfrak{F}$ as *the* field of that type. Moreover, we may prove the existence of unramified fields of any degree f over $\mathfrak{F}$. The result is given in

Theorem 13. *Let $\tau = q_{\mathfrak{F}}^f - 1$, $\mathfrak{K}$ be the root field over $\mathfrak{F}$ of the equation $x^\tau - 1 = 0$. Then $\mathfrak{K}$ is unramified of degree f over $\mathfrak{F}$, $\mathfrak{K} = \mathfrak{F}(\zeta)$ for a primitive τth root of unity ζ.*

For $x^\tau - 1 = \prod_{i=1}^{\tau}(x - \zeta_i)$ for ζ_i in $\mathfrak{K} = \mathfrak{F}(\zeta_1, \cdots, \zeta_\tau)$, $x^\tau - [1] = \prod_{i=1}^{\tau}(x - [\zeta_i])$ in $H_{\mathfrak{K}}$. The root field of $x^\tau - [1]$ is the Galois field $GF(q_{\mathfrak{F}}^f)$ of degree f over $H_{\mathfrak{F}} = GF(q_{\mathfrak{F}})$ by Theorem A7.22, and hence the residue-class degree of $\mathfrak{K}$ over $\mathfrak{F}$, which is the degree of $H_{\mathfrak{K}}$ over $H_{\mathfrak{F}}$, is $f_0 \geqq f$. By Theorem 9.11 $\mathfrak{K}$ contains a primitive τ_0th root of unity ζ_0, $\tau_0 = q_{\mathfrak{F}}^{f_0} - 1$. But $GF(q_{\mathfrak{F}}^f)$ is a subfield of $H_{\mathfrak{K}}$, f

divides f_0, τ divides $\tau_0 = \tau_1\tau$, $\zeta = \zeta_0^{\tau_1}$ is a primitive τth root of unity. Now the ζ_i are powers of ζ, $\mathfrak{K} \leqq \mathfrak{W}_1 = \mathfrak{F}(\zeta) \leqq \mathfrak{W} \leqq \mathfrak{K}$, $\mathfrak{K} = \mathfrak{F}(\zeta) = \mathfrak{W}$ is unramified of degree f over $\mathfrak{F}$ as desired.

If $\tau = q_{\mathfrak{F}}^f - 1$ and $\tau_1 = q_{\mathfrak{F}}^{f_1} - 1$ for f_1 a divisor of $f = f_0 f_1$, then $\tau = \tau_1\tau_2$ where $\tau_2 = q_{\mathfrak{F}}^{f_1(f_0-1)} + q_{\mathfrak{F}}^{f_1(f_0-2)} + \cdots + 1$. Then if ζ is a primitive τth root of unity generating an unramified field $\mathfrak{W}$ of degree f over $\mathfrak{F}$ the quantity ζ^{τ_2} is a primitive τ_1th root of unity, and Theorem 9.13 implies that $\mathfrak{W}_1 = \mathfrak{F}(\zeta^{\tau_2})$ has degree f_1 over $\mathfrak{F}$. We have proved

Theorem 14. *Every unramified field of degree f over a p-adic field $\mathfrak{F}$ has a unique subfield $\mathfrak{W}_1$ of degree f_1 over $\mathfrak{F}$ for every divisor f_1 of f.*

We may also easily prove

Theorem 15. *The composite $\mathfrak{K}$ of any finite number of unramified fields over $\mathfrak{F}$ is unramified over $\mathfrak{F}$.*

For let $\mathfrak{W}_i = \mathfrak{F}(\zeta_i)$ be unramified of degree f_i over $\mathfrak{F}$, $i = 1, \cdots, t$. By Theorem 9.9 the residue-class degree f of $\mathfrak{K}$ over $\mathfrak{F}$ is divisible by every f_i. Now $\mathfrak{K} \geqq \mathfrak{W} = \mathfrak{F}(\zeta)$ for ζ a primitive τth root of unity such that $\tau = q_{\mathfrak{F}}^f - 1$ is divisible by $\tau_i = q_{\mathfrak{F}}^{f_i} - 1$. Hence $\tau = \tau_i\sigma_i$, ζ^{σ_i} is a primitive τ_ith root of unity. By Theorems 9.13, 9.12 the field $\mathfrak{F}(\zeta^{\sigma_i}) \cong \mathfrak{F}(\zeta_i)$ and we must have $\mathfrak{F}(\zeta^{\sigma_i}) = \mathfrak{F}(\zeta_i)$ in $\mathfrak{K}$, $\mathfrak{W}_i \leqq \mathfrak{W}$. Hence the composite $\mathfrak{K} \leqq \mathfrak{W}$, $\mathfrak{K} = \mathfrak{W}$ as desired.

Observe that both the $\mathfrak{W}_i$ and their composite $\mathfrak{K}$ are generated by roots of unity and in fact in such a way that the generating root of unity ζ_i of $\mathfrak{W}_i$ may be taken to be a power of ζ generating $\mathfrak{K}$. Thus a possible alternative proof of Theorem 9.15 would consist of showing that conversely ζ is a product of properly chosen powers of the ζ_i and to prove our theorem as a consequence of

Theorem 16. *Let ζ be a primitive mth root of unity where m is prime to p, and let $\mathfrak{F}$ be a field of finite degree over $\mathfrak{R}_p$. Then $\mathfrak{F}(\zeta)$ is unramified over $\mathfrak{F}$.*

For $q_{\mathfrak{F}}$ is a power of p and is prime to m, $q_{\mathfrak{F}}^f \equiv 1 \pmod{m}$ where f is the value of the Euler ϕ-function for m. Then $\tau = q_{\mathfrak{F}}^f - 1 = mm_0$, $\zeta^\tau = \zeta^{mm_0} = 1$, $\mathfrak{F}(\zeta)$ is contained in the root field $\mathfrak{W}$ of $x^\tau - 1$. By Theorem 9.13 $\mathfrak{W}$ is unramified over $\mathfrak{F}$ and so is $\mathfrak{F}(\zeta)$.

Observe however that the result above is not necessarily true if p divides m. As an example showing this let $\mathfrak{F} = \mathfrak{R}_p$ and use the result of Section 9.3 where we showed that if ζ is a primitive pth root of unity the field $\mathfrak{K} = \mathfrak{R}_p(\zeta)$ has degree $n > 1$ over $\mathfrak{R}_p$. But $H_{\mathfrak{K}} = H_{\mathfrak{R}_p}([\zeta])$, $[\zeta]^p = [1]$ implies that $[\zeta] = [1]$ in $H_{\mathfrak{R}_p}$ of characteristic p, $H_{\mathfrak{K}} = H_{\mathfrak{R}_p}$, $\mathfrak{K}$ has residue-class degree one, ramification order n, and is completely ramified over $\mathfrak{R}_p$.

We close our list of elementary results with a consequence of Theorem 9.9 which we state as

Theorem 17. *Let $\mathfrak{W}$ be unramified of degree f over $\mathfrak{F}$, $\mathfrak{Y}$ be completely ramified of degree e over $\mathfrak{F}$. Then the composite $\mathfrak{K}$ of $\mathfrak{W}$ and $\mathfrak{Y}$ has ramification order e and residue-class degree f, and is the direct product $\mathfrak{K} = \mathfrak{W} \times \mathfrak{Y}$.*

For by Theorem 9.9 the ramification order of $\mathfrak{K}$ over $\mathfrak{F}$ is $e_0 \geqq e$ and the residue-class degree is $f_0 \geqq f$; the degree of $\mathfrak{K}$ over $\mathfrak{F}$ is $n = e_0 f_0 \geqq ef$. But by Theorem A7.9 the degree of $\mathfrak{K}$ is $n \leqq ef$ so that $n = e_0 f_0 = ef$, $e_0 = e$, $f_0 = f$, $\mathfrak{K}$ is the direct product of $\mathfrak{W}$ and $\mathfrak{Y}$.

8. The automorphism group of an unramified field. Let $\mathfrak{F}$ be a p-adic field and $\mathfrak{W} = \mathfrak{F}(\zeta)$ be unramified of degree f over $\mathfrak{F}$, where ζ is a primitive τth root of unity, $\tau = q_{\mathfrak{F}}^f - 1$. The finite field $H_{\mathfrak{W}}$ is perfect of characteristic p, and the correspondence

$$S_0: \qquad a \leftrightarrow a^{q_{\mathfrak{F}}}$$

is an automorphism of $H_{\mathfrak{W}}$ since $q_{\mathfrak{F}}$ is a power of p. But S_0 leaves every quantity of $H_{\mathfrak{F}}$ unaltered and is an automorphism over $H_{\mathfrak{F}}$ of $H_{\mathfrak{W}}$. If $\Gamma(x)$ is the minimum function of ζ over $\mathfrak{F}$, then $\bar{\Gamma}([\zeta]) = 0$, $\bar{\Gamma}([\zeta^{q_{\mathfrak{F}}}]) = 0$ so that, since the classes $[\zeta^t]$ are distinct for $t = 0, \cdots, \tau - 1$ and the ζ^t are the distinct roots of $x^\tau - 1 = \Gamma(x)\Delta(x)$, we must have $\Gamma(\zeta^{q_{\mathfrak{F}}}) = 0$. Thus the correspondence

$$S: \qquad w = w(\zeta) \leftrightarrow [w(\zeta)]^S = w(\zeta^{q_{\mathfrak{F}}}) \qquad (w \text{ in } \mathfrak{W}),$$

is an automorphism over $\mathfrak{F}$ of $\mathfrak{W}$. Now $\zeta^{q_{\mathfrak{F}}^f} = \zeta = \zeta^{S^f}$, so that S^f is the identity automorphism of the automorphism group $\mathfrak{G}$ of $\mathfrak{W}$ over $\mathfrak{F}$, and the order f_0 of S divides f. Then

$$\zeta^{S^{f_0}} = \zeta^{q_{\mathfrak{F}}^{f_0}} = \zeta, \qquad \tau_0 = q_{\mathfrak{F}}^{f_0} - 1 \leqq \tau = q_{\mathfrak{F}}^f - 1, \qquad \zeta^{\tau_0} = 1,$$

whereas ζ is a primitive τth root of unity. Hence $f_0 = f$, $\mathfrak{G}$ is generated by S, and we have proved

Theorem 18. *An unramified field $\mathfrak{W}$ of degree f over $\mathfrak{F}$ is cyclic over $\mathfrak{F}$. Moreover if we write $\mathfrak{W} = \mathfrak{F}(\zeta)$, where ζ is a primitive τth root of unity, the correspondence S above is a generating automorphism of $\mathfrak{W}$ over $\mathfrak{F}$.*

We complete our theory of algebraic extension of p-adic fields by deriving

Theorem 19. *A quantity $\alpha \neq 0$ of $\mathfrak{F}$ is the norm $N_{\mathfrak{W}|\mathfrak{F}}(w)$, for w in an unramified field $\mathfrak{W}$ of degree f over $\mathfrak{F}$, if and only if the order λ of $\alpha = \pi^\lambda u$ is divisible by f, where u is a unit of $J_{\mathfrak{F}}$, π is its prime quantity. In particular every unit u of $\mathfrak{F}$ is the norm $u = N_{\mathfrak{W}|\mathfrak{F}}(U)$, U a unit of $\mathfrak{W}$.*

For every $A \neq 0$ of $\mathfrak{W}$ has the form $A = \pi^l U$, U a unit of $\mathfrak{W}$. It follows that $N_{\mathfrak{W}|\mathfrak{F}}(A) = \pi^{fl} u$ where $u = N_{\mathfrak{W}|\mathfrak{F}}(U)$ is a unit of $\mathfrak{F}$, $fl = \lambda$ is divisible by f as desired. Conversely let $\alpha = \pi^{fl} u$ where u is a unit of $\mathfrak{F}$ so that $\alpha = u N_{\mathfrak{W}|\mathfrak{F}}(\pi^l) = N_{\mathfrak{W}|\mathfrak{F}}(w)$ if and only if $u = N_{\mathfrak{W}|\mathfrak{F}}(U)$, where $U = w\pi^{-l}$ is necessarily a unit of $\mathfrak{W}$. We have thus reduced our proof to the case where $\alpha = u$ is a unit of $\mathfrak{F}$.

We let $\mathfrak{W} = \mathfrak{F}(\zeta)$ as usual, so that $N_{\mathfrak{W}|\mathfrak{F}}(\zeta) = \zeta^s$ where it is clear that $s = 1 + q_{\mathfrak{F}} + q_{\mathfrak{F}}^2 + \cdots + q_{\mathfrak{F}}^{f-1} = (q_{\mathfrak{F}}^f - 1)[(q_{\mathfrak{F}} - 1)^{-1}]$. Hence $\zeta_0 = \zeta^s$ is a primitive $(q_{\mathfrak{F}} - 1)$th root of unity, and the maximal unramified subfield $\mathfrak{W}_0$ over $\mathfrak{R}_p$ of $\mathfrak{F}$ is thus $\mathfrak{R}_p(\zeta_0)$. Then every unit u of $\mathfrak{F}$ is congruent modulo $\mathfrak{p}_{\mathfrak{F}}$ to a power of ζ^s, and it follows that $u = \zeta^{st} v$ with $v \equiv 1 \pmod{\mathfrak{p}_{\mathfrak{F}}}$. We have shown that $u = N_{\mathfrak{W}|\mathfrak{F}}(\zeta^t) v$ is the norm of a quantity of $\mathfrak{W}$ if and only if v is such a norm.

The trace of integer α of $\mathfrak{W}$ is $T_{\mathfrak{W}|\mathfrak{F}}(\alpha) = \alpha + \alpha^S + \cdots + \alpha^{S^{f-1}}$. Now $\mathfrak{p}_{\mathfrak{W}} = (\pi)$ for π a prime quantity of $\mathfrak{F}$, since $\mathfrak{W}$ is unramified over $\mathfrak{F}$, $H_{\mathfrak{W}}$ consists of the residue classes of $J_{\mathfrak{W}}$ defined by powers of ζ modulo (π). But then if $\alpha \neq 0$ is in $J_{\mathfrak{W}}$, so that $\alpha \equiv \zeta^t \pmod{\mathfrak{p}_{\mathfrak{W}}}$, we have $\alpha^S \equiv (\zeta^t)^S \equiv \zeta^{tq_{\mathfrak{F}}} \equiv \alpha^{q_{\mathfrak{F}}} \pmod{\mathfrak{p}_{\mathfrak{W}}}$. Hence $T_{\mathfrak{W}|\mathfrak{F}}(\alpha) \equiv \alpha + \alpha^{q_{\mathfrak{F}}} + \cdots + \alpha^{q_{\mathfrak{F}}^{f-1}} \pmod{\mathfrak{p}_{\mathfrak{W}}}$. The equation $x + x^{q_{\mathfrak{F}}} + \cdots + x^{q_{\mathfrak{F}}^{f-1}} = 0$ cannot have more than $q_{\mathfrak{F}}^{f-1}$ roots in the field $H_{\mathfrak{W}}$ consisting of precisely $q_{\mathfrak{F}}^f$ distinct quantities, and hence $a = T_{\mathfrak{W}|\mathfrak{F}}(\alpha) \not\equiv 0 \pmod{\mathfrak{p}_{\mathfrak{F}}}$ for an integer α existing in $\mathfrak{W}$. Then $a \neq 0$ is in $J_{\mathfrak{F}}$, and if c is in $J_{\mathfrak{F}}$ we have $T_{\mathfrak{W}|\mathfrak{F}}(ca^{-1}\alpha) = ca^{-1}a = c$. But a is not divisible by $\mathfrak{p}_{\mathfrak{F}}$ and is a unit of $J_{\mathfrak{F}}$ and thus of $J_{\mathfrak{W}}$; $ca^{-1}\alpha$ is in $J_{\mathfrak{W}}$. We have proved that *every integer of* $\mathfrak{F}$ *is the trace of an integer of* $\mathfrak{W}$.

We shall use this result in the proof of the existence of an infinite sequence of quantities α_i of $\mathfrak{W}$ and a corresponding infinite sequence of quantities

$$\beta_0 = 1, \quad \beta_j = \beta_{j-1}(1 + \alpha_j\pi^j) \quad j > 0. \tag{19}$$

such that our given unit v of $J_{\mathfrak{F}}$ has the property $v \equiv N_{\mathfrak{W}|\mathfrak{F}}(\beta_j) \pmod{\mathfrak{p}_{\mathfrak{F}}^{j+1}}$. The sequence $\{\beta_j\}$ will converge on a quantity β of $\mathfrak{W}$, and our result will imply that $v \equiv N_{\mathfrak{W}|\mathfrak{F}}(\beta) \pmod{\mathfrak{p}_{\mathfrak{F}}^{j+1}}$ for every j. But then $v = N_{\mathfrak{W}|\mathfrak{F}}(\beta)$, and our proof will be complete as soon as we have constructed the α_i. We make the construction by an induction on j. We begin with $j = 0$ since then $v \equiv 1 \pmod{\mathfrak{p}_{\mathfrak{F}}}$ is the corresponding condition, and *this is our hypothesis* on v. Assume then that $v \equiv N_{\mathfrak{W}|\mathfrak{F}}(\beta_{j-1}) \pmod{\mathfrak{p}_{\mathfrak{F}}^j}$. The quantity $w = N_{\mathfrak{W}|\mathfrak{F}}(\beta_{j-1}) \not\equiv 0 \pmod{\mathfrak{p}_{\mathfrak{F}}}$ by our hypothesis on v, and w is an integer of $\mathfrak{W}$. By the usual congruence theory there exists an integer w_0 of $\mathfrak{W}$ such that $ww_0 \equiv 1 \pmod{\mathfrak{p}_{\mathfrak{F}}^{j+1}}$. Then $w_0v \equiv 1 \pmod{\mathfrak{p}_{\mathfrak{F}}^j}$, $w_0v = 1 + \gamma\pi^j$ where $\gamma = \pi^{-j}(w_0v - 1)$ is an integer of $\mathfrak{F}$. Then by the proof above there exists an integer α_j in $\mathfrak{W}$ such that $\gamma = T_{\mathfrak{W}|\mathfrak{F}}(\alpha_j)$. Hence $\gamma\pi^j = w_0v - 1 = \pi^jT_{\mathfrak{W}|\mathfrak{F}}(\alpha_j)$, $v \equiv w[1 + \pi^jT_{\mathfrak{W}|\mathfrak{F}}(\alpha_j)] \pmod{\mathfrak{p}_{\mathfrak{F}}^{j+1}}$. But the norm $N_{\mathfrak{W}|\mathfrak{F}}(1 + \alpha_j\pi^j) \equiv 1 + \pi^jT_{\mathfrak{W}|\mathfrak{F}}(\alpha_j) \pmod{\mathfrak{p}_{\mathfrak{F}}^{j+1}}$ since $\alpha_j^S \equiv \alpha_j^{q_{\mathfrak{F}}} \pmod{\mathfrak{p}_{\mathfrak{W}}}$, $\pi^{2j} \equiv 0 \pmod{\mathfrak{p}_{\mathfrak{F}}^{j+1}}$. Hence $v \equiv N_{\mathfrak{W}|\mathfrak{F}}[\beta_{j-1}(1 + \alpha_j\pi^j)] = N_{\mathfrak{W}|\mathfrak{F}}(\beta_j) \pmod{\mathfrak{p}_{\mathfrak{F}}^{j+1}}$ as desired, and we have completed our proof.

9. p-adic normal simple algebras. The theorems we have derived on algebraic extensions of finite degree over a p-adic field $\mathfrak{F}$ show such fields to be of an exceptionally simple structure. They imply a correspondingly simple structure for normal simple algebras over $\mathfrak{F}$. We shall derive this structure, proving first

Theorem 20. *The ramification order e and residue-class degree f of a normal division algebra* $\mathfrak{D}$ *of degree m over a p-adic field* $\mathfrak{F}$ *are both equal to m. If* $\tau = q_{\mathfrak{F}}^f - 1$ *and* $[\zeta]$ *is any primitive* τ*th root of unity of* $H_{\mathfrak{D}}$ *defined by an integer* ζ *of* $\mathfrak{D}$, *the field* $\mathfrak{W} = \mathfrak{F}(\zeta)$ *is unramified of degree m over* $\mathfrak{F}$. *Moreover,* $\mathfrak{F}(P)$ *is completely ramified of degree m over* $\mathfrak{F}$ *for every prime quantity P of* $\mathfrak{D}$.

For $\mathfrak{p}_{\mathfrak{F}} = (\pi) = (P)^e$, where $\pi = UP^e$ for a unit U of $J_{\mathfrak{D}}$, $U = \pi P^{-e}$ is in $\mathfrak{K} = \mathfrak{F}(P)$. This field has degree $m_0 \leq m$ over $\mathfrak{F}$ since the maximal subfields of

$\mathfrak{D}$ have degree m and ramification order e_0 over $\mathfrak{F}$. But in $\mathfrak{K}$ we have $(\pi) = (P)^e$, $e \leqq e_0 \leqq m_0 \leqq m$. The field $\mathfrak{W} = \mathfrak{F}(\zeta)$ has degree $m_1 \leqq m$ and residue-class degree $f_1 \leqq m_1$. Also the incongruence of the residue classes $[0]$, $[\zeta]$, $\cdots$, $[\zeta^{r-1}]$ modulo $\mathfrak{p}_{\mathfrak{D}}$ implies their incongruence modulo $\mathfrak{p}_{\mathfrak{W}}$, $f \leqq f_1 \leqq m_1 \leqq m$. Hence Theorem 9.8 implies that $m^2 = ef \leqq m_0 m_1 \leqq m^2$, $e = e_0 = m_0 = m$, $f = f_1 = m_1 = m$, $e = f = m$ as desired. This proves our theorem.

We apply this result in the proof of

Theorem 21. *Every normal simple algebra $\mathfrak{A}$ of degree n over a p-adic field $\mathfrak{F}$ is a cyclic algebra*

$$\mathfrak{A} = (\mathfrak{W}, S, \pi^{\mu}), \tag{20}$$

where $0 \leqq \mu < n$, π is any fixed prime quantity of $\mathfrak{F}$, $\mathfrak{W}$ is the unramified field of degree n over $\mathfrak{F}$, and S is the automorphism of Theorem 9.8.

For by Theorem 9.20, $\mathfrak{A}$ has a splitting field $\mathfrak{W}_0$ which is unramified of degree m over $\mathfrak{F}$, m the index of $\mathfrak{A}$. Then the unramified field $\mathfrak{W}$ of degree n over $\mathfrak{F}$ of Theorem 9.13 has $\mathfrak{W}_0$ as a subfield, is cyclic over $\mathfrak{F}$, and splits $\mathfrak{A}$. By Theorem 5.9, $\mathfrak{A} = (\mathfrak{W}, S, \gamma)$ for γ in $\mathfrak{F}$. Hence $\gamma = \pi^{\lambda}u$ for u a unit of $J_{\mathfrak{F}}$, $\lambda = nq + \mu$ for $0 \leqq \mu < n$. By Theorem 9.19 there exists an A in $\mathfrak{W}$ such that $N_{\mathfrak{W}|\mathfrak{F}}(A) = u\pi^{nq}$. Hence by Theorem 5.13, $\mathfrak{A} = (\mathfrak{W}, S, \pi^{\mu})$.

We next prove

Theorem 22. *Let $\mathfrak{A}$ and $\mathfrak{A}'$ be normal simple algebras over a p-adic field $\mathfrak{F}$ so that $\mathfrak{A} = (\mathfrak{W}, S, \pi^{\mu})$, $\mathfrak{A}' = (\mathfrak{W}', S', \pi^{\mu'})$ as in Theorem* 9.21 *for $0 \leqq \mu < n$ $0 \leqq \mu' < n'$. Then $\mathfrak{A}$ and $\mathfrak{A}'$ are equivalent if and only if $n = n'$ and $\mu = \mu'$.*

For if $n = n'$ then we may assume that $\mathfrak{W} = \mathfrak{W}'$ and by Theorem 5.13 see that $\mathfrak{A} \cong \mathfrak{A}'$ if and only if $\pi^{\mu'-\mu} = N_{\mathfrak{W}|\mathfrak{F}}(A)$ for A in $\mathfrak{W}$. But by Theorem 9.19 this occurs if and only if n divides $\mu' - \mu$, that is, $\mu = \mu'$.

The integral exponent μ of (20) has now been shown to be an invariant of $\mathfrak{A}$ which we shall signify by writing

$$\mu = \mu(\mathfrak{A}). \tag{21}$$

Thus two normal simple algebras over $\mathfrak{F}$ are equivalent if and only if they have the same degree and invariant μ.

Evidently if $\mathfrak{B}$ is the fixed algebra $(\mathfrak{W}, S, \pi)$ then we have $\mathfrak{A} \sim \mathfrak{B}^{\mu}$ for every $\mathfrak{A}$. Observe also that if π_0 is any other prime integer of $\mathfrak{F}$, then $\pi_0^{\mu} = \pi^{\mu}u$, where u is a unit of $\mathfrak{F}$, and hence

$$\mathfrak{A} = (\mathfrak{W}, S, \pi^{\mu}) = (\mathfrak{W}, S, \pi_0^{\mu}). \tag{22}$$

This result is of course included in our statement of Theorems 9.21 and 9.22.

If a normal simple algebra $\mathfrak{A}$ of degree n has index m, we have $n = mq$, $\mathfrak{A} = \mathfrak{M}_q \times \mathfrak{D}$, where $\mathfrak{D}$ is a normal division algebra of degree m, $\mathfrak{M}_q \sim 1$. By Theorem 9.21 $\mathfrak{D} = (\mathfrak{W}_0, S_0, \pi^{\mu_0})$ with $\mathfrak{W}_0$ unramified of degree m over $\mathfrak{F}$. Then $\mathfrak{A}$ has an unramified splitting field of degree m over $\mathfrak{F}$. But in fact we may prove

Theorem 23. *A field $\mathfrak{K}$ of degree t over a p-adic field $\mathfrak{F}$ splits a normal simple algebra $\mathfrak{A}$ of index m over $\mathfrak{F}$ if and only if m divides t.*

For if $\mathfrak{K}$ splits $\mathfrak{A}$ then m divides t by Theorem 4.21. Conversely, let $\mathfrak{K}$ have degree $t = mq$ over $\mathfrak{F}$, and thus $t = ef$, where e and f are the integers of Theorem 9.11 for $\mathfrak{K}$. Then $\mathfrak{K} = \mathfrak{W}_0(P_0)$, where $\mathfrak{W}_0$ is unramified of degree f over $\mathfrak{F}$, $m = m_1m_2$ such that m_1 divides e, m_2 divides f. The field $\mathfrak{K}$ contains a subfield $\mathfrak{W}_1$ which is unramified of degree m_2 over $\mathfrak{F}$, and hence the subfield $\mathfrak{K}_1 = \mathfrak{W}_1(P_0)$ which is completely ramified of degree e over $\mathfrak{W}_1$. By Theorem 9.20, $\mathfrak{A} \sim \mathfrak{D} = (\mathfrak{W}, S, \pi^\mu)$, where $\mathfrak{W}$ is unramified of degree m over $\mathfrak{F}$ and contains a subfield equivalent to $\mathfrak{W}_1$. Now π is a prime quantity of $\mathfrak{W}_1$ and by Theorem 5.8 $\mathfrak{D}_{\mathfrak{W}_1} \sim (\overline{\mathfrak{W}}, S, \pi^\mu)$ of degree m_1 over $\mathfrak{W}_1$, $\overline{\mathfrak{W}}$ is unramified of degree m_1 over $\mathfrak{W}_1$ and equivalent over $\mathfrak{F}$ to $\mathfrak{W}$. Now $\mathfrak{D}_{\mathfrak{K}_1} = (\overline{\mathfrak{W}}_{\mathfrak{K}_1}, S, \pi^\mu)$, where $\overline{\mathfrak{W}}_{\mathfrak{K}_1}$ is an unramified field of degree m_1 over $\mathfrak{K}_1$ by Theorem 9.17, $\pi = UP^e$ for a unit U of $\mathfrak{K}$, e is divisible by m_1. Hence by Theorem 9.19 π^μ is the norm of a quantity of $\overline{\mathfrak{W}}_{\mathfrak{K}_1}$, $\mathfrak{D}_{\mathfrak{K}_1} \sim 1 \sim \mathfrak{A}_{\mathfrak{K}_1}$, $\mathfrak{K}_1$ splits $\mathfrak{A}$, and so does $\mathfrak{K}$.

The form we have obtained for all normal simple algebras over $\mathfrak{F}$ yields a very explicit criterion that our algebras be division algebras. The result is given in

Theorem 24. *A normal simple algebra $\mathfrak{A}$ of degree n over a p-adic field $\mathfrak{F}$ is a division algebra if and only if its invariant $\mu(\mathfrak{A})$ is prime to n.*

For if $\mu = d\mu_0$, $n = dn_0$, the index of $\mathfrak{A} \sim (\mathfrak{W}, S, \pi)^\mu$ is at most n_0 by Theorem 5.16. Hence if $\mathfrak{A}$ is a division algebra μ is prime to n. Conversely, let μ be prime to n and $n = mq$ where m is the index of $\mathfrak{A}$. By Theorem 9.23 $\mathfrak{A}$ has an unramified splitting field $\mathfrak{W}_0$ of degree m over $\mathfrak{F}$. Also the $\mathfrak{A}$-commutator of $\mathfrak{W}_{01}$ is $\mathfrak{A}^{\mathfrak{W}_{01}} = (\mathfrak{W}, S^m, \pi^\mu)$ over $\mathfrak{W}_{01}$, where $\mathfrak{W}_{01}$ is the subfield of $\mathfrak{W}$ equivalent over $\mathfrak{F}$ to $\mathfrak{W}_0$, $\mathfrak{A}^{\mathfrak{W}_{01}}$ is equivalent over $\mathfrak{F}$ to an algebra $\mathfrak{B}$ over $\mathfrak{W}_0$ such that $\mathfrak{B} \sim \mathfrak{A}_{\mathfrak{W}_0}$. Hence $\mathfrak{A}^{\mathfrak{W}_{01}} \sim 1$, $\pi^\mu = N_{\mathfrak{W}|\mathfrak{W}_{01}}(A)$. But $\mathfrak{W}_{01}$ is unramified, π is a prime quantity of $\mathfrak{W}_{01}$, $\mathfrak{W}$ has degree q over $\mathfrak{W}_{01}$, q divides μ by Theorem 9.19. This is a contradiction unless $q = 1$, $n = m$ as desired.

We may also easily determine the exponent of every p-adic normal simple algebra, proving

Theorem 25. *The exponent of a normal simple algebra over a p-adic field is its index, and is in fact the quotient of its degree n by the greatest common divisor of n and the invariant μ.*

For let $\mathfrak{A}$ have degree n and invariant μ and write $n = md$, $\mu = \alpha d$ with α prime to m. We let $\mathfrak{W}$ be unramified of degree n over $\mathfrak{F}$ so that $\mathfrak{A} \sim \mathfrak{B}^\mu$, $\mathfrak{B} = (\mathfrak{W}, S, \pi)$. But by Theorem 7.14 $\mathfrak{B}^d \sim (\mathfrak{W}_{01}, S, \pi)$, where $\mathfrak{W}_{01}$ is the subfield of degree m over $\mathfrak{F}$ of $\mathfrak{W}$, $\mathfrak{A} \sim \mathfrak{B}^{\alpha d} \sim \mathfrak{D} = (\mathfrak{W}_{01}, S, \pi^\alpha)$. The algebra $\mathfrak{D}$ is a division algebra by Theorem 9.24 since α is prime to the degree m of $\mathfrak{D}$. Hence m is the index of $\mathfrak{A}$. If ρ is the exponent of $\mathfrak{A}$ we have $\mathfrak{A}^\rho \sim \mathfrak{D}^\rho \sim (\mathfrak{W}_{01}, S, \pi^{\alpha\rho}) \sim 1$ if and only if $\alpha\rho$ is divisible by m. Since m is prime to α, it must divide ρ, and by the minimal property of the exponent ρ we have $\rho = m$.

10. Quaternion algebras. Our determination of all rational division algebras will involve a study not only of algebras over p-adic extensions of our reference field but also over its completion with respect to its *archimedean* valuations. The normal simple algebras over such completions have index one or two. For this reason we shall begin this study by making some elementary remarks on the structure of normal simple algebras of degree two over any field.

Let $\mathfrak{A}$ be normal simple of degree two over an arbitrary field $\mathfrak{F}$. Then the index of $\mathfrak{A}$ divides the prime two and is either one or two, $\mathfrak{A}$ is either a total matric algebra or a normal division algebra over $\mathfrak{F}$. We shall obtain a special form for such an algebra and first require

LEMMA 8. *Every separable quadratic field over $\mathfrak{F}$ is a cyclic field $\mathfrak{Z}$ with generating automorphism S over $\mathfrak{F}$ such that $\mathfrak{Z}$ is the algebra*

$$(1, x) \text{ over } \mathfrak{F}, \quad x^2 = x + \alpha, \quad -4\alpha \neq 1, \tag{23}$$

where α is in $\mathfrak{F}$ and S is induced by

$$x^S = 1 - x. \tag{24}$$

Conversely, the algebra $(1, x)$ of (23) is a cyclic semi-field with generating automorphism S induced by (24) and is a field if and only if the equation $\lambda^2 - \lambda - \alpha = 0$ has no root in $\mathfrak{F}$.

For $\mathfrak{Z} = \mathfrak{F}(x_0)$ such that $x_0^2 = \alpha_1 x_0 + \alpha_2$ with α_1 and α_2 in $\mathfrak{F}$. If $\alpha_1 = 0$ the separability of $\mathfrak{Z}$ over $\mathfrak{F}$ implies that the characteristic of $\mathfrak{F}$ is not two, $x_0^2 = \alpha_2$, $\mathfrak{Z} = \mathfrak{F}(x)$, where $x = x_0 + \frac{1}{2}$, $x^2 = \alpha_2 + \frac{1}{4} + x - \frac{1}{2}$ and $x^2 = x + \alpha$ as desired. When $\alpha_1 \neq 0$ we put $x = \alpha_1^{-1} x_0$ and have $\mathfrak{Z} = \mathfrak{F}(x)$, $x^2 = \alpha_1^{-2}(\alpha_1 x_0 + \alpha_2) = x + \alpha_1^{-2}\alpha_2$ as desired. In either case $-4\alpha = 1$ implies that the characteristic of $\mathfrak{F}$ is not two, $(x - \frac{1}{2})^2 = x^2 - x + \frac{1}{4} = (x - \frac{1}{4}) - x + \frac{1}{4} = 0$ which is impossible in a quadratic field $\mathfrak{F}(x)$. Now $(1 - x)^2 = 1 - 2x + x + \alpha = (1 - x) + \alpha$ so that $x \leftrightarrow 1 - x$ induces a non-identical automorphism S over $\mathfrak{F}$ of $\mathfrak{Z}$, $S^2 = I$ as desired. Conversely, let $-4\alpha \neq 1$. Then $f(\lambda) = \lambda^2 - \lambda - \alpha$ is irreducible in $\mathfrak{F}$ if and only if it has no root in $\mathfrak{F}$, $f(x) = 0$, and the irreducibility of $f(\lambda)$ implies that the algebra $(1, x)$ over $\mathfrak{F}$ is a cyclic field of degree two over $\mathfrak{F}$ with $x^S = 1 - x$. When there exists an η in $\mathfrak{F}$ such that $\eta^2 = \eta + \alpha$ we consider the diagonal algebra (e_1, e_2) over $\mathfrak{F}$, $e_1^2 = e_1$, $e_2^2 = e_2$, $e_1 e_2 = e_2 e_1 = 0$. We put $x = \eta e_1 + (1 - \eta) e_2$ and have $x^2 = (\eta + \alpha) e_1 + (1 - \eta + \alpha) e_2 = x + \alpha$, so that $x - \eta = (1 - 2\eta) e_2$, where $2\eta = 1$ implies that $\eta^2 - \eta = -\frac{1}{4}$ contrary to hypothesis. It follows that e_2 and $e_1 = 1 - e_2$ are in $(1, x)$, $(1, x) = (e_1, e_2)$ over $\mathfrak{F}$. Now $e_2^S = (1 - 2\eta)^{-1}(1 - x - \eta) = (1 - 2\eta)^{-1}(\eta - x + 1 - 2\eta) = -e_2 + 1 = e_1$, $e_1^S = e_2^S = e_2$, and $(1, x)$ is the desired cyclic semi-field. This completes our proof.

We shall make the

DEFINITION. *The algebras*

$$\mathfrak{Q} = (1, x, y, yx), \quad x^2 = x + \beta, \quad xy = y(1 - x), \quad y^2 = \gamma, \tag{25}$$

where β and $\gamma \neq 0$ are in $\mathfrak{F}$, $-4\beta \neq 1$, are called quaternion algebras.

By Lemma 9.8 every quaternion algebra is merely a generalized cyclic algebra of degree two over $\mathfrak{F}$. By Theorem 7.1 such an algebra is normal simple and is either a total matric algebra or a division algebra, and the former conclusion occurs if and only if $\gamma = N_{\mathfrak{Z}|\mathfrak{F}}(z)$. But $z = \lambda + \mu x$, $z^S = \lambda + \mu(1 - x)$, $zz^S = \lambda^2 + \lambda\mu - \beta\mu^2$. Conversely, every normal simple algebra of degree two either is the generalized cyclic total matric algebra defined above or has a separable quadratic subfield $\mathfrak{Z}$ and hence is $(\mathfrak{Z}, S, \gamma)$. Observe that the reducibility in $\mathfrak{F}$ of $\lambda^2 - \lambda - \beta = 0$ implies that $\mathfrak{Z} = (e_1, e_2)$, $z = \gamma e_1 + e_2$, $z^S = \gamma e_2 + e_1$, $zz^S = \gamma$, so that the assumption $\gamma \neq zz^S$ for any z of $\mathfrak{Z}$ implies in particular that $\mathfrak{Z}$ is a field. We have proved

Theorem 26. *Every quaternion algebra* (25) *is normal simple of degree two over* $\mathfrak{F}$ *and is a division algebra if and only if there exist no* λ *and* μ *in* $\mathfrak{F}$ *such that* $\gamma = \lambda^2 + \lambda\mu - \beta\mu^2$. *Conversely, every normal simple algebra of degree two over* $\mathfrak{F}$ *is a quaternion algebra.*

We now have the special result obtained by replacing x by $u = x - \frac{1}{2}$ above.

Theorem 27. *Let* $\mathfrak{F}$ *have characteristic not two. Then every normal simple algebra* $\mathfrak{A}$ *of degree two over* $\mathfrak{F}$ *is a quaternion algebra*

$$(26) \qquad \mathfrak{Q} = (1, u, y, yu), \qquad u^2 = \alpha, \qquad y^2 = \gamma, \qquad uy = -yu$$

where $\alpha \neq 0$ *and* $\gamma \neq 0$ *are in* $\mathfrak{F}$. *Moreover* $\mathfrak{Q}$ *is a division algebra if and only if there exist no* λ *and* μ *in* $\mathfrak{F}$ *such that* $\lambda^2 - \alpha\mu^2 = \gamma$.

11. Simple algebras over an ordered closed field. We shall designate by $\mathcal{R}$ any ordered closed field as defined in Section A5.10, by $\mathcal{C}$ the algebraic closure of $\mathcal{R}$. Then $\mathcal{C} = \mathcal{R}(i)$, $i^2 = -1$. If $\mathfrak{K}$ is any field algebraic over $\mathcal{R}$ the composite of $\mathfrak{K}$ and $\mathcal{C}$ over $\mathcal{R}$ is $\mathcal{C}$ and $\mathfrak{K} \leqq \mathcal{C}$, $\mathfrak{K} = \mathcal{C}$ or $\mathcal{R}$. We now let $\mathfrak{A}$ be any simple algebra over $\mathcal{R}$. Then the centrum of $\mathfrak{A}$ is either $\mathcal{R}$ or $\mathcal{C}$. In the latter case $\mathfrak{A} \sim 1$, since Theorem 4.18 implies that there are no division algebras of degree greater than unity over $\mathcal{C}$. In the former case the maximum possible degree of the normal division algebra $\mathfrak{D} \sim \mathfrak{A}$ is two, and if $\mathfrak{Z}$ is a maximal subfield of such an algebra $\mathfrak{D}$ then $\mathfrak{Z} \cong \mathcal{C}$. By Theorem 9.27 $\mathfrak{D} = (1, x, y, yx)$, $x^2 = -1$, $xy = -yx$, $y^2 = \gamma$. If $\gamma > 0$ in $\mathcal{R}$ then $\gamma = \lambda^2$, a contradiction. Hence $\gamma = -\delta^2 = -N_{\mathfrak{Z}|\mathfrak{F}}(\delta)$, and $\mathfrak{D} = (\mathfrak{Z}, S, -1)$. Conversely, $-1 \neq \lambda^2 + \mu^2$ in $\mathcal{R}$, and we have proved

Theorem 28. *The algebra* $(1, i, j, ji)$, *over any ordered closed field* $\mathcal{R}$, $i^2 = j^2 = -1$, $ij = -ji$, *is a normal division algebra, and is the only such normal division algebra. This algebra of* **real quaternions** *and the field* $\mathcal{C}$ *are the only division algebras over* $\mathcal{R}$, *and* $\mathcal{C}$ *is the only division algebra over* $\mathcal{C}$.

We now return to the case where $\mathfrak{F}$ is any algebraic number field of finite degree over the field $\mathfrak{R}$ of all rational numbers. If V is any valuation of $\mathfrak{F}$, the derived field of $\mathfrak{F}$ with respect to V is uniquely determined by V in the sense of analytic equivalence, and we shall write $\mathfrak{F}_V$ for this field. If V is archimedean, then $\mathfrak{F}_V$ is analytically equivalent to either the field of all real numbers or the field of all complex numbers, and we have

Theorem 29. *Let $\mathfrak{A}$ be normal simple of degree n and index m over $\mathfrak{F}$, $\mathfrak{K}$ a field of finite degree over $\mathfrak{F}$. Then if m is odd the algebra $\mathfrak{A}_{\mathfrak{F}_V} \sim 1$ for every archimedean valuation V of $\mathfrak{F}$. Otherwise there is only a finite number of archimedean valuations $V_1, \cdots, V_s$ of $\mathfrak{F}$ such that the index of $\mathfrak{A}_{\mathfrak{F}_j}$ over $\mathfrak{F}_{V_j}$ is two, and the composites $(\mathfrak{K}, \mathfrak{F}_V)$ split $\mathfrak{A}_{\mathfrak{F}_V}$ for every archimedean V of $\mathfrak{F}$ if and only if each $(\mathfrak{K}, \mathfrak{F}_{V_j})$ is analytically equivalent to the field of all complex numbers, that is $(\mathfrak{K}, \mathfrak{F}_{V_j})$ has degree two over $\mathfrak{F}_{V_j}$.*

For by Theorem 9.28 the index of $\mathfrak{A}_{\mathfrak{F}_V}$ over $\mathfrak{F}_V$ is either one or two and divides m and hence n. It can be two only when m is even. Only a finite number of archimedean valuations exist, and if $\mathfrak{A}_{\mathfrak{F}_{V_j}}$ has index two, $\mathfrak{F}_{V_j}$ must be real. The only algebraic field $\mathfrak{L} > \mathfrak{F}_{V_j}$ is then $\mathfrak{F}_{V_j}(i)$, $i^2 = -1$, and the composites $(\mathfrak{K}, \mathfrak{F}_{V_j})$ can split $\mathfrak{A}_{\mathfrak{F}_{V_j}}$ of index two if and only if $(\mathfrak{K}, \mathfrak{F}_{V_j}) > \mathfrak{F}_{V_j}$.

Note that the final part of our result is equivalent to the following statement. Let V be an archimedean valuation of $\mathfrak{F}$ over the field $\mathfrak{R}$ of all rational numbers defined by a real conjugate field $\mathfrak{F}_0$ over $\mathfrak{R}$, and hence $\mathfrak{F}_V \sim \mathcal{R}$. Assume also that $\mathfrak{A}_{\mathcal{R}} \sim \mathfrak{Q}$, a normal division algebra of degree two. Then necessarily all relative conjugates of $\mathfrak{K}$ over $\mathfrak{F}_0$ must be imaginary so that $(\mathfrak{K}, \mathfrak{F}_V) \cong \mathcal{C}$.

12. Lemmas from the theory of algebraic numbers. Every valuation V of an algebraic number field $\mathfrak{F}$ (of finite degree over $\mathfrak{R}$) is either archimedean or non-archimedean, and in either case the field $\mathfrak{F}_V$ is uniquely determined in the sense of analytic equivalence. Now let $\mathfrak{K}$ be a field of degree ν over $\mathfrak{F}$. The composites $\mathfrak{K}_{V_{0i}} = (\mathfrak{K}_i, \mathfrak{F}_V)$ are uniquely determined by V where $\mathfrak{K}_i$ ranges over the ν algebraic number field conjugate over $\mathfrak{F}$ to $\mathfrak{K}$, and in Chapter AXII we saw that these fields are the derived fields of $\mathfrak{K}$ with respect to certain valuations V_{0i} uniquely determined by V. Let us write $\mathfrak{K}_{V_0}$ for these fields where V_0 is a ν-valued function of V. If $\mathfrak{K}$ is normal over $\mathfrak{F}$ the $\mathfrak{K}_{V_{0i}}$ are all analytically equivalent, and we write $\mathfrak{K}_V$ for this unique (in the sense of analytic equivalence) field. The importance of considering all valuations of $\mathfrak{F}$ is then first indicated for us in Hasse's

LEMMA 9. *Let $\mathfrak{Z}$ be a cyclic field of prime degree over an algebraic number field $\mathfrak{F}$, and let γ be in $\mathfrak{F}$. Then*

$$\gamma = N_{\mathfrak{Z}|\mathfrak{F}}(z),$$

for some z in $\mathfrak{Z}$, if and only if there exist quantities z_V in $\mathfrak{Z}_V$, for every valuation V of $\mathfrak{F}$, such that

$$\gamma = N_{\mathfrak{Z}_V|\mathfrak{F}_V}(z_V).$$

This result proved in [174] is equivalent, in view of Theorem 5.13, to the statement that if $\mathfrak{A} = (\mathfrak{Z}, S, \gamma)$ of prime degree then $\mathfrak{A} \sim 1$ if and only if $\mathfrak{A}_{\mathfrak{F}_V} \sim 1$ for every V. This theorem requires an extensive amount of algebraic number theory for its proof and is the first of such presupposed results. We shall also assume another arithmetic theorem, stating it partially in terms of algebras. This is a weak form of the third theorem of Grunwald [170] on the existence of cyclic fields $\mathfrak{Z}$ of degree n with $\mathfrak{Z}_V$ having a given structure for a finite number of valuations V. Grunwald showed that we may construct $\mathfrak{Z}$ such that if V is archimedean and $\mathfrak{F}_V$ is real, $\mathfrak{Z}_V$ has degree two over $\mathfrak{F}_V$ providing two divides n. By Theorem 9.29 the field $\mathfrak{Z}_V$ then splits every $\mathfrak{A}_{\mathfrak{F}_V}$ of degree n. Using the Grunwald results we then have

LEMMA 10. *Let $V_1, \cdots, V_r$ be a finite number of non-archimedean valuations of $\mathfrak{F}$, n a fixed rational integer. Then there exists a field $\mathfrak{Z}$ which is cyclic of degree n over $\mathfrak{F}$ such that $\mathfrak{Z}_{V_i}$ has degree n over $\mathfrak{F}_{V_i}$ for $i = 1, \cdots, r$, and if V is any archimedean valuation of $\mathfrak{F}$ the field $\mathfrak{Z}_V$ splits every normal simple algebra $\mathfrak{A}_{\mathfrak{F}_V}$ of degree n over $\mathfrak{F}_V$.*

We finally assume an elementary result, the proof of which may be found on page 146 of [190].

LEMMA 11. *Let $\mathfrak{K}$ be algebraic of finite degree over $\mathfrak{F}$ so that the discriminant $\mathfrak{d}$ of $\mathfrak{K}$ is an ideal of $\mathfrak{F}$. Then the only prime ideals $\mathfrak{p}$ of $\mathfrak{F}$ divisible by the square of a prime ideal of $\mathfrak{K}$ are the divisors of $\mathfrak{d}$.*

13. The p-adic extensions of algebraic number fields. Let $\mathfrak{F}$ be an algebraic number field and V be any non-archimedean valuation of $\mathfrak{F}$. Then in Section A12.13 we showed that there exists a prime ideal $\mathfrak{p}$ in $\mathfrak{F}$ uniquely determined by V. Let a be in $\mathfrak{F}$, so that $a = \pi^\lambda b$, where π is an integer of $\mathfrak{F}$ divisible by $\mathfrak{p}$ but not by $\mathfrak{p}^2$, b has numerator and denominator prime to $\mathfrak{p}$. Then $V(a) = \rho^\lambda$ for some real positive $\rho < 1$. The integer π is a prime integer of the field $\mathfrak{F}_V$, and (π) is the corresponding prime ideal. Hence, while $\mathfrak{p}$ may not be a principal ideal, it becomes one when we extend the set of integers of $\mathfrak{F}$ to that of $\mathfrak{F}_V$.

Now let $\mathfrak{K}$ be normal of degree n over $\mathfrak{F}$, $\mathfrak{P}$ be a prime ideal of $\mathfrak{K}$ dividing $\mathfrak{p}$. Then in $\mathfrak{K}_V$ we have $\mathfrak{P} = (P)$, $\mathfrak{P}^e = \mathfrak{p}$. Let $\mathfrak{p} = \mathfrak{P}^\alpha\mathfrak{Q}$ with $\mathfrak{Q}$ an ideal of $\mathfrak{K}$ prime to $\mathfrak{P}$. Then $\mathfrak{Q}$ is prime to $\mathfrak{P}$ in $\mathfrak{K}_V$. It follows that the ramification order e of $\mathfrak{K}_V$ over $\mathfrak{F}_V$ is precisely the integer α. This integer has in fact been called the *ramification order of $\mathfrak{P}$ with respect to $\mathfrak{K}$ over $\mathfrak{F}$*. We then apply Lemma 9.11 to obtain

LEMMA 12. *Let $\mathfrak{K}$ be normal over $\mathfrak{F}$ and $\mathfrak{p}$ be a prime ideal of $\mathfrak{F}$ not dividing the discriminant of $\mathfrak{K}$ over $\mathfrak{F}$. Then if V is the non-archimedean valuation of $\mathfrak{F}$ defined by $\mathfrak{p}$, the field $\mathfrak{K}_V$ is unramified over $\mathfrak{F}_V$.*

We use the result above to prove

Theorem 30. *Let $\mathfrak{A}$ be a normal simple algebra over an algebraic number field $\mathfrak{F}$ of finite degree. Then the index of $\mathfrak{A}_{\mathfrak{F}_V}$ over $\mathfrak{F}_V$ is unity except for a finite number of valuations V of $\mathfrak{F}$.*

For there is only a finite number of archimedean valuations of $\mathfrak{F}$ and hence all others are non-archimedean valuations defined by prime ideals $\mathfrak{p}$ of $\mathfrak{F}$. The algebra $\mathfrak{A} \sim (\mathfrak{Z}, a)$ by Theorem 5.1, where $\mathfrak{Z}$ is normal of degree n over $\mathfrak{F}$, $a = \{a_{S,T}\}$ for the $a_{S,T}$ in $\mathfrak{Z}$. The n^2 quantities $a_{S,T}$ contain only a finite number of prime ideal factors in their numerators and denominators and we let $\mathfrak{p}$ be any prime ideal distinct from these prime ideals as well as the prime ideal divisors of the discriminant of $\mathfrak{Z}$ over $\mathfrak{F}$. If V is the valuation defined by $\mathfrak{p}$ we have $\mathfrak{Z}_V$ unramified of degree m over $\mathfrak{F}_V$ by Lemma 9.12, the field $\mathfrak{Z}_V$ is cyclic over $\mathfrak{F}_V$, and by Theorem 5.8 $\mathfrak{A}_{\mathfrak{F}_V} \sim (\mathfrak{Z}_V, \mathfrak{b})$ where $\mathfrak{b} = \{a_{S,T}\}$, the automorphisms S and T range over the cyclic group $[S]$ of $\mathfrak{Z}_V$ over $\mathfrak{F}_V$. Now $(\mathfrak{Z}_V, \mathfrak{b}) = u_I\mathfrak{Z}_V + u_S\mathfrak{Z}_V + \cdots + u_{S^{m-1}}\mathfrak{Z}_V$ and the quantities $a_{S,T}$ clearly have order zero, the $a_{S,T}$ are units of $\mathfrak{Z}_V$. But $u_Su_S = u_{S^2}a_{S,S}$, $u_S^3 = u_{S^3}a_{S,S}a_{S,S^2}, \cdots$, $u_S^m = u_IU$, where U is a unit of $\mathfrak{Z}_V$. By (5.9) $u_Ia_{I,I}^{-1} = 1$, $u_S^m = a_{I,I}U$ is a unit of $\mathfrak{Z}_V$. But u_S^m is in $\mathfrak{F}_V$ and is a unit u of $\mathfrak{F}_V$, $(\mathfrak{Z}_V, \mathfrak{b}) = (\mathfrak{Z}_V, S, u) = (\mathfrak{Z}_V, S, 1) \sim 1$. Hence $\mathfrak{A}_{\mathfrak{F}_V} \sim 1$ for all such ideals $\mathfrak{p}$. These include all but a finite number of prime ideals of $\mathfrak{F}$ as desired.

We may now derive quickly

Theorem 31. *Let $\mathfrak{A}$ be a normal simple algebra over an algebraic number field $\mathfrak{F}$ of finite degree over $\mathfrak{R}$ and $\mathfrak{A}_{\mathfrak{F}_V} \sim 1$ for every valuation V of $\mathfrak{F}$. Then $\mathfrak{A} \sim 1$.*

For let $\mathfrak{A}$ have index $m > 1$, q be a prime divisor of m, and apply Theorem 4.30 to obtain a field $\mathfrak{K}$ of finite degree over $\mathfrak{F}$ such that $\mathfrak{A}_{\mathfrak{K}} \sim \mathfrak{D}$, where $\mathfrak{D}$ is a cyclic division algebra of degree q over $\mathfrak{K}$. Then $\mathfrak{D} = (\mathfrak{Z}, S, \gamma)$ over $\mathfrak{K}$ and if V is any valuation of $\mathfrak{F}$ and V_0 an extension of V to $\mathfrak{Z}$ we have $\mathfrak{A}_{\mathfrak{F}_V} \sim 1$, $\mathfrak{Z}_{V_0}$ is a composite of $\mathfrak{Z}$ and $\mathfrak{K}_{V_0}$, $\mathfrak{D}_{\mathfrak{K}_{V_0}} \sim (\mathfrak{A}_{\mathfrak{K}})_{\mathfrak{K}_{V_0}} = (\mathfrak{A}_{\mathfrak{F}_V})_{\mathfrak{K}_{V_0}} \sim 1$. Hence $\mathfrak{D}_{\mathfrak{K}_{V_0}} \sim (\mathfrak{Z}_{V_0}, S^r, \gamma)$, by Theorem 5.8, for an integer r such that mr^{-1} is the degree of $\mathfrak{Z}_{V_0}$ over $\mathfrak{K}_{V_0}$. It follows that $\gamma = N_{\mathfrak{Z}_{V_0}|\mathfrak{K}_{V_0}}(a_V)$ for a_V in $\mathfrak{Z}_{V_0}$ and every V_0 of $\mathfrak{K}$, every V of $\mathfrak{F}$. But then by Lemma 9.9 $\gamma = N_{\mathfrak{Z}|\mathfrak{K}}(a)$ for a in $\mathfrak{Z}$, $\mathfrak{D} \sim 1$, a contradiction. Hence $m = 1$ as desired.

14. Determination of all rational division algebras. We shall show that every rational simple algebra is a cyclic algebra over its centrum and thus, in particular, determine all rational division algebras. We state the result as

Theorem 32. *Let $\mathfrak{F}$ be a field of finite degree over $\mathfrak{R}$, $\mathfrak{A}$ be a normal simple algebra over $\mathfrak{F}$. Then $\mathfrak{A}$ is a cyclic algebra over $\mathfrak{F}$, and the exponent of $\mathfrak{A}$ is its index.*

For let $\mathfrak{A}$ have degree n and $V_1, \cdots, V_r$ be the (finite number of) non-archimedean valuations V of $\mathfrak{F}$ such that $\mathfrak{A}_{\mathfrak{F}_V}$ is not a total matric algebra. We construct a cyclic field $\mathfrak{Z}$ of degree n as in Lemma 9.10, and consider $\mathfrak{B} = \mathfrak{A}_{\mathfrak{Z}}$ over $\mathfrak{Z}$. The degree of $\mathfrak{B}$ is n, and by the choice of $\mathfrak{Z}$ in Lemma 9.10 $\mathfrak{B}_{\mathfrak{Z}_V} \sim 1$ for every archimedean V of $\mathfrak{F}$. By Theorem 9.23 $\mathfrak{B}_{\mathfrak{Z}_{V_i}} \sim 1$ for every V_i $(i = 1, \cdots, r)$. However if $V \neq V_i$ is non-archimedean then $\mathfrak{B}_{\mathfrak{Z}_V} = (\mathfrak{A}_{\mathfrak{F}_V})_{\mathfrak{Z}_V} \sim 1$ since $\mathfrak{A}_{\mathfrak{F}_V} \sim 1$. By Theorem 9.31 $\mathfrak{A}_{\mathfrak{Z}} = \mathfrak{B} \sim 1$, $\mathfrak{Z}$ splits $\mathfrak{A}$. Since $\mathfrak{Z}$ and $\mathfrak{A}$ both have degree n, $\mathfrak{A}$ is a cyclic algebra.

To prove that the index m of $\mathfrak{A}$ is its exponent we apply Theorem 5.19 to reduce our proof to the case where $m = c^e$ for c a rational prime. Let the V_i be defined as above and m_i be the index of $\mathfrak{A}_{\mathfrak{F}_{V_i}}$. Then m_i divides m, $m_i = c^{e_i}$. We let e_0 be the maximum of the e_i. By the proof above, with the integer n of Lemma 9.10 taken to be c^{e_0}, there exists a field $\mathfrak{Z}$ of degree c^{e_0} over $\mathfrak{F}$ such that $\mathfrak{Z}$ splits $\mathfrak{A}$. By Theorem 4.21 $c^{e_0} \geqq m$. However c^{e_0} divides m, $c^{e_0} = m$. Hence there is a valuation V of $\mathfrak{F}$ such that $\mathfrak{A}_{\mathfrak{F}_V}$ has index m. By Theorem 9.25 the exponent of $\mathfrak{A}_{\mathfrak{F}_V}$ is m. But the exponent of $\mathfrak{A}_{\mathfrak{F}_V}$ is clearly at most the exponent ρ of $\mathfrak{A}$, and $\rho \leqq m$. Hence $\rho = m$.

The index of $\mathfrak{A}_{\mathfrak{F}_V}$ may be called the *V-index* of $\mathfrak{A}$. As a corollary of our proof above we then have

Theorem 33. *The index (= exponent) of a normal simple algebra over an algebraic number field of finite degree is the least common multiple of its V-indices.*

For $\mathfrak{A} \sim \mathfrak{D}_1 \times \cdots \times \mathfrak{D}_t$ where $\mathfrak{D}_i$ is a normal division algebra of degree $c_i^{e_i}$ for distinct primes c_i, $m = c_1^{e_1} \cdots c_t^{e_t}$ is the index of $\mathfrak{A}$. By our proof there exists a valuation V_i of $\mathfrak{F}$ such that $\mathfrak{D}_i$ has V_i-index $c_i^{e_i}$. But every V-index m_V divides m, the least common multiple m_0 of the m_V is divisible by every $c_i^{e_i}$ and hence by m. Thus $m_0 = m$.

15. The equivalence of normal simple algebras over an algebraic number field. If $\mathfrak{A}$ is any normal simple algebra over an algebraic number field $\mathfrak{F}$ the valuations V such that $\mathfrak{A}_{\mathfrak{F}_V}$ is not a total matric algebra are a finite number of invariants V_i $(i = 1, \cdots, r)$ of $\mathfrak{A}$. With each non-archimedean V_i we associate the index m_i of $\mathfrak{A}_{\mathfrak{F}_{V_i}}$ and the integer μ_i of Theorem 9.21. If V_i is archimedean then necessarily $\mathfrak{F}_{V_i}$ is analytically equivalent to the field of all real numbers, the only field over $\mathfrak{F}_{V_i}$ is the field $\mathfrak{F}_{V_i}(z)$, $z^2 = -1$. Hence $\mathfrak{A}_{V_i} \sim \mathfrak{Q} = (1, z, y, yz)$, where we may clearly take $y^2 = -1$. This shows that $\mathfrak{Q}$ is the algebra of real quaternions, $\mathfrak{A}_{\mathfrak{F}_{V_i}}$ has index two, and we put $\mu_i = 1$ in this case. We then call the complete set V_i, m_i, μ_i $(i = 1, \cdots, r)$ *the invariants* of $\mathfrak{A}$ and prove

Theorem 34. *Two normal simple algebras of the same degree n over $\mathfrak{F}$ are equivalent if and only if they have the same invariants V_i, m_i, μ_i.*

For let $\mathfrak{A}$ and $\mathfrak{B}$ have the same invariants. Then clearly $\mathfrak{A}_{\mathfrak{F}_V}$ and $\mathfrak{B}_{\mathfrak{F}_V}$ are equivalent for every non-archimedean V_i by Theorem 9.22, they are trivially equivalent for every archimedean V_i and every $V \neq V_i$. It follows that $(\mathfrak{A} \times \mathfrak{B}^{-1})_{\mathfrak{F}_V} \sim 1$ for every V. By Theorem 9.31 $\mathfrak{A} \times \mathfrak{B}^{-1} \sim 1$, $\mathfrak{A}$ and $\mathfrak{B}$ are equivalent. The converse is trivial.

Algebras over an algebraic number field not of finite degree over $\mathfrak{R}$ will be considered in our Chapter XI on special results and we shall indicate, with reference to a paper of the author, how their properties are derivable out of those we have already obtained above.

CHAPTER X

INVOLUTIONS OF ALGEBRAS

1. Definition and elementary properties of involutions. Let $\mathfrak{A}$ be an algebra over $\mathfrak{F}$, and let us assume throughout our discussion that $\mathfrak{A}$ has a unity quantity 1. Then as usual we may and shall assume also that $\mathfrak{A}$ contains $\mathfrak{F}$, $\mathfrak{F} = (1)$ over $\mathfrak{F}$ is a subalgebra over $\mathfrak{F}$ of $\mathfrak{A}$.

The set $\mathfrak{G}_{\mathfrak{F}}$ of all non-singular linear transformations over $\mathfrak{F}$ of $\mathfrak{A}$ is a group whose identity element is the identity automorphism I of $\mathfrak{A}$. We have already defined the subgroup of $\mathfrak{G}_{\mathfrak{F}}$ consisting of all automorphisms over $\mathfrak{F}$ of $\mathfrak{A}$, with complete results in the case where $\mathfrak{A}$ is normal simple over $\mathfrak{F}$, and shall now study certain other interesting transformations.

A non-singular linear transformation J over $\mathfrak{F}$ of $\mathfrak{A}$ is called an *$\mathfrak{F}$-involution* or, briefly, an *involution* of $\mathfrak{A}$, if

$$J^2 = I, \qquad (ab)^J = b^J a^J \tag{1}$$

for every a and b of $\mathfrak{A}$, and if such a J exists $\mathfrak{A}$ is then called a *J-involutorial algebra.* Such linear transformations were studied in Chapter AV where $\mathfrak{A}$ was a total matric algebra, and we shall proceed to obtain the general properties which inspired that treatment. We first have the elementary

Theorem 1. *An involution J of an algebra $\mathfrak{A}$ is an automorphism over $\mathfrak{F}$ of $\mathfrak{A}$ if and only if $\mathfrak{A}$ is a commutative algebra. In this case J has order one or two.*

For $(ab)^J = b^J a^J = a^J b^J$ for every a and b of $\mathfrak{A}$ if and only if $ab = ba$, $\mathfrak{A}$ is a commutative algebra.

We next obtain

Theorem 2. *Let $\mathfrak{C}$ be the centrum of an algebra $\mathfrak{A}$. Then the correspondence*

$$c \leftrightarrow c^J \tag{2}$$

is an involution (automorphism over $\mathfrak{F}$) of $\mathfrak{C}$.

For $\mathfrak{C}$ is the set of all quantities c such that $ca = ac$ for every a of $\mathfrak{A}$. Hence $ca^J = a^J c$, $ac^J = c^J a$, c^J is in $\mathfrak{C}$ for every c of $\mathfrak{C}$. Clearly (2) is an automorphism over $\mathfrak{F}$ of $\mathfrak{C}$.

Theorem 3. *The product $S = TJ$ of any two involutions of $\mathfrak{A}$ is an automorphism over $\mathfrak{F}$ of $\mathfrak{A}$.*

For TJ is in $\mathfrak{G}_{\mathfrak{F}}$, $(ab)^{TJ} = (b^T a^T)^J = a^{TJ} b^{TJ}$.

2. The J-symmetric and J-skew quantities. A quantity s of $\mathfrak{A}$ is called *J-symmetric* if $s = s^J$. It is clear that the set $\mathfrak{S}_J(\mathfrak{A})$ consisting of all J-symmetric quantities of $\mathfrak{A}$ is a linear subset over $\mathfrak{F}$ of $\mathfrak{A}$. This set contains $a + a^J$,

aa^J, a^Ja for every a of $\mathfrak{A}$. It also contains the product ab of every a and b in it if and only if $ab = ba$. For, $(ab)^J = b^Ja^J = ba$. We state these results as

Theorem 4. *The set $\mathfrak{S}_J(\mathfrak{A})$ is a subalgebra of $\mathfrak{A}$ if and only if all J-symmetric quantities of $\mathfrak{A}$ are commutative with one another.*

As immediate corollaries we have

Theorem 5. *If $\mathfrak{A} = \mathfrak{S}_J(\mathfrak{A})$ the algebra $\mathfrak{A}$ is a commutative algebra.*

Theorem 6. *Let $\mathfrak{C}$ be the centrum of $\mathfrak{A}$. Then $\mathfrak{S}_J(\mathfrak{C})$ is a subalgebra of $\mathfrak{C}$.*

A quantity t of $\mathfrak{A}$ is called *J-skew* if $t^J = -t$. The set $\mathfrak{T}_J(\mathfrak{A})$ of all J-skew quantities of $\mathfrak{A}$ is a linear subset over $\mathfrak{F}$ of $\mathfrak{A}$. It contains $a - a^J$ for every a of $\mathfrak{A}$. If a and b are J-symmetric then $ab + ba$ is J-symmetric, $ab - ba$ is J-skew. If $a = a^J$, $b = -b^J$, then $ab + ba$ is J-skew, $ab - ba$ is J-symmetric. Similarly $a = -a^J$, $b = -b^J$ imply that $ab + ba$ is J-symmetric, $ab - ba$ is J-skew. Our definitions also imply

Theorem 7. *The product ab of commutative quantities a and b is J-symmetric if a and b are either both J-symmetric or both J-skew. Also $ab = ba$ is J-skew if one of a and b is J-symmetric, the other J-skew.*

If the characteristic of $\mathfrak{F}$ is two the sets $\mathfrak{S}_J(\mathfrak{A})$ and $\mathfrak{T}_J(\mathfrak{A})$ are the same since $-1 = 1$. In the contrary case we have

Theorem 8. *Let $\mathfrak{F}$ have characteristic not two and $\mathfrak{A}$ be a J-involutorial algebra over $\mathfrak{F}$. Then $\mathfrak{A}$ is the supplementary sum*

$$\mathfrak{A} = \mathfrak{S}_J(\mathfrak{A}) + \mathfrak{T}_J(\mathfrak{A}). \tag{3}$$

For $a = \frac{1}{2}(a + a^J) + \frac{1}{2}(a - a^J) = s + t$ with $s = \frac{1}{2}(a + a^J)$ in $\mathfrak{S}_J(\mathfrak{A})$, t in $\mathfrak{T}_J(\mathfrak{A})$. The sum is supplementary since if s is in $\mathfrak{S}_J(\mathfrak{A})$ and in $\mathfrak{T}_J(\mathfrak{A})$ we have $s^J = s = -s$, $2s = 0$, $s = 0$.

The result above may be regarded as a generalization of the theorem stating that every matrix A over $\mathfrak{F}$ of characteristic not two is uniquely expressible as the sum of a symmetric and a skew matrix. There is an analogous theorem stating that every complex matrix is uniquely expressible as the sum $A_1 + A_2i$ with $i^2 = -1$, A_1 and A_2 Hermitian matrices, and we shall obtain the following generalization of that result.

Theorem 9. *Let $\mathfrak{A}$ be J-involutorial over $\mathfrak{F}$ of characteristic not two and $\mathfrak{C}$ be the centrum of $\mathfrak{A}$. Then if $\mathfrak{C}$ contains a regular J-skew quantity $q = -q^J$ the set $\mathfrak{T}_J(\mathfrak{A}) = q\mathfrak{S}_J(\mathfrak{A})$, q^2 is in $\mathfrak{S}_J(\mathfrak{A})$,*

$$\mathfrak{A} = \mathfrak{S}_J(\mathfrak{A}) + q\mathfrak{S}_J(\mathfrak{A}). \tag{4}$$

For $(q^{-1})^J = (q^J)^{-1} = (-q)^{-1} = -q^{-1}$, q^{-1} is J-skew. Now q^{-1} is in $\mathfrak{C}$, $q^{-1}b = bq^{-1}$. By Theorem 10.7 if $b^J = -b$ then $a = q^{-1}b = a^J$, $b = qa$ is in $q\mathfrak{S}_J(\mathfrak{A})$. Conversely if $a = a^J$ and $b = qa$ then $b^J = aq^J = -qa = -b$ is in $\mathfrak{T}_J(\mathfrak{A})$. Also $(q^2)^J = (-q)^2 = q^2$. We obtain (4) from (3).

The result of Theorem 10.9 has no apparent analogue in the case where two is the characteristic of $\mathfrak{F}$. But we shall obtain a generalization of Theorem 10.9 for $\mathfrak{F}$ of arbitrary characteristic in the case where $\mathfrak{C}$ is a field.

3. The two types of involutions. Involutions fall into two distinct types which we define as follows.

DEFINITION. *Let $\mathfrak{A}$ have an involution J and $\mathfrak{C}$ be the centrum of $\mathfrak{A}$. Then we shall say that $\mathfrak{A}$ is J-involutorial of the first or second kind (J is an involution of the first or second kind) according as $\mathfrak{S}_J(\mathfrak{C}) = \mathfrak{C}$ or $\mathfrak{S}_J(\mathfrak{C}) \neq \mathfrak{C}$.*

We shall be interested particularly in the case where $\mathfrak{A}$ is a simple algebra and so shall take a step in the direction of such algebras by assuming that henceforth the centrum of $\mathfrak{A}$ is a field $\mathfrak{K}$.

Every quantity $q \neq 0$ of a field $\mathfrak{K}$ is regular. Moreover the subset $\mathfrak{S}$ of all J-symmetric quantities of $\mathfrak{K}$ is a subfield over $\mathfrak{F}$ of $\mathfrak{K}$ by Theorem 10.6. Hence either $\mathfrak{A}$ is J-involutorial of the first kind or, if $\mathfrak{F}$ has characteristic not two, $\mathfrak{K}$ contains a quantity q as in Theorem 10.9. We may then prove

Theorem 10. *Let the centrum of a J-involutorial algebra $\mathfrak{A}$ of the second kind be a field $\mathfrak{K}$ so that $\mathfrak{K} > \mathfrak{S}$, $\mathfrak{S}$ is the subfield over $\mathfrak{F}$ of all J-symmetric quantities of $\mathfrak{K}$. Then $\mathfrak{K} = \mathfrak{S}(\theta)$ is a separable quadratic field over $\mathfrak{S}$ such that*

$$\theta^J = 1 - \theta, \qquad \theta^2 - \theta = \beta \text{ in } \mathfrak{S}, \tag{5}$$

$$\mathfrak{A} = \mathfrak{S}_J(\mathfrak{A}) + \theta\mathfrak{S}_J(\mathfrak{A}) = (u_1, \cdots, u_n) \text{ over } \mathfrak{K} \tag{6}$$

with $u_i = u_i^J$ in $\mathfrak{S}_J(\mathfrak{A})$. Moreover, if $\mathfrak{F}$ has characteristic not two we may replace θ in (6) by $q = \theta - \frac{1}{2}$ and obtain $\mathfrak{K} = \mathfrak{S}(q)$,*

$$q^J = -q, \qquad q^2 = \alpha \text{ in } \mathfrak{S}. \tag{7}$$

If k is in $\mathfrak{K}$ and not in $\mathfrak{S}$ the equation $f(\lambda) = \lambda^2 - (k + k^J)\lambda + kk^J$ has coefficients in $\mathfrak{S}$ and $\mathfrak{S}(k)$ has degree two over $\mathfrak{S}$. Now $f(\lambda)$ is inseparable only if $\mathfrak{F}$ has characteristic two, $k = -k^J = k^J$ is in $\mathfrak{S}$, a contradiction. Hence $\mathfrak{K}$ is separable over $\mathfrak{S}$, $\mathfrak{K} = \mathfrak{S}(k)$ for some k, $\mathfrak{K}$ is quadratic. By Lemma 9.8 we have $\mathfrak{K} = \mathfrak{S}(\theta)$ with (5) holding. If a is in $\mathfrak{A}$ we define

$$s = (a - a^J)(2\theta - 1)^{-1}, \tag{8}$$

which exists for every a of $\mathfrak{A}$ since θ is not in $\mathfrak{S}$, $2\theta - 1 \neq 0$. Then $a^J = a - s(2\theta - 1)$, $s^J = (a^J - a)(2 - 2\theta - 1)^{-1} = s$, $s_0 = a - s\theta$, $s_0^J = a^J - s(1 - \theta) = a - s(2\theta - 1) - s(1 - \theta) = a - s\theta = s_0$. Hence $\mathfrak{A} = (\mathfrak{S}_J(\mathfrak{A}), \theta\mathfrak{S}_J(\mathfrak{A}))$. If $s_2 = \theta s_1 = s_2^J$, $s_1 = s_1^J$, then $\theta s_1 = (\theta s_1)^J = (1 - \theta)s_1$, $(1 - 2\theta)s_1 = 0$, $s_1 = 0 = s_2$. Hence $\mathfrak{S}_J(\mathfrak{A})$ and $\theta\mathfrak{S}_J(\mathfrak{A})$ are supplementary

* While it is desirable, in general, to obtain forms of our results not distinguishing the case where $\mathfrak{F}$ has characteristic two from that where it does not, there are some proofs where it is much better to use the simplifications possible in each of the two cases. We therefore make this and similar additions.

and we have (6). The result (7) is derivable either as a restatement of Theorem 10.9 or by putting $q = \theta - \frac{1}{2}$.

4. Involutions over $\mathfrak{S}$ of a simple algebra. We assume henceforth that $\mathfrak{A}$ is a simple algebra over $\mathfrak{F}$. Then the centrum $\mathfrak{K}$ of $\mathfrak{A}$ is always a field and the subfield $\mathfrak{S}_J(\mathfrak{K})$ of all J-symmetric quantities of $\mathfrak{K}$ is uniquely determined by J. We now make the

DEFINITION. *Let $\mathfrak{S}$ be a subfield over $\mathfrak{F}$ of the centrum $\mathfrak{K}$ of a simple algebra $\mathfrak{A}$. Then we call an involution J of $\mathfrak{A}$ an involution over $\mathfrak{S}$ of $\mathfrak{A}$ if $\mathfrak{S}_J(\mathfrak{K}) = \mathfrak{S}$, that is, $k = k^J$ for k in $\mathfrak{K}$ if and only if k is in $\mathfrak{S}$.*

The result of Theorem 10.10 now implies that we may limit our study of the existence of involutions over $\mathfrak{S}$ of a normal simple algebra $\mathfrak{A}$ over $\mathfrak{K}$ to the discussion of the case where $\mathfrak{K}$ has an automorphism C over $\mathfrak{S}$ as follows: Either $\mathfrak{K} = \mathfrak{S}$, C is the identity automorphism I, or $\mathfrak{K} = \mathfrak{S}(\theta)$ is a separable quadratic extension over the set $\mathfrak{S}$ consisting of all quantities of $\mathfrak{K}$ unaltered by C, $C^2 = I$, $\theta^2 - \theta$ is in $\mathfrak{S}$,

$$\theta^C = 1 - \theta = \theta^J. \tag{9}$$

We shall adopt these notations here and henceforth.

If T is an involution over $\mathfrak{S}$ of $\mathfrak{A}$ then we have seen that necessarily $k^T = k^C$ for every k of $\mathfrak{K}$. If also J is an involution over $\mathfrak{S}$ we have $k^J = k^C = k^T$, $C^2 = I$ gives $k^{TJ} = k$. Combining this result with that of Theorem 10.3 we have

LEMMA 1. *Let T and J be involutions over $\mathfrak{S}$ of $\mathfrak{A}$ over $\mathfrak{F}$. Then TJ is an automorphism over the centrum $\mathfrak{K}$ of $\mathfrak{A}$.*

Complete information on the relation between any two involutions over the same $\mathfrak{S}$ of $\mathfrak{A}$ is now given by

Theorem 11. *Let $\mathfrak{K}$ be the centrum of a simple algebra $\mathfrak{A}$ and T be an involution over $\mathfrak{S}$ of $\mathfrak{A}$. Then a self-correspondence $a \to a^J$ is an involution J over $\mathfrak{S}$ of $\mathfrak{A}$ if and only if there exists a regular quantity $y = \pm y^T$ in $\mathfrak{A}$ such that*

$$a^J = y^{-1}a^Ty \qquad (a \text{ in } \mathfrak{A}). \tag{10}$$

For the correspondence S given by $a \leftrightarrow y^{-1}ay$ is an automorphism of $\mathfrak{A}$ over $\mathfrak{K}$. If $y = \pm y^T$ then the resulting J of (10) is clearly the product $J = TS$ and hence is a non-singular linear transformation over $\mathfrak{S}$ of $\mathfrak{A}$. Now $a^{S^{-1}} = yay^{-1}$, $a^{ST} = (y^{-1}ay)^T = y^Ta^T(y^T)^{-1} = ya^Ty^{-1} = a^{TS^{-1}}$, $ST = TS^{-1}$. Then $J^2 = TSTS = T^2S^{-1}S = I$. Also $(ab)^J = (ab)^{ST} = (a^Sb^S)^T = b^{ST}a^{ST}$, so that J is an involution of $\mathfrak{A}$. It is an involution over $\mathfrak{S}$ since k^T is in $\mathfrak{K}$, $k^{TS} = (k^T)^S = k^T$ for every k of $\mathfrak{K}$, S is an automorphism over $\mathfrak{K}$.

Conversely let J be an involution over $\mathfrak{S}$. By Lemma 10.1 and Theorem 4.5 $S = TJ$ is an inner automorphism of $\mathfrak{A}$, $a^S = g_0^{-1}ag_0$ for g_0 a regular quantity of $\mathfrak{A}$. Then $a^T = a^{SJ} = (g_0^J)a^J(g_0^J)^{-1}$, $a^J = g^{-1}a^Tg$ where $g = g_0^J$ is regular. We apply J to $a^J = g^{-1}a^Tg$ and obtain $a = g^J(a^T)^J(g^J)^{-1} = g^{-1}g^Tgg^{-1}a^{T^2}g(g^{-1}g^Tg)^{-1} =$

$(g^{-1}g^T)a(g^{-1}g^T)^{-1}$ for every a of $\mathfrak{A}$. Then $\gamma = g^{-1}g^T$ is in the centrum $\mathfrak{K}$ of $\mathfrak{A}$, $g^T = \gamma g$. If $\gamma = -1$ the quantity $y = g = -g^T$ has the property we desire. Otherwise $y = g + g^T = (1 + \gamma)g$ is a T-symmetric quantity and is regular, $y^{-1}a^Ty = g^{-1}a^Tg = a^J$ as desired.

It is clear that if J is any involution over $\mathfrak{S}$ of $\mathfrak{A}$ defined by a quantity y of $\mathfrak{A}$ satisfying (10) the multiples αy of y by non-zero quantities α of $\mathfrak{K}$ have the property $(\alpha y)^{-1}a^T(\alpha y) = a^J$ and define the same involution J of $\mathfrak{A}$ as y. Conversely if $a^J = y^{-1}a^Ty = y_0^{-1}a^Ty_0$ for every a of $\mathfrak{A}$ then $y_0y^{-1}a^T = a^Ty_0y^{-1}$ for every a of $\mathfrak{A}$. But $a = (a^T)^T$, $y_0y^{-1}a = ay_0y^{-1}$ for every a of $\mathfrak{A}$, $y_0y^{-1} = \alpha$ in $\mathfrak{K}$, $y_0 = \alpha y$. We have shown that the quantity y of (10) is uniquely determined by J up to a non-zero factor in $\mathfrak{K}$.

Two involutions J and J_0 are called *cogredient* if there exists an automorphism S over $\mathfrak{K}$ of $\mathfrak{A}$ such that $J_0 = S^{-1}JS$. Then S is an inner automorphism of $\mathfrak{A}$ over $\mathfrak{K}$ and $a^S = z^{-1}az$ for a regular quantity z of $\mathfrak{A}$. But then $a^{S^{-1}} = zaz^{-1}$,

$$a^{J_0} = z^{-1}y^{-1}(zaz^{-1})^Tyz = (z^Tyz)^{-1}a^T(z^Tyz) = y_0^{-1}a^Ty_0\,, \tag{11}$$

where $y_0 = z^Tyz$. The argument above shows that then two involutions J and J_0 over $\mathfrak{S}$ are cogredient if and only if the defining y_0 is T-congruent to a multiple of y by a quantity in the centrum.

The automorphisms S of an algebra may be thought of as replacing any fixed representation a of its abstract arbitrary quantity by another representation a^S. Now J_0 is the involution $a^S \leftrightarrow (a^S)^{J_0} = a^{SS^{-1}JS} = (a^J)^S$. Thus *cogredient involutions are essentially merely different representations of the same abstract involution.*

5. Involutions of a direct product. The reader may verify the truth of the following result as in Section 6.6.

LEMMA 2. *Let $\mathfrak{B}$ and $\mathfrak{C}$ be normal simple algebras over $\mathfrak{K}$ with respective involutions J and J_0 over $\mathfrak{S}$. Then the correspondence in $\mathfrak{A} = \mathfrak{B} \times \mathfrak{C}$ induced by*

$$b \leftrightarrow b^J, \qquad c \leftrightarrow c^{J_0}, \tag{12}$$

for every b of $\mathfrak{B}$ and c of $\mathfrak{C}$, is an involution over $\mathfrak{S}$ of $\mathfrak{A}$.

The converse of the lemma above stating that the existence of involutions over $\mathfrak{S}$ of $\mathfrak{A}$ implies the existence of involutions in its direct factors $\mathfrak{B}$ and $\mathfrak{C}$ is also true in certain cases which we now consider.

Let $\mathfrak{M} = (e_{ij}\,;\, i, j = 1, \cdots, m)$ be a total matric algebra over a field $\mathfrak{K}$ with the notations we assumed at the beginning of Section 10.4. Then $\mathfrak{M}$ has an involution T over $\mathfrak{S}$ given by

$$\sum_{i,j=1}^{m} k_{ij}e_{ij} \leftrightarrow \sum_{i,j=1}^{m} k_{ij}^C e_{ji}\,, \tag{13}$$

for all k_{ij} of $\mathfrak{K}$. Now ($\mathfrak{M}$ over $\mathfrak{K}$) $= \mathfrak{K} \times$ ($\mathfrak{M}$ over $\mathfrak{S}$), and $e_{ij}^T = e_{ji}$. The involution T in $\mathfrak{M}$ over $\mathfrak{S}$ may thus be represented as the usual operation of transposition on the set of all m-rowed square matrices with elements in $\mathfrak{S}$. We now prove

Theorem 12. *Let $\mathfrak{A}$ be normal simple over $\mathfrak{K}$ so that $\mathfrak{A} = \mathfrak{M} \times \mathfrak{D}$ where $\mathfrak{D}$ is a normal division algebra, $\mathfrak{M}$ is a total matric algebra. Then $\mathfrak{A}$ has an involution T over $\mathfrak{S}$ if and only if there exists an involution over $\mathfrak{S}$ of $\mathfrak{D}$. In this case $\mathfrak{A}$ has an involution J over $\mathfrak{S}$ such that $\mathfrak{D}$ is J-involutorial and $e_{ij}^J = e_{ji}$.*

For let $f_{ij} = e_{ji}^T$ and obtain an algebra $\mathfrak{M}_0 = (f_{ij}\,; i, j = 1, \cdots, m)$ equivalent over $\mathfrak{K}$ to $\mathfrak{M}$ under the correspondence $e_{ij} \leftrightarrow f_{ij}$. Then $\mathfrak{A} = \mathfrak{M}_0 \times \mathfrak{A}^{\mathfrak{M}_0}$ and by Theorem 4.14 there exists an inner automorphism S of $\mathfrak{A}$ such that $f_{ij} = e_{ij}^S$. Hence there exists a regular quantity g of $\mathfrak{A}$ such that $e_{ij} = g^{-1}f_{ij}g$. Then $e_{ji} = g^{-1}e_{ij}^T g$, $e_{ij}^T = g e_{ji} g^{-1}$, $e_{ij} = (g^T)^{-1}e_{ji}^T g^T = (g^T)^{-1}g e_{ij}(g^{-1}g^T)$. It follows that $d = g^{-1}g^T$ is in the $\mathfrak{A}$-commutator $\mathfrak{D}$ of $\mathfrak{M}$ and that either $g = -g^T$ and we write $y = g$, or else $y = g + g^T = g(1 + d) = y^T$ is a regular quantity of $\mathfrak{A}$. For $\mathfrak{D}$ is a division algebra, $1 + d \neq 0$ is regular in $\mathfrak{D}$, g is regular. But in either case $y e_{ji} y^{-1} = g e_{ji} g^{-1} = e_{ij}^T$ since $(1 + d)e_{ji}(1 + d)^{-1} = e_{ji}$. Hence $e_{ji} = y^{-1}e_{ij}^T y$, and the involution J of Theorem 10.11 defined by $a^J = y^{-1}a^T y$ has the property $e_{ij}^J = e_{ji}$ as desired. If d is in $\mathfrak{D}$ then $de_{ji} = e_{ji}d$, $de_{ij}^J = e_{ij}^J d$, $e_{ij}d^J = d^J e_{ij}$ for all e_{ij}, d^J is in the $\mathfrak{A}$-commutator $\mathfrak{D}$ of $\mathfrak{M}$. Hence J induces the involution $d \leftrightarrow d^J$ of $\mathfrak{D}$.

The result above may be used to prove the following generalization.

Theorem 13. *Let $\mathfrak{M} = (e_{ij}\,; i, j = 1, \cdots, m)$ be a total matric subalgebra of a normal simple algebra $\mathfrak{A}$ over $\mathfrak{K}$ with the same unity quantity as $\mathfrak{A}$ so that $\mathfrak{A} = \mathfrak{M} \times \mathfrak{B}$, $\mathfrak{B} = \mathfrak{A}^{\mathfrak{M}}$. Then if $\mathfrak{A}$ has an involution over $\mathfrak{S}$ it has an involution J such that $e_{ij}^J = e_{ji}$ and $\mathfrak{B}$ is J-involutorial. Thus $\mathfrak{A}$ has an involution over $\mathfrak{S}$ if and only if $\mathfrak{B}$ does for any $\mathfrak{B} \sim \mathfrak{A}$.*

For $\mathfrak{A} = \mathfrak{M}_0 \times \mathfrak{D}$, $\mathfrak{B} = \mathfrak{M}_1 \times \mathfrak{D}$, $\mathfrak{M}_0 = \mathfrak{M} \times \mathfrak{M}_1$. Now $\mathfrak{D}$ has an involution J_0 by Theorem 10.12 and hence by Lemma 10.2 so does $\mathfrak{B}$. Using (13) and Lemma 10.2 we obtain our result.

We may also prove

Theorem 14. *Let $\mathfrak{D} = \mathfrak{D}_1 \times \mathfrak{D}_2$ be a normal division algebra over $\mathfrak{K}$ where $\mathfrak{D}_1$ and $\mathfrak{D}_2$ have relatively prime degrees. Then $\mathfrak{D}$ has an involution over $\mathfrak{S}$ if and only if $\mathfrak{D}_1$ and $\mathfrak{D}_2$ have involutions over $\mathfrak{S}$.*

For let ρ_1 and ρ_2 be the respective exponents of $\mathfrak{D}_1$ and $\mathfrak{D}_2$ so that, by Theorem 5.17, they are relatively prime integers. Then $a\rho_1 + b\rho_2 = 1$ for integers a and b, $q = b\rho_2 \equiv 1 \pmod{\rho_1}$. Now $\mathfrak{D}^q = \mathfrak{D}_1^q \times \mathfrak{D}_2^q = \mathfrak{D}_1 \times \mathfrak{M}$ where $\mathfrak{M} \sim 1$. By Lemma 10.2 if $\mathfrak{D}$ has an involution over $\mathfrak{S}$, so does $\mathfrak{D}^q$ and hence, by Theorem 10.12, so does $\mathfrak{D}_1$. Similarly $\mathfrak{D}_2$ has an involution over $\mathfrak{S}$. The converse follows from Lemma 10.2.

Theorem 10.12 reduces the problem of determining all involutorial simple algebras to that of determining all involutorial division algebras $\mathfrak{D}$, and Theorem 10.14 reduces the latter problem to that where $\mathfrak{D}$ has prime-power degree. This latter reduction will not be of great importance in our study although the theorem itself will be useful. Conversely, if a division algebra $\mathfrak{D}$ over its centrum $\mathfrak{K}$ is given, the problem of the existence of an involution over $\mathfrak{S}$ of $\mathfrak{D}$

is reduced to the solution of the same question for any desired $\mathfrak{A} \sim \mathfrak{D}$. We shall find this property of great use in our determination theory.

6. The construction of involutions. The question arises as to how an involution may be constructed under the hypothesis of its existence. We shall consider a normal simple algebra $\mathfrak{A}$ of degree n over $\mathfrak{K}$ and may easily answer this question in terms of the regular representation of $\mathfrak{A}$. Then $\mathfrak{A}$ is a subalgebra of the algebra $\mathfrak{M}_{n^2} = (E_{ij}; i, j = 1, \cdots, n^2)$ over $\mathfrak{K}$, where E_{ij} is an n^2-rowed square matrix with elements in $\mathfrak{K}$, and

$$\mathfrak{M}_{n^2} = \mathfrak{A} \times \mathfrak{A}^{-1}. \tag{14}$$

The algebra $\mathfrak{M}_{n^2}$ has the involution

$$T: \qquad a = \sum_{i,j=1}^{n^2} k_{ij} E_{ij} \leftrightarrow a^T = \sum_{i,j=1}^{n^2} k_{ij}^C E_{ji}.$$

If $\mathfrak{A}$ has an involution J over $\mathfrak{S}$ it is clear that $\mathfrak{A}^{-1}$ has a corresponding involution J_0 over $\mathfrak{S}$ and, by Lemma 10.2, $\mathfrak{M}_{n^2}$ has an involution J_1 over $\mathfrak{S}$ such that

$$a^{J_1} = a^J, \qquad b^{J_1} = b^{J_0} \tag{15}$$

for every a of $\mathfrak{A}$ and b of $\mathfrak{A}^{-1}$. By Theorem 10.11 there exists a non-singular matrix

$$y = \pm y^T, \tag{16}$$

such that

$$a^J = y^{-1} a^T y. \tag{17}$$

The quantity a^T is either the transpose or the conjugate (with respect to C) transpose of the matrix a, and we have now demonstrated that *an involution over $\mathfrak{S}$ of $\mathfrak{A}$ exists if and only if there exists a T-symmetric (symmetric or Hermitian in the respective cases) or a T-skew (skew or skew-Hermitian) matrix y carrying a^T into a quantity of $\mathfrak{A}$, where a ranges over all quantities of $\mathfrak{A}$.*

7. J-symmetric subfields. In considering normal simple algebras $\mathfrak{A}$ over $\mathfrak{K}$ we discussed the structure of $\mathfrak{A}$ in terms of properties defined relative to maximal separable subfields $\mathfrak{Z}$ over $\mathfrak{K}$ of special types. We shall do this also for involutorial algebras, but shall now require also that $\mathfrak{Z} = \mathfrak{K}(x)$ with x J-symmetric. The fundamental result which will allow us to make such a restriction on x is given by

Theorem 15. *Let $\mathfrak{A}$ be normal simple over $\mathfrak{K}$, $\mathfrak{Z} = \mathfrak{K}(x)$ be a separable subfield of $\mathfrak{A}$, and $\mathfrak{A}$ have an involution over $\mathfrak{S}$. Then x is J-symmetric with respect to some involution J over $\mathfrak{S}$ of $\mathfrak{A}$ if and only if the minimum function $\phi(\lambda)$ of x has coefficients in $\mathfrak{S}$.*

For $\phi(\lambda) = \lambda^m + k_1\lambda^{m-1} + \cdots + k_m$ with k_i in $\mathfrak{K}$ and $\phi(x) = 0 = [\phi(x)]^J$. Now $k^J = k^C$ for every k of $\mathfrak{K}$, and if $x = x^J$ then

$$\psi(x) = \phi(x) - [\phi(x)]^J = \sum_{i=1}^{m} (k_i - k_i^J) x^{m-i} = 0.$$

The definition of $\phi(\lambda)$ implies that $\psi(\lambda) \equiv 0$, the $k_i = k_i^J$ are in $\mathfrak{S}$. Conversely, let the k_i be all in $\mathfrak{S}$ so that $[\phi(x)]^J = \phi(x^J)$. The $\mathfrak{A}$-commutator of $\mathfrak{Z}$ has the form $\mathfrak{M}_0 \times \mathfrak{G}$ where $\mathfrak{G}$ is a normal division algebra over $\mathfrak{Z}$, and $\mathfrak{A} = \mathfrak{M}_0 \times \mathfrak{M}_1 \times \mathfrak{D}$ with $\mathfrak{M}_0 \sim \mathfrak{M}_1 \sim 1$, $\mathfrak{B} = \mathfrak{M}_1 \times \mathfrak{D}$ contains $\mathfrak{Z}$, $\mathfrak{D}$ is a normal division algebra over $\mathfrak{K}$, $\mathfrak{G}$ is the $\mathfrak{B}$-commutator of $\mathfrak{Z}$. By Theorem 10.13 $\mathfrak{B}$ has an involution T over $\mathfrak{S}$. Now $\phi(x) = \phi(x^T) = 0$ and $\mathfrak{K}(x)$ is equivalent over $\mathfrak{K}$ to $\mathfrak{K}(x^T)$ such that $x \leftrightarrow x^T$. By Theorem 4.14 there exists a regular quantity g of $\mathfrak{B}$ such that $x^T = gxg^{-1}$, $x = (g^T)^{-1}gxg^{-1}g^T = \beta^{-1}x\beta$, where $\beta = g^{-1}g^T$ is then in $\mathfrak{G}$. If $\beta = -1$ then $g^T = -g$ and we write $y = g$. Otherwise $y = g + g^T = g + g\beta = g(1 + \beta)$ is a regular quantity of $\mathfrak{B}$, and since $1 + \beta$ is in $\mathfrak{G}$, $yxy^{-1} = gxg^{-1}$. We have proved the existence of a regular quantity y such that $y^T = \pm y$, $x = y^{-1}x^Ty$. By Theorem 10.11 the involution $b \leftrightarrow b^{J_0} = y^{-1}b^Ty$ is an involution over $\mathfrak{S}$ of $\mathfrak{B}$ such that $x^{J_0} = x$. The corresponding involution J in Lemma 10.2 of $\mathfrak{M}_0 \times \mathfrak{B}$ has the desired property $x^J = x$.

8. Involutorial crossed products. We have shown that every normal simple algebra $\mathfrak{A}$ has a normal splitting field $\mathfrak{Z}$, and thus that $\mathfrak{A}$ is similar to a crossed product $(\mathfrak{Z}, \mathfrak{a})$. This result is not adequate for our theory of crossed products and we shall require the existence of a splitting field which is the direct product

$$\mathfrak{Z} = \mathfrak{X} \times \mathfrak{K} \tag{18}$$

over $\mathfrak{S}$, where $\mathfrak{X}$ is normal over $\mathfrak{S}$. Note that in view of Theorem 5.1 this is no restriction for involutorial algebras of the first kind (where $\mathfrak{K} = \mathfrak{S}$). But it is an assumption otherwise. We may prove this assumption valid under very mild restrictions on the reference field $\mathfrak{F}$ basic in our discussion, and this was done in [43]. However the question as to the existence in general of such splitting fields would require a much more elaborate treatment than it seems desirable to give in these LECTURES. We shall then content ourselves with only the more important cases to be considered in connection with the source from which the theory of involutorial algebras arose.

Let $\mathfrak{G} = (S_1, S_2, \cdots, S_n)$, $S_1 = I$, be the automorphism group over $\mathfrak{S}$ of the field $\mathfrak{X}$ and we shall later use S, T, U, and so on, for these automorphisms as in Chapter V. The field $\mathfrak{Z}$ is the direct product of two normal fields $\mathfrak{X}$ and $\mathfrak{K}$ over $\mathfrak{S}$, and the group of $\mathfrak{Z}$ over $\mathfrak{S}$ is the direct product $\mathfrak{G} \times \mathfrak{H}$, where $\mathfrak{H} = (I)$ or (I, C) is the group of $\mathfrak{K}$ over $\mathfrak{S}$. If S and T range over all automorphisms of $\mathfrak{G}$, and

$$\mathfrak{a} = \{a_{S,T}\} \tag{19}$$

is a factor set of $\mathfrak{Z}$ over $\mathfrak{K}$, then we shall define a second (conjugate) factor set

$$\mathfrak{a}^C = \{a_{S,T}^C\}. \tag{20}$$

Evidently the product

$$\mathfrak{b} = \mathfrak{a}\mathfrak{a}^C = \{a_{S,T}a_{S,T}^C\} \tag{21}$$

is a factor set of $\mathfrak{X}$ *over* $\mathfrak{S}$. Thus the crossed product $\mathfrak{A} = (\mathfrak{Z}, \mathfrak{a})$ over $\mathfrak{K}$ defines a corresponding algebra $\mathfrak{A}^C = (\mathfrak{Z}, \mathfrak{a}^C)$ and

$$\mathfrak{A} \times \mathfrak{A}^C \sim (\mathfrak{Z}, \mathfrak{a}\mathfrak{a}^C) = \mathfrak{B} \times \mathfrak{K}, \tag{22}$$

where

$$\mathfrak{B} = (\mathfrak{X}, \mathfrak{b}) \text{ over } \mathfrak{S}. \tag{23}$$

We shall prove

Theorem 16. *The crossed product* $\mathfrak{A} = (\mathfrak{Z}, \mathfrak{a})$ *defined above has an* $\mathfrak{F}$*-involution over* $\mathfrak{S}$ *if and only if the algebra* $\mathfrak{B}$ *of* (23), (21) *is a total matric algebra, that is,*

$$\mathfrak{B} = (\mathfrak{X}, \mathfrak{b}) \sim 1. \tag{24}$$

For let $\mathfrak{A}$ have an involution over $\mathfrak{S}$. The minimum function of a properly chosen generating quantity x_1 of $\mathfrak{Z} = \mathfrak{K}(x_1)$ has coefficients in $\mathfrak{S}$ since $\mathfrak{X}_{\mathfrak{K}} = \mathfrak{X} \times \mathfrak{K} = \mathfrak{Z}$, and we choose x_1 so that $\mathfrak{X} = \mathfrak{S}(x_1)$. We may thus choose an involution J over $\mathfrak{S}$ of $\mathfrak{A}$ such that $x = x^J$ for every x of $\mathfrak{X}$. Now x^S is in $\mathfrak{X}$ for every x of $\mathfrak{X}$ and $(x^S)^J = x^S$. The algebra $\mathfrak{A}$ is the supplementary sum of its subsets $u_S\mathfrak{Z}$, S in $\mathfrak{G}$, such that $zu_S = u_Sz^S$. Then $(x_0u_S)^J = u_S^Jx_0 = (u_Sx_0^S)^J = x_0^Su_S^J$, $u_S^Jx_0^{S^{-1}} = x_0u_S^J$ for every x_0 of $\mathfrak{X}$. Then $x_0u_S^Ju_S = u_S^Jx_0^{S^{-1}}u_S = u_S^Ju_Sx_0$. Since $\mathfrak{Z} = \mathfrak{K}(x_1)$ is a maximal subfield of $\mathfrak{A}$ the quantity $u_S^Ju_S$ is in $\mathfrak{Z}$. But $u_S^Ju_S$ is J-symmetric and hence in $\mathfrak{X}$, so is $c_S = (u_S^Ju_S)^{-1} = u_S^{-1}(u_S^J)^{-1}$,

$$v_S = (u_S^J)^{-1} = u_Sc_S\,. \tag{25}$$

The condition $u_Su_T = u_{ST}a_{S,T}$ implies that $u_T^Ju_S^J = a_{S,T}^Ju_{ST}^J = a_{S,T}^Cu_{ST}^J$. Then $(u_S^J)^{-1}(u_T^J)^{-1} = (u_{ST}^J)^{-1}(a_{S,T}^C)^{-1}$, $v_Sv_T = v_{ST}(a_{S,T}^C)^{-1}$. Thus $u_Sc_Su_Tc_T = u_{ST}c_{ST}(a_{S,T}^C)^{-1} = u_Su_Tc_S^Tc_T = u_{ST}a_{S,T}c_S^Tc_T$, and we have proved that there exist quantities c_S in $\mathfrak{X}$, for every S of $\mathfrak{G}$, such that

$$b_{S,T} = a_{S,T}a_{S,T}^C = \frac{c_{ST}}{c_Tc_S^T}. \tag{26}$$

But then the factor set $\mathfrak{b} = \{b_{S,T}\}$ of $\mathfrak{X}$ is associated with the unit set, $\mathfrak{B} \sim 1$ as desired.

Conversely, if $\mathfrak{B} \sim 1$ we have (26) for quantities c_S in $\mathfrak{X}$. Define a correspondence J in $\mathfrak{A}$ by

$$(\sum_S u_Sz_S)^J = \sum_S z_S^C(u_Sc_S)^{-1} \qquad (z \text{ in } \mathfrak{Z}). \tag{27}$$

By Section 5.8 we may assume that $u_I = c_I = 1$, and then see that $z^J = z^C$ for every z of $\mathfrak{Z}$, $x^J = x$ for every x of $\mathfrak{X}$. Clearly J is a linear transformation of $\mathfrak{A}$ over $\mathfrak{F}$. Since $u_Su_T = u_{ST}a_{S,T}$ we see that

$$u_Sc_Su_Tc_T = u_{ST}c_{ST}g_{S,T},\ g_{S,T} = a_{S,T}\frac{c_S^Tc_T}{c_{ST}} = (a_{S,T}^C)^{-1}.$$

Now $(u_Sc_S)^{-1} = u_S^J$, so that the first equation above becomes $(u_S^J)^{-1}(u_T^J)^{-1} = (u_{ST}^J)^{-1}(a_{S,T}^C)^{-1}$, $u_T^Ju_S^J = a_{S,T}^Cu_{ST}^J$. For every z of $\mathfrak{Z}$ we have $(u_Sz)^J = z^C(u_Sc_S)^{-1} =$

$z^C u_S^J$, so that $(u_S u_T)^J = (u_{ST} a_{S,T})^J = a_{S,T}^C u_{ST}^J$, and we have proved $(u_S u_T)^J = u_T^J u_S^J$. That $(ab)^J = b^J a^J$ follows now by a simple computation and the fact that $zu_S c_S = u_S c_S z^S$ so that $u_S^J z = z^S u_S$. Hence $(aa^{-1})^J = (a^{-1})^J a^J = 1^J = 1$, $(a^J)^{-1} = (a^{-1})^J$ for every a of $\mathfrak{A}$. Now $u_S^{J^2} = (c_S^{-1} u_S^{-1})^J = (u_S^J)^{-1} c_S^{-1} = u_S c_S c_S^{-1} = u_S$, and it follows readily that $J^2 = I$. Thus J is an involution over $\mathfrak{S}$ of $\mathfrak{A}$.

It is interesting to determine all involutions J over $\mathfrak{S}$ of the algebra $\mathfrak{A}$ above such that $x = x^J$ for every x in $\mathfrak{X}$. The determination is given in

Theorem 17. *Let a set of quantities c_S be a fixed solution of* (26) *and J_0 be the corresponding involution over $\mathfrak{S}$ of $\mathfrak{A}$ defined by* (27), *that is, generated by $u_S^{J_0} = (u_S c_S)^{-1}$. Then every involution J over $\mathfrak{S}$ of $\mathfrak{A}$, such that the quantities of $\mathfrak{X}$ are all J-symmetric, is generated by $u_S^J = (u_S d_S)^{-1}$ where*

$$d_S = (t^S)^{-1} t c_S \tag{28}$$

is the general solution of (26), *t ranges over all quantities of $\mathfrak{X}$.*

Our theorem thus gives a determination of all solutions of the c_S by means of which a given factor set of the type in (21) is associated with the unit set in terms of a fixed solution. Observe that, in view of our proof of Theorem 10.16, the quantities d_S do satisfy (26) and we require only the derivation of (28). Now by Theorem 10.11 there exists a regular quantity $t = \pm t^J$ in $\mathfrak{A}$ such that $a^J = t^{-1} a^{J_0} t$ for every a of $\mathfrak{A}$. Then $x^J = x = t^{-1} x^{J_0} t = t^{-1} x t$ implies that t is in $\mathfrak{Z}$. Thus $t = \pm t^J = \pm t^C$. If t is not in $\mathfrak{X}$ then $t^C = -t \neq t$, the characteristic of $\mathfrak{S}$ is not two and $\mathfrak{K} = \mathfrak{S}(q)$, $q^C = -q$, $tq = (tq)^J$ is in $\mathfrak{X}$, $a^J = (tq)^{-1} a^{J_0} (tq)$. Hence we may assume that t is in $\mathfrak{X}$. But then $(u_S^J)^{-1} = u_S d_S = t^{-1} (u_S^{J_0})^{-1} t = t^{-1} u_S t c_S = u_S (t^S)^{-1} t c_S$, d_S satisfies (28).

We are naturally interested particularly in the case where the field $\mathfrak{X}$ of Theorem 10.16 is cyclic over $\mathfrak{S}$ with generating automorphism S. Then we have

Theorem 18. *Let $\mathfrak{X}$ be cyclic over $\mathfrak{S}$ with generating automorphism S, $\mathfrak{K} > \mathfrak{S}$, and $\mathfrak{Z} = \mathfrak{X} \times \mathfrak{K}$ be a field so that the algebra $\mathfrak{A}$ of Theorem* 10.16 *is the cyclic algebra*

$$\mathfrak{A} = (\mathfrak{Z}, S, \gamma) \qquad (\gamma \text{ in } \mathfrak{K}). \tag{29}$$

Then $\mathfrak{A}$ has an involution S if and only if

$$N_{\mathfrak{K}|\mathfrak{S}}(\gamma) = N_{\mathfrak{X}|\mathfrak{S}}(g) \tag{30}$$

for g in $\mathfrak{X}$.

For, the algebra $\mathfrak{B}$ of Theorem 10.16 is the algebra $(\mathfrak{X}, S, \gamma\gamma^C) \sim 1$ if and only if $\gamma\gamma^C = N_{\mathfrak{X}|\mathfrak{S}}(g)$. But the norm $N_{\mathfrak{K}|\mathfrak{S}}(\gamma) = \gamma\gamma^C$ by definition and the assumption that $\mathfrak{K} > \mathfrak{S}$. We shall treat the case $\mathfrak{K} = \mathfrak{S}$ completely in the next section.

9. Involutorial simple algebras of the first kind. If $\mathfrak{A}$ is a normal simple algebra over $\mathfrak{K}$ and $\mathfrak{A}$ has exponent one or two then $\mathfrak{A}^2 \sim 1$, that is, $\mathfrak{A}$ and $\mathfrak{A}^{-1}$

are equivalent. But then $\mathfrak{A}$ is self-reciprocal, that is, there exists a linear transformation L over $\mathfrak{K}$ of $\mathfrak{A}$ such that $(ab)^L = b^L a^L$. Our definition states that L is an involution over $\mathfrak{K}$ of $\mathfrak{A}$ if and only if L^2 is the identity automorphism of $\mathfrak{A}$. This is clearly not necessarily true of an arbitrary L. We have seen in Section 10.5 that if $\mathfrak{A}$ has exponent one it does have an involution over $\mathfrak{K}$. But we shall also show by Theorem 10.16 that the existence of an L above is equivalent to the existence of an involution J over $\mathfrak{K}$ even in the remaining case. We state the result as

Theorem 19. *A normal simple algebra $\mathfrak{A}$ over $\mathfrak{K}$ is involutorial of the first kind if and only if $\mathfrak{A}$ has exponent one or two.*

For if $\mathfrak{A}$ has an involution J over $\mathfrak{K}$ then the correspondence $a \leftrightarrow a^J$ is a reciprocal correspondence of $\mathfrak{A}$, $\mathfrak{A} \cong \mathfrak{A}^{-1}$, $\mathfrak{A}^2 \cong \mathfrak{A} \times \mathfrak{A}^{-1} \sim 1$, $\mathfrak{A}$ has exponent one or two. Conversely let $\mathfrak{A}^2 \sim 1$. Then there exists a crossed product $\mathfrak{A}_0 \sim \mathfrak{A}$ and $\mathfrak{A}_0^2 \sim 1$. But since $\mathfrak{K} = \mathfrak{S}$ the hypotheses of Theorem 10.16 are fulfilled and $\mathfrak{A}_0$ has an involution J, so does $\mathfrak{A}$ by Theorem 10.13.

This result together with the result of Theorem 9.32 on the exponent of a rational division algebra implies

Theorem 20. *Let $\mathfrak{K}$ be an algebraic field of finite degree over the field $\mathfrak{R}$ of all rational numbers. Then the involutorial normal division algebras of the first kind over $\mathfrak{K}$ are $\mathfrak{K}$ and the quaternion division algebras, that is, normal division algebras of degree one or two.*

Note that for cyclic algebras of Theorem 10.18 with $\mathfrak{K} = \mathfrak{S}$ the condition that $\mathfrak{A}$ be involutorial is γ^2 a norm, and this is simply $\gamma\gamma^C = \gamma^2 = N_{\mathfrak{X}|\mathfrak{K}}(g)$. This result holds trivially for all quaternion algebras since $\mathfrak{X}$ has degree two.

We now pass to a study of certain algebras of the second kind.

10. Involutorial quaternion algebras of the second kind. The quaternion division algebras over $\mathfrak{K} = \mathfrak{S}(\theta) \neq \mathfrak{S}$ which have an involution over $\mathfrak{S}$ have a rather remarkable, if elementary, property. We shall actually prove

Theorem 21. *A quaternion division algebra $\mathfrak{A}$ over $\mathfrak{K} > \mathfrak{S}$ has an involution J over $\mathfrak{S}$ if and only if $\mathfrak{A} = \mathfrak{Q} \times \mathfrak{K}$ where $\mathfrak{Q}$ is a quaternion division algebra over $\mathfrak{S}$.*

For if $\mathfrak{A} = \mathfrak{Q} \times \mathfrak{K}$ then $\mathfrak{Q}$ has an involution by Theorem 10.19 and so does $\mathfrak{A}$ by Lemma 10.2. Conversely, let $\mathfrak{A}$ have an involution J over $\mathfrak{S}$. By Theorem 10.10 $\mathfrak{A} = (1, u_2, u_3, u_4)$ over $\mathfrak{K}$ with $u_i = u_i^J$. Every quantity a of $\mathfrak{A}$ is a root of an equation

$$x^2 - T(a)x + N = 0,$$

where $T(a)$ is the reduced trace, that is, the trace in the matrix representation of $\mathfrak{A}$ by an algebra of two-rowed square matrices with elements in a splitting field of $\mathfrak{A}$. The function $T(a)$ is linear over $\mathfrak{K}$ and $T(\lambda_1 + \lambda_2 u_2 + \lambda_3 u_3 + \lambda_4 u_4) = \lambda_1 T(1) + \lambda_2 T(u_2) + \lambda_3 T(u_3) + \lambda_4 T(u_4)$. Now $\mathfrak{A}$ contains an inseparable field over $\mathfrak{K}$ only if $\mathfrak{K}$ has characteristic two, $T(1) = 0$. But by Theorem 4.18 $\mathfrak{A}$

contains a separable quadratic field $\mathfrak{K}(x)$ over $\mathfrak{K}$, $T(x) \neq 0$, and thus $T(u_i) \neq 0$ for some i when $T(1) = 0$. When the characteristic of $\mathfrak{K}$ is not two, each u_i generates a separable quadratic field over $\mathfrak{K}$. For any characteristic, then, a quantity $x_0 = u_i$ generates a separable field over $\mathfrak{K}$, and since $x_0 = x_0^J$ we have $\mathfrak{S}(x_0)$ separable of degree two over $\mathfrak{S}$ by Theorem 10.15. Now $\mathfrak{A}$ is a cyclic algebra $\mathfrak{A} = (1, x_0, y, yx_0)$, $y^2 = \gamma$ in $\mathfrak{K}$, $zy = yz^S$ for every z of $\mathfrak{Z} = \mathfrak{K}(x_0)$. Also x_0^S is in $\mathfrak{S}(x_0)$, $x_0^S = x_0^{SJ}$, $(x_0y)^J = y^Jx_0 = (yx_0^S)^J = x_0^Sy^J$, so that $yy^Jx_0 = x_0yy^J$, yy^J is in $\mathfrak{K}(x_0)$ and is J-symmetric. Hence $yy^J = x_2$ is in $\mathfrak{X} = \mathfrak{S}(x_0)$. Then $y^J = y\gamma^{-1}x_2$, $y_0 = y + y^J = y(1 + \gamma^{-1}x_2)$ is J-symmetric and has the property $zy_0 = y_0z^S$ for every z of $\mathfrak{Z}$. If $y_0 \neq 0$ it follows that $y_0^2 = g \neq 0$ in $\mathfrak{S}$, and that if $\mathfrak{Q} = (\mathfrak{X}, S, g)$ then $\mathfrak{A} = \mathfrak{Q} \times \mathfrak{K}$. Otherwise $y^J = -y$, $(y^J)^2 = \gamma^J = (-y)^2 = \gamma$ is in S. Hence the algebra $\mathfrak{Q} = (\mathfrak{X}, S, \gamma)$ is again an algebra over $\mathfrak{S}$, $\mathfrak{A} = \mathfrak{Q} \times \mathfrak{K}$.

11. Involutorial simple algebras over an algebraic number field. The problem of determining all involutorial simple algebras $\mathfrak{A}$ over an algebraic field of finite degree over $\mathfrak{R}$ was reduced in Theorem 10.12 to the case where $\mathfrak{A}$ is a normal division algebra $\mathfrak{D}$ over $\mathfrak{K}$ of finite degree over $\mathfrak{R}$. When $\mathfrak{D}$ is of the first kind we proved that $\mathfrak{D}$ has degree two over $\mathfrak{K}$ in Theorem 10.20, and conversely that every normal division algebra $\mathfrak{D}$ of degree two over $\mathfrak{K}$ has an involution J over $\mathfrak{K}$. There remains the case where $\mathfrak{D}$ is of the second kind, and we complete our determination by proving

Theorem 22. *A normal division algebra $\mathfrak{D}$ of degree n over a field $\mathfrak{K} = \mathfrak{S}(\theta)$ of degree two over an algebraic extension $\mathfrak{S}$ of finite degree over $\mathfrak{R}$ has an involution over $\mathfrak{S}$ if and only if $\mathfrak{D}$ is a cyclic algebra of Theorem* 10.18 *satisfying* (30).

For we saw in Chapter IX that there is no loss of generality if we take $\mathfrak{K}$ to be a field of ordinary algebraic numbers and that the index of $\mathfrak{D}_{\mathfrak{K}_V}$ over $\mathfrak{K}_V$ is unity for all except a finite number of valuations $V_1, \cdots, V_r$ of $\mathfrak{S}$. Apply Lemma 9.10 to construct a cyclic field $\mathfrak{X}$ of degree n over $\mathfrak{S}$ such that the degree of $\mathfrak{X}_{V_i}$ over $\mathfrak{S}_{V_i}$ is n for every non-archimedean V_i and that either n is odd or $\mathfrak{X}_{V_i}$ is analytically equivalent to the field of all complex numbers for every archimedean V_i. We let $\mathfrak{Z}$ be the composite of $\mathfrak{X}$ and $\mathfrak{K}$. If n is odd then $\mathfrak{D}_{\mathfrak{K}_V} \sim 1$ for every archimedean V, the field $\mathfrak{Z} = \mathfrak{X} \times \mathfrak{K}$, $\mathfrak{Z}_{V_i} = \mathfrak{X}_{V_i} \times \mathfrak{K}_{V_i}$ has degree n over $\mathfrak{K}_{V_i}$ for every non-archimedean V_i and splits $\mathfrak{D}_{\mathfrak{K}_{V_i}}$, $\mathfrak{Z}_V$ splits $\mathfrak{D}_{\mathfrak{K}_V}$ for every V of $\mathfrak{S}$. Hence $\mathfrak{Z}$ splits $\mathfrak{D}$. Now let n be even, $n = 2^e n_0 = 2\nu$, where n_0 is odd. If the index of $\mathfrak{D}_{\mathfrak{K}_{V_i}}$ over $\mathfrak{K}_{V_i}$ divides ν then $\mathfrak{Z}_{V_i}$ splits $\mathfrak{D}_{\mathfrak{K}_{V_i}}$ since the degree of $\mathfrak{Z}_{V_i}$ over $\mathfrak{K}_{V_i}$ is either n or ν. Otherwise the index of $\mathfrak{D}_{\mathfrak{K}_{V_i}}$ over $\mathfrak{K}_{V_i}$ is $2^e m_0$ where m_0 divides n_0, the index of $(\mathfrak{D}^\nu)_{\mathfrak{K}_{V_i}}$ is two. However $\mathfrak{D}^\nu \sim \mathfrak{B}$, a normal division algebra of degree two over $\mathfrak{K}$ with an involution J over $\mathfrak{S}$ by Theorem 10.13 if $\mathfrak{D}$ has an involution over $\mathfrak{S}$. Then by Theorem 10.21 $\mathfrak{B} = \mathfrak{K} \times \mathfrak{Q}$, where $\mathfrak{Q}$ is a normal division algebra over $\mathfrak{S}$, $\mathfrak{B}_{\mathfrak{K}_{V_i}} = \mathfrak{K}_{V_i} \times \mathfrak{Q}_{\mathfrak{S}_{V_i}}$ is a division algebra since $(\mathfrak{D}^\nu)_{\mathfrak{K}_{V_i}} \sim \mathfrak{B}_{\mathfrak{K}_{V_i}}$. The degree of $\mathfrak{K}_{V_i}$ over $\mathfrak{S}_{V_i}$ is one or two and cannot be two since then $\mathfrak{K}_{V_i}$ would split $\mathfrak{Q}_{\mathfrak{S}_{V_i}}$ by

Theorem 9.29. Hence $\mathfrak{K}_{V_i} = \mathfrak{S}_{V_i}$, $\mathfrak{X}_{V_i}$ of degree n over $\mathfrak{S}_{V_i}$ is the composite $\mathfrak{Z}_{V_i}$, $\mathfrak{Z}_{V_i}$ splits $\mathfrak{D}_{\mathfrak{K}_{V_i}}$. Then $\mathfrak{Z}_V$ splits $\mathfrak{D}_V$ for every V, $\mathfrak{Z}$ splits $\mathfrak{D}$. The degree ν or n of $\mathfrak{Z}$ over $\mathfrak{K}$ cannot be less than the index n of $\mathfrak{D}$ over $\mathfrak{K}$ so that $\mathfrak{Z}$ has degree n over $\mathfrak{K}$, $\mathfrak{Z} = \mathfrak{X} \times \mathfrak{K}$. Our result follows from Theorem 10.18.

12. Total real and pure imaginary fields. We return now to the fields $\mathfrak{F}$, $\mathcal{R}$ and $\mathcal{C}$ of Section 8.13 where we defined generalized Riemann matrices Ω, and, to avoid confusion in notation, designate by $\mathfrak{F}_0$ the (real) subfield of $\mathfrak{F}$ contained in $\mathcal{R}$. Before considering the problem of determining the multiplication algebra $\mathfrak{A}$ of Ω we shall consider some properties of certain types of fields occurring in this theory.

Since $\mathfrak{F}_0$ is real it is non-modular and every algebraic extension $\mathfrak{Z}$ of finite degree of $\mathfrak{F}_0$ is a simple extension $\mathfrak{F}_0(A)$. We shall call such an extension *total real* if the minimum function

$$g(\lambda) = \lambda^t + a_1\lambda^{t-1} + \cdots + a_t \qquad (a_i \text{ in } \mathfrak{F}_0) \tag{31}$$

of A factors into linear factors,

$$g(\lambda) = (\lambda - \alpha_1) \cdots (\lambda - \alpha_t) \tag{32}$$

in $\mathcal{R}$, that is, with the α_i in $\mathcal{R}$. This property is clearly independent of the particular generating quantity A used in its definition. We shall be particularly interested in the case where A is a ν-rowed square matrix with elements in $\mathfrak{F}$ and clearly have, by Theorem A5.17,

LEMMA 3. *Let A be similar to a Hermitian matrix. Then $\mathfrak{F}_0(A)$ is total real.*

A quantity $\mu(A)$ of a total real field $\mathfrak{F}_0(A)$ is called *total positive* if $\mu(\alpha_i) > 0$ for every root α_i of the (irreducible) minimum function $g(\lambda)$ of A. We call $\mu(A)$ *total negative* if $\mu(\alpha_i) < 0$ for $i = 1, \cdots, t$. Then we call $\mathfrak{F}_0(A)$ a *total pure imaginary* field over $\mathfrak{F}_0$ if $\mathfrak{F}_0(A^2)$ is total real and A^2 is a total negative quantity of $\mathfrak{F}_0(A^2)$. Evidently, by Theorem A5.20 we have

LEMMA 4. *Let A be similar to a skew-Hermitian matrix. Then $\mathfrak{F}_0(A)$ is total pure imaginary.*

Let A now be any ν-rowed square matrix with elements in $\mathfrak{F}$ and with irreducible minimum function $g(\lambda)$ of (31) such that (32) holds with α_i now in $\mathcal{C}$. Then $\nu = tq$ and, by Exercise 4 of page A82, we may take

$$A = \begin{pmatrix} 0 & I_q & 0 & \cdots & 0 \\ 0 & 0 & I_q & \cdots & 0 \\ \cdot & \cdot & \cdot & \cdots & \cdot \\ -a_tI_q & -a_{t-1}I_q & -a_{t-2}I_q & \cdots & -a_1I_q \end{pmatrix}. \tag{33}$$

Moreover the computation of Section A10.8 shows that if

$$V = (\alpha_j^{i-1}I_q) \qquad (i, j = 1, \cdots, t). \tag{34}$$

then

$$V^{-1}AV = A_0 = \text{diag}\,\{\alpha_1 I_q, \cdots, \alpha_t I_q\}. \tag{35}$$

We now prove

LEMMA 5. *Let $\psi(A) \neq 0$ be in $\mathfrak{F}[A]$. Then $\mathfrak{F}[\lambda, \mu]$ contains a polynomial $\Psi(\lambda, \mu)$ of degree at most $t - 1$ in each of the indeterminates λ, μ such that*

$$\Psi(\alpha_i, \alpha_i) = \psi(\alpha_i), \qquad \Psi(\alpha_i, \alpha_j) = 0 \qquad (i \neq j;\ i, j = 1, \cdots, t). \tag{36}$$

For $g(\lambda) = (\lambda - \alpha_j)\gamma(\lambda, \alpha_j)$ where $\gamma(\lambda, \alpha_j)$ is a polynomial of $\mathfrak{F}[\lambda, \alpha_j]$ of degree $t - 1$ in λ such that $\gamma(\alpha_j, \alpha_j) = \gamma_1(\alpha_j) \neq 0$ is in $\mathfrak{F}(\alpha_j)$. We put

$$\Psi(\lambda, \alpha_j) = [\gamma_1(\alpha_j)]^{-1}[\psi(\alpha_j)][\gamma(\lambda, \alpha_j)].$$

The degree of $\Psi(\lambda, \alpha_j)$ in λ is $t - 1$ and its coefficients are in $\mathfrak{F}[\alpha_j]$ of degree t over $\mathfrak{F}$ so that we may write $\Psi(\lambda, \alpha_j) = \sum_{i,k=1}^{t} \lambda^{i-1} a_{ik} \alpha_j^{k-1}$ such that the a_{ik} are in $\mathfrak{F}$ and define $\Psi(\lambda, \mu) = \sum_{i,k=1}^{t} \lambda^{i} a_{ik} \mu^{k-1}$. But then $\gamma(\alpha_i, \alpha_j) = 0$ for $i \neq j$, $\Psi(\alpha_i, \alpha_j) = 0$ for $i \neq j$,

$$\Psi(\alpha_j, \alpha_j) = [\gamma_1(\alpha_j)]^{-1}\gamma_1(\alpha_j)\psi(\alpha_j) = \psi(\alpha_j)$$

as desired.

We now let $\mathfrak{F}_0(A)$ be total pure imaginary of degree $t = 2t_0$ over $\mathfrak{F}_0$, $A^2 = B$ be total negative. Then the characteristic roots α_i of (32), (34) have the form $\alpha_j^2 = -\beta_j$ where $\beta_j > 0$ is in $\mathfrak{R}$ and the β_j are the characteristic roots of $-B$. But then we may take the characteristic roots of A to be $\alpha_1 \cdots \alpha_{t_0}, \bar{\alpha}_1, \cdots, \bar{\alpha}_{t_0}$, and write $\alpha_{j+t_0} = \bar{\alpha}_j$, $\alpha_{t+1} = \alpha_1$. In this case we prove

LEMMA 6. *Let $\psi(A) \neq 0$ be in $\mathfrak{F}[A]$. Then there exists a polynomial $\Psi(\lambda, \mu)$ of degree at most $t - 1$ in the indeterminates λ, μ and with coefficients in $\mathfrak{F}$ such that*

$$\Psi(\alpha_j, \bar{\alpha}_j) = \psi(\alpha_j), \qquad \Psi(\alpha_j, \alpha_k) = 0 \tag{37}$$

for $j, k = 1, \cdots, t$ and every $\alpha_k \neq \bar{\alpha}_j$.

For we define $\gamma(\lambda, \alpha_k)$ as in the proof of Lemma 10.5 and use the facts that the a_j of (31) are in $\mathfrak{F}_0$, $\bar{\alpha}_k = -\alpha_k$ in $\mathfrak{F}_0(\alpha_k)$. Put

$$\Psi(\lambda, \bar{\alpha}_k) = [\gamma_1(\alpha_k)]^{-1}[\psi(\alpha_k)][\gamma(\lambda, \alpha_k)], \tag{38}$$

where we clearly have replaced α_k by $-\bar{\alpha}_k$ in the corresponding polynomial of Lemma 10.5. This gives the desired result.

As a consequence of our two lemmas we have

LEMMA 7. *Let $\mathfrak{F}_0(A)$ be either total real or total pure imaginary so that A may be taken to have the form (33), and let G be a matrix with elements in $\mathfrak{F}$,*

$$\bar{V}'GV = (G_{jk}) \qquad (j, k = 1, \cdots, t) \tag{39}$$

for q-rowed square matrices G_{jk}. Then there exists a matrix G_0 with elements in $\mathfrak{F}$ such that $\bar{V}'G_0V = \text{diag}\,\{G_{11}, \cdots, G_{tt}\}$.

For we clearly have $G_{jk} = \psi(\alpha_j, \bar{\alpha}_k)$ for a polynomial $\psi(\lambda, \mu)$ of degree at most $t - 1$ in each of λ and μ and coefficients q-rowed square matrices. By

Lemmas 10.5 and 10.6 there exists a $\Psi(\lambda, \mu)$ such that $\Psi(\alpha_j, \bar{\alpha}_k) = 0$ if $k \neq j$, $\Psi(\alpha_j, \bar{\alpha}_j) = \psi(\alpha_j, \bar{\alpha}_j)$. But then if $\Psi(\lambda, \mu) = \sum_{k,l=1}^{t} \lambda^{k-1} H_{kl} \mu^{l-1}$ the matrix $G_0 = (H_{kl})$ is the desired matrix.

If $\mathfrak{F}_0(A)$ is as above the matrix

$$(40) \qquad T = V\bar{V}'$$

has symmetric functions of $\alpha_1, \cdots, \alpha_t$ for elements. Hence the elements of T are in $\mathfrak{F}_0$, since the coefficients of (32) are in $\mathfrak{F}_0$. Then for any matrix H we have

$$(41) \qquad V^{-1}HV = \bar{V}'GV, \qquad G = T^{-1}H,$$

so that G has elements in $\mathfrak{F}$ if and only if H does. Moreover it follows that $V^{-1}HV = (G_{jk})$ for $G_{jk} = \psi(\alpha_j, \bar{\alpha}_k)$. Now $HA = AH$ if and only if $V^{-1}HVA_0 = A_0V^{-1}HV$, $G_{jk}\alpha_k = \alpha_j G_{jk}$. Since the α_j are distinct this gives $G_{jk} = 0$ for $j \neq k$. In particular this implies

Lemma 8. *Let A be as above and in the multiplication algebra of a GR-matrix Ω over $\mathfrak{F}$ with principal matrix $C = \epsilon\bar{C}'$. Then*

$$(42) \qquad V^{-1}\Omega V = \text{diag}\{\Omega_1, \cdots, \Omega_t\}$$

or GR-matrices Ω_j over $\mathfrak{F}(\alpha_j)$ such that Ω_j has the same type number ϵ as Ω.

For $V^{-1}\Omega C(\bar{V}')^{-1}$ is positive definite if and only if $\Gamma = \Omega C$ is positive definite. But $V^{-1}\Omega V$ is commutative with A_0 and we have (42). It follows that $V^{-1}\Gamma(\bar{V}')^{-1} = V^{-1}\Omega V V^{-1} C(\bar{V}')^{-1} = (\Omega_j C_{jk})$, where

$$(43) \qquad V^{-1}C(\bar{V}')^{-1} = \bar{V}'(T^{-1}CT^{-1})V = (C_{jk}).$$

The principal minors $\Omega_j C_{jj}$ are positive definite, $C_{jj} = \epsilon\bar{C}'_{jj}$ has elements in $\mathfrak{F}(\alpha_j)$ and we have our desired result.

As a corollary of the above we may prove

Lemma 9. *Let A, C, and Ω be as in Lemma 10.8. Then there exists a principal matrix $C_0 = \bar{C}_0'$ of Ω such that $\bar{A}' = \delta C_0^{-1}AC_0$ where $\delta = \pm 1$ according as $\mathfrak{F}_0(A)$ is total real or total pure imaginary.*

For by the proof of Lemma 10.8 the matrices C_{jj} of (43) are non-singular, the matrix diag $\{C_{11}^{-1}, \cdots, C_{tt}^{-1}\}$ has the property $C_{jj}^{-1} = \psi(\alpha_j, \alpha_j)$ for a polynomial $\psi(x, y)$ with coefficients matrices with elements in $\mathfrak{F}$. But by Lemma 10.7 there exists a matrix G_0 with elements in $\mathfrak{F}$ such that $\bar{V}'G_0V = \text{diag}\{C_{11}^{-1}, \cdots, C_{tt}^{-1}\}$, G_0 is non-singular. Then $(\bar{V}'G_0V)^{-1} = V^{-1}C_0(\bar{V}')^{-1}$ for $C_0^{-1} = G_0$, $V^{-1}C_0(\bar{V}')^{-1} = \text{diag}\{C_{11}, \cdots, C_{tt}\}$, $V^{-1}\Omega C_0(\bar{V}')^{-1} = \text{diag}\{\Omega_1 C_{11}, \cdots, \Omega_t C_{tt}\}$ is positive definite. Hence so is ΩC_0, C_0 is a principal matrix of Ω. But $\bar{C}'_{jj} = \epsilon C_{jj}$, $\bar{C}_0' = \epsilon C_0$. Also $V^{-1}C_0(\bar{V}')^{-1}$ is commutative with $A_0 = \delta\bar{A}_0'$, where δ is defined as above. Then $V^{-1}C_0(\bar{V}')^{-1}A_0 = A_0V^{-1}C_0(\bar{V}')^{-1}$, $C_0(\bar{V}')^{-1}A_0 = AC_0(\bar{V}')^{-1}$. However $\bar{A}_0' = \bar{V}'\bar{A}'(\bar{V}')^{-1} = \delta A_0$, $C_0\delta(\bar{V}')^{-1}\bar{V}'(\bar{A}')(\bar{V}')^{-1} = AC_0(\bar{V}')^{-1}$, and thus $\delta C_0\bar{A}'C_0^{-1} = A$ as desired.

We shall use the property derived above in the proofs of two fundamental results for our investigation of the structure of multiplication algebras.

13. Special subfields of multiplication algebras. The first of the results just referred to will be stated as

Theorem 23. *Let* $\mathfrak{A}$ *be the multiplication algebra of a* ν*-rowed GR-matrix* Ω *over* $\mathfrak{F}$ *with principal matrix* $C = \epsilon\bar{C}'$, $\epsilon = \pm 1$. *Then the correspondence* J *defined by*

$$A^J = C\bar{A}'C^{-1} \qquad (A \text{ in } \mathfrak{A}) \tag{44}$$

is an involution of $\mathfrak{A}$ *and has the property that if* $A = A^J$ *the characteristic roots of* A *are all real, if* $A = -A^J$ *the characteristic roots of* A *are all pure imaginary. Moreover* $\alpha^J = \bar{\alpha}$ *for every* α *of* $\mathfrak{F}$.

For $A\Omega = \Omega A$, $\Gamma = \Omega C$ is positive definite. Then $\Omega C = \epsilon C\bar{\Omega}'$, $\bar{\Omega}' = \epsilon C^{-1}\Omega C$, $\bar{\Omega}'\bar{A}' = \bar{A}'\bar{\Omega}'$, $C^{-1}\Omega C\bar{A}' = \bar{A}'C^{-1}\Omega C$, and we have shown that $A^J\Omega = \Omega A^J$, where A^J is defined by (44) and is in $\mathfrak{A}$. The algebra $\mathfrak{A} \leq \mathfrak{M}_\nu$ and the correspondence $A \leftrightarrow A^J$ of (44) was shown in Theorem 10.11 to be an involution of $\mathfrak{M}_\nu$. But then A^J in $\mathfrak{A}$ implies that J is an involution of $\mathfrak{A}$. If $A = \alpha I_\nu$ for α in $\mathfrak{F}$ then $A^J = \bar{\alpha}I_\nu$, and if we agree as usual to identify $\mathfrak{F}$ with $\mathfrak{F}(I_\nu)$ we have $\bar{\alpha} = \alpha^J$. Now by Section A5.12 we have $\Gamma = \Delta\bar{\Delta}'$ for a non-singular complex matrix Δ. Then $C\bar{A}'C^{-1} = \delta A$ for $\delta = \pm 1$ implies that $\bar{A}' = \delta C^{-1}AC$, $A\Omega = A\Gamma C^{-1} = \Gamma C^{-1}A$, $A\Gamma = \Gamma C^{-1}AC = \delta\Gamma\bar{A}'$. We substitute $\Delta\bar{\Delta}'$ for Γ and have

$$A_0 = \Delta^{-1}A\Delta = \delta\bar{\Delta}'\bar{A}'(\bar{\Delta}')^{-1} = \delta\bar{A}_0'. \tag{45}$$

If $\delta = 1$ the matrix A_0 is Hermitian and similar to A, the characteristic roots of A are all real by Theorem A5.17. If $\delta = -1$ then iA_0 is Hermitian, A_0 and $A = \Delta A_0\Delta^{-1}$ have characteristic roots which are all pure imaginary.

The second of our results is a partial converse of Theorem 10.23. We state it as

Theorem 24. *Let* $\mathfrak{A}$ *and* Ω *be as in Theorem* 10.23, *and* A *be in* $\mathfrak{A}$ *such that* $\mathfrak{F}_0(A)$ *is either a total real or a total pure imaginary field over* $\mathfrak{F}_0$. *Then* Ω *has a principal matrix* $C^{(0)}$ *such that*

$$A^J = C^{(0)}\bar{A}'(C^{(0)})^{-1} = \pm A \tag{46}$$

in the respective cases.

For if $\mathfrak{Z} = \mathfrak{F}_0(A)$ is any subfield of $\mathfrak{A}$ there exists a non-singular matrix B with elements in $\mathfrak{F}$ such that $A_1 = BAB^{-1}$ has the form (33). Then $\Omega_0 = B\Omega B^{-1}$ is isomorphic to Ω and has A_1 in its multiplication algebra $\mathfrak{A}_0$ which is a (1-1) representation of an abstract algebra equivalent (by means of B) to $\mathfrak{A}$. Then by Lemma 10.9 Ω_0 has a principal matrix C_0 such that $C_0\bar{A}_1'C_0^{-1} = \delta A_1$. Hence

$$\delta B^{-1}A_1B = \delta A = [B^{-1}C_0(\bar{B}')^{-1}][\overline{(B^{-1}A_1B)}'][B^{-1}C_0(\bar{B}')^{-1}]^{-1} = C^{(0)}\bar{A}'(C^{(0)})^{-1},$$

where $C^{(0)} = B^{-1}C_0(\bar{B}')^{-1}$. Also $B^{-1}\Omega_0C_0(\bar{B}')^{-1} = \Omega C^{(0)}$ is positive definite. Then $C^{(0)}$ is a principal matrix of Ω, so that if G is in the multiplication algebra

$\mathfrak{A}$ of Ω the correspondence $G \leftrightarrow G^J = C^{(0)}\bar{G}'(C^{(0)})^{-1}$ is an involution J of $\mathfrak{A}$. But then $A^J = C^{(0)}\bar{A}'(C^{(0)})^{-1} = \delta A$ as desired. This proves our theorem.

We saw in Section 8.13 that the only case which we need consider is that where the multiplication algebra $\mathfrak{A}$ of Ω is a simple algebra over $\mathfrak{F}$. Let us then examine the restrictions imposed on the centrum $\mathfrak{K}$ of $\mathfrak{A}$ by the theorem above. We have already proved that $\mathfrak{K} = \mathfrak{S}$ or $\mathfrak{S}(\theta)$, $\mathfrak{S}$ consists of all quantities k of $\mathfrak{K}$ such that $k = k^J$, and, since $\mathfrak{K}$ is non-modular, $\theta^J = -\theta$. Then $\mathfrak{S}$ is a total real field over $\mathfrak{F}_0$, and if $\mathfrak{K} = \mathfrak{S}$ we must have $\mathfrak{F}_0 = \mathfrak{F}$. Thus the only case where $\mathfrak{A}$ is an algebra of the first kind is that where $\mathfrak{F}$ is a real field. When $\mathfrak{F}$ is not real then we may take $\mathfrak{F} = \mathfrak{F}_0(\rho^{\frac{1}{2}})$ for ρ a negative quantity of $\mathfrak{F}_0$ and then clearly $\theta\rho^{\frac{1}{2}}$ is in $\mathfrak{S}$. Hence $\mathfrak{K} = \mathfrak{S}(\rho^{\frac{1}{2}})$. However we may have $\mathfrak{F} = \mathfrak{F}_0$ real and still $\mathfrak{K} = \mathfrak{S}(\theta) > \mathfrak{S}$, $\rho = \theta^2$ a total negative quantity of $\mathfrak{S}$, $\mathfrak{K}$ total pure imaginary over $\mathfrak{F}_0$. We have proved

Theorem 25. *Let the multiplication algebra $\mathfrak{A}$ of a ν-rowed GR-matrix Ω over $\mathfrak{F}$ be a simple algebra with centrum $\mathfrak{K}$, and let $\mathfrak{F}_0$ be the subfield of all real quantities of $\mathfrak{F}$. Then $\mathfrak{A}$ is J-involutorial of the first kind with respect to the involution J of Theorem* 10.23 *if and only if $\mathfrak{F} = \mathfrak{F}_0$, $\mathfrak{K}$ is a total real extension of $\mathfrak{F}$. When $\mathfrak{A}$ is J-involutorial of the second kind the field $\mathfrak{K}$ is a total pure imaginary extension of $\mathfrak{F}_0$.*

14. The structure of multiplication algebras. The problem of determining the structure of the multiplication algebra $\mathfrak{A}$ over $\mathfrak{F}$ of any generalized Riemann matrix Ω over $\mathfrak{F}$ was reduced to the case where Ω is pure and $\mathfrak{A}$ is a division algebra $\mathfrak{D}$, and we saw in Theorem 10.25 that Theorem 10.23 imposes an additional condition on (the centrum of) $\mathfrak{D}$. We shall continue our investigations and shall determine a complete set of necessary structural conditions on $\mathfrak{D}$ (actually in terms of a crossed product similar to $\mathfrak{D}$). That our set is complete would be an obvious consequence of the result that conversely if $\mathfrak{D}$ is given such that the derived necessary conditions are satisfied then there exists a GR-matrix Ω with $\mathfrak{D}$ as multiplication algebra. The proof of this latter result has no place in our present study of the structure of algebras and we refer the reader to [43]. Let us observe that the partial converse, Theorem 10.24, of Theorem 10.23 is obviously a tool of importance in the proofs of these existence theorems.

However we shall see that Theorem 10.24 is of importance for our immediate problem of determining the structure of the multiplication algebra $\mathfrak{D}$ over $\mathfrak{F}$ of a pure GR-matrix Ω over $\mathfrak{F}$. To observe this let us begin with the study of the case where $\mathfrak{D}$ is of the second kind. Then the algebra $\mathfrak{D}$ is a normal division algebra of degree m, order $M = m^2$, over its centrum $\mathfrak{K} = \mathfrak{S}(\theta)$, where $\theta^2 = \rho$ is a total negative quantity of the total real field $\mathfrak{S}$. By Theorem 10.10 $\mathfrak{D} = (u_1, \cdots, u_M)$ over $\mathfrak{K}$ with $u_i = u_i^J$. We apply Theorem 1.15 and the comment following it to obtain quantities $\sigma_1, \cdots, \sigma_M$ in $\mathfrak{S}$ such that the subfield $\mathfrak{K}(s)$ generated by $s = \sigma_1 u_1 + \cdots + \sigma_M u_M$ is a field of degree m over $\mathfrak{K}$. Then clearly $s = s^J$, $\mathfrak{S}(s)$ is total real over $\mathfrak{S}$. But then we may choose s so that $\mathfrak{S}(s) = \mathfrak{F}_0(s)$ is total real over $\mathfrak{F}_0$. There exists a field $\mathfrak{X} = \mathfrak{S}(s_1, \cdots, s_m)$ in which the minimum function of s over $\mathfrak{K}$ factors into linear factors $\lambda - s_i$,

$\mathfrak{X} = \mathfrak{S}(x)$ is normal of degree $n = mq$ over $\mathfrak{S}$ and is clearly total real over $\mathfrak{F}_0$. Then $\mathfrak{Z} = \mathfrak{K}(x)$ splits $\mathfrak{D}$ and $\mathfrak{D} \sim \mathfrak{M}_q \times \mathfrak{D} = (\mathfrak{Z}, \mathfrak{a}) = \mathfrak{A}$ over $\mathfrak{K}$. Moreover $\mathfrak{A}$ is the multiplication algebra of the GR-matrix $\Omega_1 = \text{diag}\ \{\Omega, \cdots, \Omega\}$ with q terms Ω. Now we wish to be able to state that $\mathfrak{A}$ is J-involutorial with respect to an involution J of (44) defined by a principal matrix C_1 of Ω_1 and such that $x^J = x$. This statement is true in view of Theorem 10.24 and we shall prove

Theorem 26. *Let $\mathfrak{D}$ be the multiplication algebra of a pure GR-matrix Ω over $\mathfrak{F}$ such that $\mathfrak{D}$ is J-involutorial of the second kind. Then $\mathfrak{D}$ is similar to a crossed product $\mathfrak{A} = (\mathfrak{Z}, \mathfrak{a})$ of Theorem* 10.16 *where $\mathfrak{Z} = \mathfrak{X} \times \mathfrak{K}$, $\mathfrak{X}$ is a total real field over $\mathfrak{F}_0$, the solutions c_S of* (26) *are total positive.*

For we have already shown that $\mathfrak{D} \sim \mathfrak{A} = (\mathfrak{Z}, \mathfrak{a})$ where $\mathfrak{Z} = \mathfrak{K}(x)$ for a field $\mathfrak{X} = \mathfrak{F}_0(x)$ normal over $\mathfrak{S}$ and total real over both $\mathfrak{S}$ and $\mathfrak{F}_0$, and moreover that $\mathfrak{A}$ is the multiplication algebra of a GR-matrix Ω_1 with principal matrix C_1 such that $A \leftrightarrow A^J = C_1\bar{A}'C_1^{-1}$ is an involution of $\mathfrak{A}$, $x^J = x$, $\mathfrak{K} = \mathfrak{S}(\theta)$, $\theta^J = -\theta$, $s^J = s$ for every s of $\mathfrak{S}$. Since $\mathfrak{S}(\theta)$ is total pure imaginary the field $\mathfrak{X}$ over $\mathfrak{S}$ cannot contain $\mathfrak{K}$ as a subfield, $\mathfrak{Z} = \mathfrak{S}(\theta, x) = \mathfrak{K} \times \mathfrak{X}$. Now we showed in Theorem 10.16 that $\mathfrak{A} = u_{S_1}\mathfrak{Z} + \cdots + u_{S_n}\mathfrak{Z}$ such that $(u_S^J)^{-1} = u_S c_S$, $u_S u_T = u_{ST} a_{S,T}$ with $a_{S,T}$ in $\mathfrak{Z}$. We write $a_{S,T} = a_{S,T}^{(1)} + a_{S,T}^{(2)}\theta$ for quantities $a_{S,T}^{(j)}$ in $\mathfrak{X}$, and see that $a_{S,T}\bar{a}_{S,T} = [a_{S,T}^{(1)}]^2 - [a_{S,T}^{(2)}\theta]^2$ is a total positive quantity of $\mathfrak{X}$. Now $\Gamma_1 = \Omega_1 C_1$ is positive definite and so is $u_S\Gamma_1\bar{u}_S'$. By Theorem 10.16 and our definition of J we have $u_S^J = (u_S c_S)^{-1}$ for c_S in $\mathfrak{X}$, $u_S^J = (c_S^{S^{-1}} u_S)^{-1} = u_S^{-1}(c_S^{S^{-1}})^{-1} = C_1\bar{u}_S'C_1^{-1}$. Then $u_S u_S^J = (c_S^{S^{-1}})^{-1}$, $u_S\Gamma_1\bar{u}_S' = u_S\Omega_1(C_1\bar{u}_S'C_1^{-1})C_1 = u_S\Omega_1 u_S^J C_1 = u_S u_S^J \Gamma$ since u_S^J is in the multiplication algebra of Ω_1. We now write $\Gamma_1 = \Delta\bar{\Delta}'$ for a non-singular matrix Δ and have $G_S = \Delta^{-1}(u_S\Gamma_1\bar{u}_S')(\bar{\Delta}')^{-1} = \Delta^{-1}u_S u_S^J \Delta\bar{\Delta}'(\bar{\Delta}')^{-1} = \Delta^{-1}u_S u_S^J\Delta$, a positive definite Hermitian matrix. The characteristic roots of G_S are now all positive and this is then true of those of $u_S u_S^J = (c_S^{S^{-1}})^{-1}$. Hence $(c_S^{S^{-1}})^{-1}$ is a total positive quantity of $\mathfrak{X}$ and this must be true of $(c_S)^{-1} = [(c_S^{S^{-1}})^{-1}]^S$ and of c_S. We have proved our theorem.

We next study algebras of the first kind so that $\mathfrak{F}$ is real. We then have two results the first of which is

Theorem 27. *Let the multiplication algebra of a GR-matrix Ω over $\mathfrak{F}$ be a crossed product $\mathfrak{A} = (\mathfrak{Z}, \mathfrak{a})$ over its total real centrum $\mathfrak{S}$ over $\mathfrak{F}$, where $\mathfrak{Z}$ is total real over $\mathfrak{F}$. Then $\mathfrak{A}^2 \sim 1$, $\mathfrak{a} = \{a_{S,T}\}$ for quantities a_{ST} of $\mathfrak{Z}$ such that the equations*

$$a_{S,T}^2 = c_{ST}(c_S^T c_T)^{-1} \tag{47}$$

have total positive solutions c_S in $\mathfrak{Z}$.

The proof is exactly as in the case of Theorem 10.26 and we may thus turn to our final result on GR-matrices over an arbitrary real $\mathfrak{F}$ which we state as

Theorem 28. *Let $\mathfrak{D}$ be the multiplication algebra of a pure GR-matrix Ω over $\mathfrak{F}$ and let $\mathfrak{D}$ be J-involutorial of the first kind with respect to J of* (44) *so that $\mathfrak{F}$ is real. Then if $\mathfrak{D}$ is not similar to an algebra of Theorem* 10.27 *it has the form $\mathfrak{D} \sim \mathfrak{A} =$*

$\mathfrak{A}_1 \times \mathfrak{Q}$, *where* $\mathfrak{A}_1$ *is an algebra of Theorem* 10.27, *and*

$$(48) \qquad \mathfrak{Q} = (1, v, j, jv), \; vj = -jv, \; v^2 = j^2 = -1,$$

over $\mathfrak{F}$.

For let $\mathfrak{W} = \mathfrak{S}(x)$ be a J-symmetric subfield of maximal possible degree over the centrum $\mathfrak{S}$ of $\mathfrak{D}$, and let $\mathfrak{B}$ be the $\mathfrak{D}$-commutator of $\mathfrak{W}$, a normal division algebra over $\mathfrak{W}$. If b is in $\mathfrak{B}$ then $bx = xb$, $b^J x = xb^J$, b^J is in $\mathfrak{B}$. Now $\lambda^2 - (b + b^J)\lambda + bb^J$ has J-symmetric coefficients in $\mathfrak{D}^{\mathfrak{W}}$ and if any $b + b^J$ or bb^J were not in $\mathfrak{W}$ the corresponding field $\mathfrak{W}(b + b^J) > \mathfrak{W}$ or $\mathfrak{W}(bb^J) > \mathfrak{W}$ would be J-symmetric contrary to our definition of $\mathfrak{W}$. Hence either $\mathfrak{B} = \mathfrak{W}$ or $\mathfrak{B}$ has degree two over $\mathfrak{W}$. In the former case $\mathfrak{W}$ has degree m over $\mathfrak{S}$ where m is the degree of $\mathfrak{D}$ over $\mathfrak{S}$, $\mathfrak{W}$ is total real over $\mathfrak{F}$, the corresponding normal field $\mathfrak{Z} = \mathfrak{F}(x) = \mathfrak{S}(x)$ is total real over $\mathfrak{F}$ and $\mathfrak{D} \sim \mathfrak{A} = (\mathfrak{Z}, \mathfrak{a})$. In this case $\mathfrak{D} \sim \mathfrak{A}$ of Theorem 10.27. In the contrary case $\mathfrak{B}$ has degree two over $\mathfrak{W}$ and is a quaternion algebra

$$\mathfrak{B} = (1, v_1, j_1, j_1 v_1), \qquad v_1 j_1 = -j_1 v_1, \qquad v_1^2 = \alpha, \qquad j_1^2 = \beta.$$

Moreover the only J-symmetric quantities of $\mathfrak{B}$ are those of $\mathfrak{W}$ so that we may take $v_1 = x - x^J$ for some x of $\mathfrak{B}$, $v_1 = -v_1^J$. Also $j_1^J v_1^J = -v_1^J j_1^J$, $v_1 j_1^J = -j_1^J v_1$ and we may replace j_1 by $j_1 - j_1^J$ if necessary and hence take $j_1 = -j_1^J$. Then $\alpha = v_1^2$ and $\beta = j_1^2$ must be total negative quantities of $\mathfrak{W}$. We now define $\mathfrak{Q}$ as in our theorem and form $\mathfrak{A}_0 = \mathfrak{Q} \times \mathfrak{D}$. Let $\mathfrak{Z}$ be the field which is normal over $\mathfrak{F}$, total real over $\mathfrak{S}$, and contains a subfield $\mathfrak{Y} = \mathfrak{W}_0(\xi, \eta)$ such that $\mathfrak{W}_0$ is equivalent over $\mathfrak{S}$ to $\mathfrak{W}$ under a correspondence such that $\alpha_0 \leftrightarrow \alpha$, $\beta_0 \leftrightarrow \beta$ for α_0 and β_0 in $\mathfrak{W}_0$, $\xi^2 = -\alpha_0$, $\eta^2 = -\beta_0$. Consequently, we have $(\mathfrak{A}_0)_{\mathfrak{Z}} \geqq (\mathfrak{A}_0)_{\mathfrak{Y}} \sim \mathfrak{Q} \times \mathfrak{B}_0 \times \mathfrak{W}_0(\xi, \eta)$, where $\mathfrak{B}_0$ over $\mathfrak{W}_0$ is equivalent over $\mathfrak{S}$ to $\mathfrak{B}$ over $\mathfrak{W}$ But then if $v_1 \leftrightarrow v_0$ in $\mathfrak{B}_0$, $j_1 \leftrightarrow j_0$ in $\mathfrak{B}_0$ we have $v_0^2 = \alpha_0$, $j_0^2 = \beta_0$. It follows that the set $(\mathfrak{B}_0)_{\mathfrak{Y}} = (1, \xi^{-1} v_0, \eta^{-1} j_0, \xi^{-1}\eta^{-1} j_0 v_0)$ such that $(\xi^{-1} v_0)^2 = -1$, $(\eta^{-1} v_0)^2 = -1$. But then $\mathfrak{Q} \times \mathfrak{B}_{0\mathfrak{Z}} \sim 1$, $\mathfrak{Z}$ splits $\mathfrak{A}_0$. It follows that $\mathfrak{A}_0 \sim \mathfrak{A}_1 = (\mathfrak{Z}, \mathfrak{a})$, $\mathfrak{D} \sim \mathfrak{Q} \times \mathfrak{A}_1$. By Theorem 10.24 $\mathfrak{Q} \times \mathfrak{A}_1$ is the multiplication algebra of a GR-matrix Ω_1 with a principal matrix C_1 such that J is an involution $A \leftrightarrow A^J = C_1 \bar{A}' C_1^{-1}$ of $\mathfrak{Q} \times \mathfrak{A}_1$ with the property $u^J = -u$, $z^J = z$ for every z of $\mathfrak{Z}$. The field $\mathfrak{Z}(u)$ is clearly normal over $\mathfrak{S}$ and if L is the automorphism of $\mathfrak{Z}(u)$ replacing v by $-v$ we have $\mathfrak{Z}(u) = \mathfrak{S}(v) \times \mathfrak{Z}$, L commutative with all automorphisms S of $\mathfrak{Z}(v)$ induced by the automorphisms of $\mathfrak{Z}$ over $\mathfrak{S}$. Now we may take $u_L = j$, $u_L u_S = u_S u_L$, where clearly the u_S are in $\mathfrak{A}_1$. The algebra $\mathfrak{D}$ has exponent one or two and so does $\mathfrak{Q}$, so that so does $\mathfrak{A}_1$. Hence $u_S = (u_S c_S)^{-1}$ for c_S in $\mathfrak{Z}$ and satisfying (47), and S ranging over the automorphisms of $\mathfrak{Z}$. Now $\mathfrak{Q} \times \mathfrak{A}_1$ evidently has an involution T over $\mathfrak{S}$ with $v^T = -v$, $j^T = -j$, $z^T = z$, $u_S^T = (u_S c_S)^{-1}$. By Theorem 10.15 and Theorem 10.11,

$$A^J = g^{-1} A^T g,$$

where $g^{-1} z g = z$, $g^{-1} v g = v$ so that g is in $\mathfrak{Z}(v)$. Also $g^J = \pm g$, and if $g^J = -g$ then $g = g_0 v$ with g_0 in $\mathfrak{Z}$, $j^J = v^{-1} g_0^{-1}(-j) g_0 v = -v^{-1} j v = j$, which is impossible since $\mathfrak{S}(j)$ is not total real. Hence $g^J = g$ is in $\mathfrak{Z}$ and hence in $\mathfrak{A}_1$, $A^J = g^{-1}A^Tg$ is in $\mathfrak{A}_1$ for every A of $\mathfrak{A}_1$, $\mathfrak{A}_1$ is J-involutorial. Our proof of Theorem 10.27 then implies that $u_S^J = (u_S c_S)^{-1}$ for c_S total positive as desired.

We have completed our results on the structure of multiplication algebras over an arbitrary real $\mathfrak{F}$ and wish to make some remarks in closing on the restrictions which must be made on the degree of $\mathfrak{A}$ relative to the order ν of Ω. We are of course interested only in the case of pure GR-matrices Ω of ν rows and let $\mathfrak{A}$ then be a normal division algebra $\mathfrak{D}$ of degree m over $\mathfrak{K}$, $\mathfrak{K}$ of degree t over $\mathfrak{F}$. By Theorem 8.2 we have $\nu = m^2tq$. The existence proofs of [43] then imply that if mq is sufficiently large and $\mathfrak{D}$ is an algebra of Theorems 10.26, 10.27, 10.28 there exists a corresponding pure Ω with $\mathfrak{D}$ as multiplication algebra. However in the limiting cases where mq takes on the values 1 or 2 it is known that certain additional restrictions must be made.

15. Multiplication algebras over an algebraic number field. Let $\mathfrak{F}$ be an algebraic extension of finite degree over the field $\mathfrak{R}$ of all rational numbers. This is the case actually arising in the original theory of Riemann matrices where in fact $\mathfrak{F} = \mathfrak{R}$. If Ω is pure with $\mathfrak{D}$ of the first kind then $\mathfrak{D} = (1, v, j, jv)$, $v^2 = \alpha$ in $\mathfrak{S}$, $j^2 = \beta$ in $\mathfrak{S}$, $vj = -jv$ and the cases of Theorems 10.28 and 10.27 arise respectively according as α and β are or are not both total negative. We leave the details of the verification of this result to the reader. The more interesting case is that of algebras $\mathfrak{D}$ of the second kind where we have $\mathfrak{F}_0$ the real subfield of $\mathfrak{F}$, $\mathfrak{K}$ is the centrum of $\mathfrak{D}$, $\mathfrak{K} = \mathfrak{S}(\theta) > \mathfrak{S}$, and we prove

Theorem 29. *The multiplication algebra $\mathfrak{D}$ of a pure GR-matrix Ω over an algebraic number field $\mathfrak{F}$ is an algebra of the second kind if and only if $\mathfrak{D}$ is a cyclic algebra $(\mathfrak{Z}, S, \gamma)$ of Theorem* 10.18 *where $\mathfrak{Z} = \mathfrak{X} \times \mathfrak{K}$, the centrum $\mathfrak{K}$ of $\mathfrak{D}$ is a total pure imaginary field over $\mathfrak{F}_0$, $\mathfrak{X}$ is a total real cyclic field over $\mathfrak{S}$, and* (30) *holds for g a total positive quantity of $\mathfrak{X}$.*

In view of Theorems 10.18 and 10.26 it is sufficient merely to prove the existence of a total real field $\mathfrak{X}$ over $\mathfrak{F}_0$ such that $\mathfrak{X}$ is cyclic over $\mathfrak{S}$, $\mathfrak{X}(\theta)$ splits $\mathfrak{D}$. For evidently since $\mathfrak{K}$ is total pure imaginary $\mathfrak{Z} = \mathfrak{X}(\theta) = \mathfrak{X} \times \mathfrak{K}$. We let m be the degree of $\mathfrak{D}$ over $\mathfrak{K}$ and use Lemma 9.10 to construct a cyclic field $\mathfrak{W}$ of degree $2m$ over $\mathfrak{S}$ such that the field $\mathfrak{W}_{\mathfrak{S}_V}$ has degree $2m$ over $\mathfrak{S}_V$ for every non-archimedean valuation V of $\mathfrak{S}$ such that the index of $\mathfrak{D}_{\mathfrak{K}_V}$ is not unity. The field $\mathfrak{W}$ has automorphism group $[S_0]$ where $[S_0]$ has order $2m$, S_0^m leaves unaltered a subfield $\mathfrak{X}$ which is cyclic of degree m over $\mathfrak{S}$ with generating automorphism S induced by S_0. If $\mathfrak{W}$ is total real over $\mathfrak{F}_0$ then so is $\mathfrak{X}$. Otherwise there is a field $\mathfrak{S}_i$ over $\mathfrak{F}_0$ contained in $\mathcal{R}$ and such that $\mathfrak{S}_i$ is equivalent over $\mathfrak{F}_0$ to $\mathfrak{S}$ over $\mathfrak{F}_0$, the corresponding $\mathfrak{W}_i$ over $\mathfrak{S}_i$ is imaginary. But $\mathfrak{W}_i = \mathfrak{S}_i(\xi_i)$, $\bar{\xi}_i$ is a root of the minimum function over $\mathfrak{S}_i$ of ξ_i, $\bar{\xi}_i$ is in $\mathfrak{W}_i$, $\bar{\xi}_i = \xi_i^{S^t}$ for some integer $t < 2m$, $\xi_i = \bar{\xi}_i^{S^t} = \xi_i^{S^{2t}}$. Then $2t < 4m$ is divisible by $2m$, $t = m$, S^m carries each quantity of $\mathfrak{W}_i$ into its complex conjugate and leaves unaltered the subfield $\mathfrak{X}$. Hence $\mathfrak{X}$ is total real over $\mathfrak{F}_0$. Now $\mathfrak{K}$ is total pure imaginary, $\mathfrak{Z} = \mathfrak{X}(\theta) = \mathfrak{X} \times \mathfrak{K}$. By the proof of Theorem 10.22 $\mathfrak{Z}$ splits $\mathfrak{D}$. This completes our proof.

CHAPTER XI

SPECIAL RESULTS

1. Remarks on the structure of arbitrary algebras. There are certain theorems on the structure of algebras which are both general and of sufficient interest to be included in these LECTURES. They are somewhat isolated results, however, and their insertion in the main body of our theory would have disrupted its continuity. We shall thus give an exposition of a selected group of these results in this final chapter, shall state others without proof but with reference to sources, and shall merely refer to the articles in which still others are to be found. We shall also indicate supplementary reading on the subject matter of these LECTURES.

Let us begin with a discussion of the general problem of the structure of any algebra $\mathfrak{A}$ over a field $\mathfrak{F}$. The main theorems of the resulting theory are the Wedderburn structure theorems given in Chapter III as Theorem 3.9 on the structure of a simple algebra, Theorem 3.8 on the structure of a semi-simple algebra, Theorem 3.23 stating that if $\mathfrak{A} - \mathfrak{N}$ is separable then $\mathfrak{A} = \mathfrak{S} + \mathfrak{N}$ where $\mathfrak{S} \cong \mathfrak{A} - \mathfrak{N}$. These results were first proved for the case where $\mathfrak{F}$ is non-modular in [425], and are partly given in [350]. More detailed expositions of these proofs appear in [136], [142], and form the basis of all subsequent proofs. Observe that the hypothesis that $\mathfrak{A} - \mathfrak{N}$ is separable is always satisfied if $\mathfrak{F}$ is non-modular, and so does not appear in the statement of the early forms of Theorem 3.23. The theorem was first proved in its present form in [119], [414], and an example was given in the former book to show the hypothesis necessary if $\mathfrak{F}$ is modular. A somewhat different proof was also given in [432], and shown to hold for modular fields in [433].

In connection with these results let us mention their extensions to rings in [54], [55], and [119]. See also the theory of algebras of infinite order in [428], [200], [229]. Our bibliography also contains many other references to somewhat related articles on the structure of rings, and ideal theory. Here we have made conspicuous omission of the many interesting papers of W. Krull. The subject matter of these papers and others like them seem to be too far removed from our theory to be given the space which references to such a great body of theory would require, and we prefer rather to do them justice by the present mention. We shall not make any more explicit discussion of the papers on rings in our bibliography, as their titles adequately indicate their contents. Note that we have listed a number of papers as samples of the types of material contained in such work.

The Wedderburn proof of the property that $\mathfrak{A} - \mathfrak{N}$ is separable if $\mathfrak{F}$ is non-modular uses a property which we certainly wish to state here as it is not even indicated in our present proof. This is the criterion for properly nilpotent quantities which is proved on pages 108–110 of [142]. It involves the trace function $T(a)$ defined for the first regular representation and we state the result without proof as

Theorem 1. *A quantity x of an algebra $\mathfrak{A}$ over a non-modular field $\mathfrak{F}$ is in the radical of $\mathfrak{A}$ if and only if $T(xy) = 0$ for every y of $\mathfrak{A}$.*

This result and consequent method are false if $\mathfrak{F}$ is modular since we may have $T(a) \equiv 0$ in this case. The theorem is used in [142] to prove that if r is the rank of the discriminant matrix of $\mathfrak{A}$ of order n over $\mathfrak{F}$ the order of the radical of $\mathfrak{A}$ is $n - r$. This is seen to imply, in particular, that if $\mathfrak{A}$ over $\mathfrak{F}$ has radical $\mathfrak{N}$, any scalar extension $\mathfrak{A}_{\mathfrak{K}}$ has radical $\mathfrak{N}_{\mathfrak{K}}$. All of these results clearly hold only when $\mathfrak{F}$ is a non-modular field.

If $\mathfrak{A}$ is any algebra for which $\mathfrak{A} - \mathfrak{N}$ is separable the results of Chapter III imply that the structure of $\mathfrak{A}$ will be fully determined if we determine the structure of all normal division algebras over $\mathfrak{F}$, the structure of all nilpotent algebras over $\mathfrak{F}$, and the multiplicative relations between $\mathfrak{S}$ and $\mathfrak{N}$ in $\mathfrak{A} = \mathfrak{S} + \mathfrak{N}$. Here $\mathfrak{N}$ is an ideal of $\mathfrak{A}$ so that $\mathfrak{S}\mathfrak{N} \leqq \mathfrak{N}$, $\mathfrak{N}\mathfrak{S} \leqq \mathfrak{N}$. The study of this final question was made the first part of the author's Chicago doctorate dissertation of 1928. It has probably been studied by many other mathematicians. However no complete solution has ever been published to the author's knowledge, and it seems likely that only fragmentary results have been obtained so far. In this connection see the complexity of the results even for the special case considered in Section 102 of [136].

The early papers on the multiplication tables of special algebras are closely allied to the topic above and are listed in the bibliographical text [362]. See also the bibliography in [350]. It is quite interesting to the research worker on algebras to work out such multiplication tables for algebras of low order which are neither semi-simple nor nilpotent, and it is not surprising that many have done so. In a letter to L. E. Dickson in 1926, J. H. M. Wedderburn gave a complete determination (without details of proof) of all such algebras over any non-modular field $\mathfrak{F}$. The results were also obtained later by the author in his master's dissertation of 1927, the work constituting an independent verification of that of Wedderburn. The results have appeared since in the literature by other authors.

One of the outstanding remaining problems on algebras is the second of those mentioned above, the classification of nilpotent algebras. The conclusions are usually based on the normalization of pages 110–111 of [142], and we refer the reader to [362], and the papers on nilpotent algebras of our bibliography, for the results themselves.

2. Division algebras over special fields. The most interesting of all the three structure problems mentioned above is that of determining all normal division algebras over any field $\mathfrak{F}$. The case where $\mathfrak{F}$ is finite was settled completely in Section 4.12 and our proof has its origin in [424], [448]. See also [124] in this connection. *Henceforth let $\mathfrak{F}$ be an infinite field.*

The theory of Chapter IX is a solution of our problem for the case where $\mathfrak{F}$ is any algebraic number field of finite degree over the field $\mathfrak{R}$ of all rational numbers. The problem was recognized to be an arithmetic and not an algebraic one early by H. Hasse, the author, and probably others, and elementary considerations using the theory of quadratic forms were made in [7], [13], [23], and [25]. See other connections of algebras with quadratic forms in [31], [49], [454]. The ultimate successful method seems to have had its origin in the papers [172], [173] of Hasse on quadratic forms, and the final proof is based on [177] and given in [89], [52]. For a complete exposition differing from ours in the proof of Theorem 9.32 see [119] which contains nearly all of the subject matter of these LECTURES and will not be referred to again in our present summary except for special references.

The results obtained for algebras over fields $\mathfrak{F}$ of finite degree over $\mathfrak{R}$ are also valid when $\mathfrak{F}$ is *any* field of algebraic numbers, and were derived in [32] by the use of the following device.

Let $\mathfrak{A} = (u_1, \cdots, u_n)$ over $\mathfrak{F}$ be an algebra, so that $u_i u_j = \sum_{k=1}^{n} \gamma_k^{(ij)} u_k$ with the $\gamma_k^{(ij)}$ in $\mathfrak{F}$. We let $\mathfrak{P}$ be the prime subfield of $\mathfrak{F}$, whence $\mathfrak{P}$ may be taken to be the field of all rational numbers when $\mathfrak{F}$ is non-modular, or the field of residue classes of integers modulo p if $\mathfrak{F}$ has characteristic p. Let $\mathfrak{F}_0 = \mathfrak{P}(\gamma_1^{(11)}, \cdots, \gamma_k^{(ij)}, \cdots, \gamma_n^{(nn)})$. Then $\mathfrak{F}_0$ is the field obtained from $\mathfrak{P}$ by the adjunction of a finite number of quantities and is what we call a *reduced field*. Clearly $\mathfrak{A} = (\mathfrak{A}_0)_{\mathfrak{F}}$, where $\mathfrak{A}_0 = (u_1, \cdots, u_n)$ over $\mathfrak{F}_0$. This argument reduces the problem of determining the structure of all division algebras over $\mathfrak{F}$ to the case where $\mathfrak{F}$ is a reduced field. Moreover it is shown in [32] that the theory of exponents is thereby simplified similarly.

Division algebras over an algebraic extension of any field $\mathfrak{F}_0 = \mathfrak{P}(\eta)$, η an indeterminate over the prime field $\mathfrak{P}$ of characteristic p, have also been shown in [451], [393] to be cyclic algebras whose degree and exponent are equal. The results are obtained as consequences of a valuation theory of such fields actually simpler than that of algebraic number fields, but it is necessary to study the question of separability and thus to use Theorem 5.18 to reduce the study to the cases of algebras of degree p^e and algebras of degree n prime to p. The former is the troublesome case in [393], but actually is the most elementary case of all, as was shown in [46]. There the author considered normal division algebras of degree p^e over an algebraic function field $\mathfrak{F}$ of one indeterminate and perfect constant field, that is, the case where $\mathfrak{F}$ is algebraic of finite degree over $\mathfrak{L}(\eta)$, $\mathfrak{L}$ perfect of characteristic p, η an indeterminate over $\mathfrak{L}$. Our results were a simple consequence of the following

LEMMA 1. *Let $\mathfrak{F}$ be as above. Then η may be chosen so that every a of $\mathfrak{F}$ has the property $a = b^p$ for b in $\mathfrak{F}(\eta^{1/p})$.*

A generalization of this latter result was obtained in [49] for algebraic function fields $\mathfrak{F}_r$ of r independent indeterminates, and perfect constant field of characteristic p, and will be applied at some future time with Theorem 7.28 to give some new theorems on the structure of p-algebras over $\mathfrak{F}_r$. We wish also to call attention to the theorem of [405], [406], where it was shown that the index of any normal simple algebra over an algebraic function field of one indeterminate and algebraically closed constant field is unity. This result is the inspiration for the results of [407], [108], [419].

3. The exponent of a normal division algebra. If $\mathfrak{D}$ is any normal division algebra of degree n over $\mathfrak{F}$ and $m = p_1 \cdots p_r$ is the product of all the distinct prime divisors p_i of n, the exponent of $\mathfrak{D}$ was shown in Theorem 5.17 to be an integral divisor ρ of n which is divisible by m. Conversely it has been shown in [82], [289] that there exist fields $\mathfrak{F}$ and normal division algebras $\mathfrak{D}$ of degree n over $\mathfrak{F}$ and exponent ρ for every ρ satisfying these two conditions.

By Section 5.10 the study of the exponent of a normal division algebra may be reduced to that of the exponent of a *primary* algebra, that is, a normal division algebra $\mathfrak{D}$ not expressible as a direct product $\mathfrak{D} = \mathfrak{D}_1 \times \mathfrak{D}_2$ where $\mathfrak{D}_i$ has lower degree than that of $\mathfrak{D}$. By Theorem 5.20 $\mathfrak{D}$ has degree p^e over $\mathfrak{F}$, p a prime. It was expected when the exponent of an algebra was first studied that then the exponent of $\mathfrak{D}$ is its degree, but this was shown in [29] to be false. A criterion for the exponent of a p-algebra was given in Section 7.9, and has its origin in [37], [35], [44], and [45]. The material of [45] was also derived in [293], [294], [398], papers which were inspired by [37], [44]. See also the conjecture in [45] on the exponent of algebras of degree p^e over $\mathfrak{F}$ of characteristic not p.

In the study of the exponent of algebras of degree 2^e one may simplify the considerations somewhat as a consequence of the following result of [12], [25].

Theorem 2. *A normal division algebra of degree four over any field $\mathfrak{F}$ is primary if and only if it has exponent four.*

The property above is used in [23] together with the theory of quadratic forms to obtain a simple proof of the fact that every normal division algebra of degree 2^e over an algebraic number field $\mathfrak{K}$ has exponent 2^e. As a first step Theorem 11.2 is seen to imply that all normal division algebras of degree four over $\mathfrak{K}$ are primary and hence of exponent four. The theory is then completed by the use of a device of considerable interest which will be given here. We begin with the

DEFINITION. *Let $\mathfrak{F}$ be a field, p a prime, e any positive integer. Then $\mathfrak{F}$ will be said to have the property $E(p, e)$ if the exponent ρ of every normal division algebra of degree p^e over any algebraic extension of finite degree over $\mathfrak{F}$ is its degree p^e.*

We next prove

LEMMA 2. *Let $\mathfrak{F}$ have the property $E(p, 2)$. Then if $\mathfrak{A}$ is any normal simple algebra of index p^e the index of $\mathfrak{A}^p$ is p^{e-1}.*

For by Theorem 5.16 the index of $\mathfrak{A}^p$ is $p^f \leqq p^{e-1}$. By Theorems 4.27, 4.18 there exists a splitting field $\mathfrak{K}$ of degree p^f over $\mathfrak{F}$ of $\mathfrak{A}^p$, and if $p^f < p^{e-1}$ then $n = p^e = p^{e-f}p^f$ where $e - f \geqq 2$. By Theorem 4.20 the index of $\mathfrak{A}_{\mathfrak{K}}$ is at least p^2. Now there exists an extension $\mathfrak{L}$ of $\mathfrak{K}$ such that the index of $\mathfrak{A}_{\mathfrak{L}}$ is p^2 by Theorems 4.31, 4.16. By our hypothesis on $\mathfrak{F}$, $\mathfrak{A}_{\mathfrak{L}}$ has exponent p^2, $(\mathfrak{A}_{\mathfrak{L}})^p$ has index not unity, whereas $(\mathfrak{A}_{\mathfrak{L}})^p = (\mathfrak{A}^p)_{\mathfrak{L}} = [(\mathfrak{A}^p)_{\mathfrak{K}}]_{\mathfrak{L}} \sim 1$. Hence $f = e - 1$.

We now prove the result desired, which we may state as

Theorem 3. *A field $\mathfrak{F}$ has the property $E(p, e)$ for every e if and only if $\mathfrak{F}$ has the property $E(p, 2)$.*

For $\mathfrak{F}$ has the property $E(p, 1)$ by Theorem 5.17, and we may assume as the basis of an induction on e that $\mathfrak{F}$ has the property $E(p, f)$ for $f \leqq e$. If $\mathfrak{A}$ has index p^e the index of $\mathfrak{A}^p$ is p^{e-1} and the hypothesis of our induction implies that $\mathfrak{A}^p$ has exponent p^{e-1}. But then the exponent of $\mathfrak{A}$ is p^e as desired. The converse is trivial.

4. Normal division algebras with a pure maximal subfield. The theory of p-algebras was obtained by the author as the result of an attempt to prove the converse of a trivial corollary of Theorem 5.9. There we showed that every cyclic normal division algebra $\mathfrak{D} = (\mathfrak{Z}, S, \gamma)$ has the form $\mathfrak{D} = \mathfrak{Z} + j\mathfrak{Z} + \cdots + j^{n-1}\mathfrak{Z}$, where $\mathfrak{Z}$ is a cyclic field of degree n over $\mathfrak{F}$ with generating automorphism S, $j^n = \gamma$ in $\mathfrak{F}$, $zj = jz^S$ for every z of $\mathfrak{Z}$. The form of $\mathfrak{D}$ implies that $\lambda^n - \gamma$ is the minimum function of j, $\mathfrak{F}(j)$ is what may be called a *pure field* of degree n over $\mathfrak{F}$. Thus the property that a normal division algebra is a cyclic algebra implies that it has a pure maximal subfield. One immediately conjectures the converse.

Theorem 7.20 clearly reduces our investigation of the truth of this conjecture to the case of algebras $\mathfrak{D}$ of degree a power p^e of a prime p. Then Theorem 7.26 states that the proposition is true if p is the characteristic of $\mathfrak{F}$. However it was proved in [48] that the proposition is false if p is not the characteristic of $\mathfrak{F}$ and $e > 1$. We refer the reader to that paper for the counter-example, a non-cyclic normal division algebra which has degree and exponent four over a non-modular field and has a pure quartic subfield. There remains the case of normal division algebras $\mathfrak{D}$ of prime degree p over $\mathfrak{F}$ of characteristic not p, and we shall prove the conjecture true for this case.

Let $\mathfrak{D}$ of prime degree p contain a subfield $\mathfrak{Y} = \mathfrak{F}(y)$, $y^p = \gamma$ in $\mathfrak{F}$. We let $\mathfrak{K} = \mathfrak{F}(\zeta)$, where ζ is a primitive pth root of unity, and showed in Section A8.11 that $\mathfrak{K}$ is a cyclic field of degree ν over $\mathfrak{F}$ such that ν divides $p - 1$. If $\mathfrak{K} = \mathfrak{F}$ then $\mathfrak{Y}$ is cyclic over $\mathfrak{F}$ and our result is trivial. Hence let $\nu > 1$, $\mathfrak{K} > \mathfrak{F}$. Then $\mathfrak{K}$ has a generating automorphism T over $\mathfrak{F}$ induced by

$$\zeta^T = \zeta^t \neq \zeta, \tag{1}$$

where t is an integer chosen so that $0 < t < p$,

$$t^\nu \equiv 1 \pmod{p}, \qquad t^k \not\equiv 1 \pmod{p} \tag{2}$$

for any positive integer $k < t$. There exist integers ν_0, s such that

$$(3) \qquad st \equiv \nu\nu_0 \equiv 1 \pmod{p}.$$

We then define integers

$$(4) \qquad s_k = \nu_0 s^k = s_{k-1}s \qquad (k = 0, 1, \cdots, \nu),$$

and have

$$(5) \qquad \sigma = \sum_{k=1}^{\nu} t^k s_k \equiv 1 \pmod{p},$$

as in Section A9.7.

The scalar extension $\mathfrak{D}_{\mathfrak{K}} = \mathfrak{D} \times \mathfrak{K}$ is a division algebra over $\mathfrak{K}$ by the corollary of Theorem 4.20. The correspondence T_0 of $\mathfrak{D}_{\mathfrak{K}}$ defined by

$$(6) \qquad d^{T_0} = d, \qquad k^{T_0} = k^T \qquad (d \text{ in } \mathfrak{D},\ k \text{ in } \mathfrak{K}),$$

is clearly an automorphism over $\mathfrak{F}$ of $\mathfrak{D} \times \mathfrak{K}$.

Theorem A8.12 states that $\mathfrak{Y} \times \mathfrak{K} = \mathfrak{W}$ is cyclic of degree p over $\mathfrak{K}$ with generating automorphism W induced by $y^W = \zeta y$. Then

$$(7) \qquad \mathfrak{D}_{\mathfrak{K}} = (\mathfrak{W}, W, \lambda) = \mathfrak{W} + j\mathfrak{W} + \cdots + j^{p-1}\mathfrak{W}$$

such that $j^p = \lambda$ in $\mathfrak{K}$, $wj = jw^W$ for every w of $\mathfrak{W}$. In particular $yj = j(\zeta y)$, and since y is in $\mathfrak{D}$ we have $y = y^{T_0}$,

$$(8) \qquad yj^{T_0} = j^{T_0}\zeta^t y = j^{T_0}y^{W^t}.$$

It follows that $j^{T_0} = j^t w_0$ for w_0 in $\mathfrak{W}$, $(j^{T_0})^p = \lambda^{T_0} = \lambda^T = \lambda^t N_{\mathfrak{W}|\mathfrak{K}}(w_0)$. Then $\lambda^{T^2} = \lambda^{t^2} N_{\mathfrak{W}|\mathfrak{K}}(w_0^t w_0^{T_0}) = \lambda^{t^2} N_{\mathfrak{W}|\mathfrak{K}}(w_{02})$ since $T_0 W = WT_0$. Similarly $\lambda^{T^k} = \lambda^{t^k} N_{\mathfrak{W}|\mathfrak{K}}(w_{0k})$. Then $(\lambda^{T^k})^{s_k} = \lambda^{t^k s_k} N_{\mathfrak{W}|\mathfrak{K}}(w_k)$ with $w_k = w_{0k}^{s_k}$, and $\lambda_0 = M(\lambda) = \prod_{k=1}^{\nu} (\lambda^{T^k})^{s_k} = \lambda^\sigma N_{\mathfrak{W}|\mathfrak{K}}(w^{(0)}) = \lambda N_{\mathfrak{W}|\mathfrak{K}}(w)$, with w in $\mathfrak{W}$, since $\sigma \equiv 1 \pmod{p}$ by (5). By Theorem A9.13 the quantity λ_0 has the property that $\lambda_0^T \lambda_0^{-t}$ is the pth power of a quantity of $\mathfrak{K}$. But then $\mathfrak{D}_{\mathfrak{K}} = (\mathfrak{W}, W, \lambda_0)$ and hence we may assume that $\lambda = \lambda_0$ has the property

$$(9) \qquad \lambda^T = \lambda^t \mu^p \qquad (\mu \text{ in } \mathfrak{K}).$$

The field $\mathfrak{K}(j)$ is cyclic over $\mathfrak{K}$ with $j^{S_1} = \zeta j$. Now $yj = \zeta jy$, $j^{-1}y = y(\zeta j^{-1})$. Hence

$$(10) \qquad \mathfrak{D}_{\mathfrak{K}} = (\mathfrak{Z}_1, S_1, \gamma) = \mathfrak{Z}_1 + y\mathfrak{Z}_1 + \cdots + y^{p-1}\mathfrak{Z}_1,$$

where $\mathfrak{Z}_1 = \mathfrak{K}(j_1)$, $j_1 = j^{-1}$, $(j_1)^{S_1} = \zeta j_1$, $j_1^p = \lambda_1 = \lambda^{-1}$ has the property $\lambda_1^T = \lambda_1^t \mu_1^p$, $\mu_1 = \mu^{-1}$ in $\mathfrak{K}$. By Theorem A9.15

$$(11) \qquad \mathfrak{Z}_1 = \mathfrak{Z} \times \mathfrak{K},$$

where $\mathfrak{Z}$ is cyclic of degree p over $\mathfrak{F}$ with generating automorphism

$$S: \qquad z \leftrightarrow z^S = z^{S_1} \qquad (z \text{ in } \mathfrak{Z}).$$

Then

$$\mathfrak{D}_{\mathfrak{K}} = (\mathfrak{Z}, S, \gamma) \times \mathfrak{K}, \tag{12}$$

so that $\mathfrak{D} \times (\mathfrak{Z}, S, \gamma^{-1}) \times \mathfrak{K} \sim 1$, $\mathfrak{K}$ of degree ν over $\mathfrak{F}$ splits $\mathfrak{D} \times (\mathfrak{Z}, S, \gamma^{-1})$ of degree p^2 over $\mathfrak{F}$. Since ν is prime to p^2 we have $\mathfrak{D} \times (\mathfrak{Z}, S, \gamma^{-1}) \sim 1$, $\mathfrak{D} = (\mathfrak{Z}, S, \gamma^{-1})^{-1} = (\mathfrak{Z}, S, \gamma)$. We have proved

Theorem 4. *A normal division algebra $\mathfrak{D}$ of prime degree p over $\mathfrak{F}$ is a cyclic algebra $\mathfrak{D} = (\mathfrak{Z}, S, \gamma)$ if and only if $\mathfrak{D}$ has a subfield $\mathfrak{Y} = \mathfrak{F}(y)$, $y^p = \gamma$.*

The proof we have given above as well as the result itself is a slight improvement of that given by the author in [39].

5. The structure of normal division algebras of degree three. In [24], [30] and the subsequent [48] the author proved the existence of non-cyclic algebras. This result naturally increases the interest in the problem of the determination of normal division algebras of low degrees, since it is now clear that no proof that all normal division algebras are cyclic can exist. We shall thus begin a determination of all normal division algebras of degree $n \leqq 4$ over an arbitrary field $\mathfrak{F}$.

If $n = 1$ then $\mathfrak{D} = \mathfrak{F}$ and the result on the structure of $\mathfrak{D}$ is trivial. If $n = 2$ we showed the algebras cyclic (quaternion algebras) in Section 9.10. The next case will be completed in

Theorem 5. *Every normal division algebra of degree three over $\mathfrak{F}$ is a cyclic algebra.*

For by Theorem 4.18 $\mathfrak{D}$ has a separable cubic subfield $\mathfrak{X} = \mathfrak{F}(x)$. If $\mathfrak{X}$ is cyclic over $\mathfrak{F}$ or a pure field as in Theorem 11.4 we see that $\mathfrak{D}$ is a cyclic algebra. Otherwise we showed in Exercise 8 of page A178 that $\mathfrak{X}$ is contained in a normal field $\mathfrak{W}$ of degree six over $\mathfrak{F}$ whose automorphism group $\mathfrak{G}$ is equivalent to the symmetric permutation group on three letters. Then $\mathfrak{G} = \mathfrak{H} + \mathfrak{H}T$, $\mathfrak{H} = (I, S, S^2)$, $S^3 = T^2 = I$, $ST = TS^2$. The group $\mathfrak{H}$ is a normal divisor of $\mathfrak{G}$, the corresponding subfield of $\mathfrak{W}$ is a cyclic field $\mathfrak{K}$ of degree two over $\mathfrak{F}$ with generating automorphism induced by T, $\mathfrak{W} = \mathfrak{X} \times \mathfrak{K} = \mathfrak{X}_{\mathfrak{K}}$, $\mathfrak{D}_{\mathfrak{K}} = \mathfrak{D} \times \mathfrak{K}$ contains $\mathfrak{W}$. The algebra $\mathfrak{D} \times \mathfrak{K}$ has an automorphism which we may designate by T and which is such that $d = d^T$ for every d of $\mathfrak{D}$. Also $\mathfrak{W}$ is cyclic over $\mathfrak{K}$, $\mathfrak{D}_{\mathfrak{K}} = (\mathfrak{W}, S, \gamma) = \mathfrak{W} + j\mathfrak{W} + j^2\mathfrak{W}$, $j^3 = \gamma$ in $\mathfrak{K}$. Here $ST = TS^2$ and $\mathfrak{D}_{\mathfrak{K}}$ is a cyclic algebra such that $wj = jw^S$ for every w of $\mathfrak{W}$. Now $\mathfrak{X} = \mathfrak{F}(x)$ with $x = x^T$ in $\mathfrak{D}$ and $xj = jx^S$, $(xj)^T = xj^T = j^Tx^{ST} = j^Tx^{TS^2} = j^Tx^{S^2}$. Then $j^T = j^2u$ for u in $\mathfrak{K}$, $(j^T)^3 = \gamma^T = \gamma^2 N_{\mathfrak{W}|\mathfrak{K}}(u)$.

$$\Gamma = \gamma^{-1}\gamma^T = \gamma N_{\mathfrak{W}|\mathfrak{K}}(u)$$

so that $\mathfrak{D}_{\mathfrak{K}} = (\mathfrak{W}, S, \Gamma)$. But $T^2 = I$ and $\Gamma^T = \Gamma^{-1}$. Hence we may assume, for convenience of notation, that $\mathfrak{D}_{\mathfrak{K}} = (\mathfrak{W}, S, \gamma)$ as above but with the property that $\gamma^T = \gamma^{-1}$.

If $\mathfrak{F}$ has characteristic three the quantity $y = (j + j^{-1})$ has the property $y^3 = \gamma + \gamma^{-1} = \gamma + \gamma^T = \delta$ in $\mathfrak{F}$, and $y = j + \gamma^{-1}j^2$ is clearly not in $\mathfrak{F}$, $\mathfrak{D}_{\mathfrak{K}}$

contains the pure field $\mathfrak{F}(y)$, $\mathfrak{D}_{\mathfrak{K}}$ has a pure splitting field $\mathfrak{K}(\delta^{\frac{1}{3}})$, so that $\mathfrak{F}(\delta^{\frac{1}{3}})$ splits $\mathfrak{D}$ and our result follows from Theorem 11.4. Hence let $\mathfrak{F}$ have characteristic not three. Then we may assume that $\mathfrak{X} = \mathfrak{F}(x)$ with $x^3 + \alpha x + \beta = 0$ for α and β in $\mathfrak{F}$, and thus have $T_{\mathfrak{X}|\mathfrak{F}}(x) = x + x^S + x^{S^2} = 0$, $N_{\mathfrak{X}|\mathfrak{F}}(x) = -\beta$. Put $z = (1 + j + j^{-1})x$, so that

$$z^2 = (1 + j + j^{-1})x(1 + j + j^{-1})x = (1 + j + j^{-1})(x + jx^S + j^{-1}x^{S^2})x$$
$$= [(x + x^{S^2} + x^S) + j(x + \gamma^{-1}x^{S^2} + x^S) + j^{-1}(x + x^{S^2} + \gamma x^S)]x.$$

Use $x + x^S + x^{S^2} = 0$ to obtain $z^2 = jxx^{S^2}(\gamma^{-1} - 1) + j^{-1}xx^S(\gamma - 1)$. Then

$$\begin{aligned} z^3 &= (1 + j + j^{-1})[jxx^Sx^{S^2}(\gamma^{-1} - 1) + j^{-1}xx^Sx^{S^2}(\gamma - 1)] \\ &= -\beta(\gamma - 1)(1 + j + j^{-1})(j^{-1} - j\gamma^{-1}) \\ &= -\beta(\gamma - 1)(j^{-1} - j\gamma^{-1} + 1 + j^{-1} + \gamma^{-1}j - \gamma^{-1}) \\ &= \beta(\gamma - 1)(\gamma^T - 1) \text{ in } \mathfrak{F}. \end{aligned}$$

Since z is not in $\mathfrak{K}$ we apply the argument above to see that $\mathfrak{D}$ has a subfield over $\mathfrak{F}$ equivalent to $\mathfrak{F}(z)$ and apply Theorem 11.4 to complete our proof.

We refer the reader to [51] for a generalization of a part of the result above to algebras of prime degree. We also refer the reader to [427] for the original Wedderburn proof of our result for the case where the field $\mathfrak{F}$ is non-modular. The proof given there begins by showing that $\mathfrak{D}$ contains a pure cubic subfield and the author showed in [37] that this part of Wedderburn's proof was valid for any $\mathfrak{F}$. The remainder of the proof is then valid for any $\mathfrak{F}$ of characteristic not three and was modified in [37] to hold for the remaining (characteristic three) case. Note that the completion may be more readily made by the use of the general result of Theorem 11.4, and that in view of the exceedingly simple proof of Theorem 4.17 this would provide essential simplification of the characteristic three case.

As an immediate corollary of Theorems 11.5, and 7.20 we have

Theorem 6. *Every normal division algebra of degree six over a field $\mathfrak{F}$ is a cyclic algebra.*

See also [10], [16] for elementary considerations on algebras of degree six the methods of which may be useful in other future studies on the determination of the structure of normal division algebras.

The result of Theorem 11.5 and the fact that every quaternion division algebra is cyclic may also be used to prove

Theorem 7. *Let $\mathfrak{D}$ be a normal division algebra of degree four or nine over $\mathfrak{F}$ of characteristic two or three respectively. Then $\mathfrak{D}$ is cyclic if and only if $\mathfrak{D}$ has a subfield which is pure inseparable over $\mathfrak{F}$.*

This result is indeed a special case of

Theorem 8. *Let $\mathfrak{F}$ be a field of finite characteristic p with the property that every normal division algebra of degree p over any algebraic extension of finite degree over $\mathfrak{F}$ is a cyclic algebra. Then a normal division algebra $\mathfrak{D}$ of degree p^2 over $\mathfrak{F}$ is a cyclic algebra if and only if $\mathfrak{D}$ has a subfield which is pure inseparable over $\mathfrak{F}$.*

For by Theorem 7.26 $\mathfrak{D}$ is cyclic if and only if it has a pure inseparable subfield of degree p^2 over $\mathfrak{F}$. Thus we require only to prove that if $\mathfrak{D}$ has a pure inseparable subfield $\mathfrak{Y}$ of degree less than p^2 over $\mathfrak{F}$ then $\mathfrak{D}$ is cyclic. By Theorem 1.6 $\mathfrak{Y}$ has degree p over $\mathfrak{F}$, $\mathfrak{Y} = \mathfrak{F}(y)$, $y^p = \gamma$ in $\mathfrak{F}$. The $\mathfrak{D}$-commutator $\mathfrak{B}$ of $\mathfrak{Y}$ is a normal division algebra of degree p over $\mathfrak{Y}$ and, by our hypothesis, $\mathfrak{B}$ is a cyclic algebra

$$\mathfrak{B} = (\mathfrak{Z}, S, g) = \mathfrak{Z} + j\mathfrak{Z} + \cdots + j^{p-1}\mathfrak{Z}$$

over $\mathfrak{Y}$, such that $j^p = g \neq 0$ in $\mathfrak{Y}$, $\mathfrak{Y}(j)$ has degree p over $\mathfrak{Y}$. If g is not in $\mathfrak{F}$ the field $\mathfrak{F}(j)$ is pure inseparable of degree p^2 over $\mathfrak{F}$ with $\mathfrak{Y}$ as proper subfield, $\mathfrak{D}$ is a cyclic algebra by Theorem 7.26. Hence let $g = \delta$ in $\mathfrak{F}$. Then clearly Theorems 2.29 and 2.32 imply that $\mathfrak{Z} = \mathfrak{Y} \times \mathfrak{Z}_0$ where $\mathfrak{Z}_0$ is cyclic of degree p over $\mathfrak{F}$, $\mathfrak{B} = (\mathfrak{Z}_{0\mathfrak{Y}}, S, \delta) = (\mathfrak{Z}_0, S, \delta) \times \mathfrak{Y}$, where $\mathfrak{B}_0 = (\mathfrak{Z}_0, S, \delta)$ is cyclic of degree p over $\mathfrak{F}$. By Theorem 4.6 $\mathfrak{D} = \mathfrak{B}_0 \times \mathfrak{D}^{\mathfrak{B}_0}$ where the $\mathfrak{D}$-commutator $\mathfrak{D}^{\mathfrak{B}_0}$ of $\mathfrak{B}_0$ is a normal division algebra of degree p over $\mathfrak{F}$. By our hypothesis $\mathfrak{D}^{\mathfrak{B}_0}$ is cyclic. By Lemma 7.13 so is $\mathfrak{D}$. This proves our theorem.

6. The structure of normal division algebras of degree four. The principle result on normal division algebras of degree four over any field $\mathfrak{F}$ may be stated as

Theorem 9. *Every normal division algebra of degree four over $\mathfrak{F}$ contains a normal quartic field*

$$\mathfrak{Z} = \mathfrak{U} \times \mathfrak{B} \tag{13}$$

where $\mathfrak{U}$ and $\mathfrak{B}$ are separable (cyclic) quadratic fields over $\mathfrak{F}$. Hence $\mathfrak{D}$ is a crossed product $(\mathfrak{Z}, \mathfrak{a})$.

We first observe that if $\mathfrak{D}$ is a cyclic algebra $(\mathfrak{W}, S, \gamma) = \mathfrak{W} + j\mathfrak{W} + j^2\mathfrak{W} + j^3\mathfrak{W}$, $wj = jw^S$ for every w of $\mathfrak{W}$, $j^4 = \gamma$ in $\mathfrak{F}$, then $\mathfrak{W}$ has a quadratic subfield $\mathfrak{U}$ every quantity of which is unaltered by S^2, $j^2u = uj^2$ for every u of $\mathfrak{U}$. Now $\mathfrak{F}(j^2)$ has degree two over $\mathfrak{F}$ and if the characteristic of $\mathfrak{F}$ is not two we may take $\mathfrak{B} = \mathfrak{F}(j^2)$, $\mathfrak{Z} = \mathfrak{U} \times \mathfrak{B}$ is the $\mathfrak{Z}$ of our theorem. We have thus shown that cyclic algebras over $\mathfrak{F}$ of characteristic not two trivially satisfy the conclusion of Theorem 11.9. This is not true however of the remaining case where $\mathfrak{F}(j^2)$ is pure inseparable and since this case arises in our general proof of Theorem 11.9 we must treat it.

We use the results of Theorem A9.3 to write $\mathfrak{W} = \mathfrak{U}(x) = \mathfrak{F}(x)$, $\mathfrak{U} = \mathfrak{F}(u)$, such that

$$u^2 - u = \alpha, \qquad x^2 - x = a, \tag{14}$$

for α in $\mathfrak{F}$, a in $\mathfrak{U}$, and

$$u^S = u + 1, \qquad x^S = x + b, \qquad b^2 - b = a^S - a, \tag{15}$$

where b is in $\mathfrak{U}$. Since a is in $\mathfrak{U}$ we have $(b^2 - b)^S = a^{S^2} - a^S = a^S - a$, $b^2 - b = \beta$ in $\mathfrak{F}$. Hence we may take b as the generating quantity u of $\mathfrak{U} = \mathfrak{F}(u)$ and still have (14). We write

$$a = \alpha_0 + \alpha_1 u, \qquad b = u \tag{16}$$

and see that the condition $T_{\mathfrak{U}|\mathfrak{F}}(b^2) = T_{\mathfrak{U}|\mathfrak{F}}(u + \alpha) = T_{\mathfrak{U}|\mathfrak{F}}(b) = 1$ is satisfied, $\alpha = b^2 - b = a^S - a = \alpha_0 + \alpha_1(u + 1) - \alpha_0 + \alpha_1 u = \alpha_1$, so that

$$a = \alpha_0 + \alpha u, \qquad x^{S^2} = x + 1. \tag{17}$$

Now the $\mathfrak{D}$-commutator of $\mathfrak{U}$ is clearly the algebra

$$\mathfrak{B} = (\mathfrak{W}, S^2, \gamma) = \mathfrak{W} + j^2\mathfrak{W} \tag{18}$$

of degree two over $\mathfrak{U}$. Every quadratic subfield $\mathfrak{Z}$ over $\mathfrak{U}$ of $\mathfrak{B}$ defines a quartic field $\mathfrak{Z}$ over $\mathfrak{F}$, and if $\mathfrak{Z} = \mathfrak{U}(v)$, where $\mathfrak{V} = \mathfrak{F}(v)$ is a separable quadratic field over $\mathfrak{F}$, then $\mathfrak{Z} = \mathfrak{U} \times \mathfrak{V}$ is the quartic field of our theorem. We thus wish to prove the existence of a quantity v in $\mathfrak{B}$ and not in $\mathfrak{F}$ such that

$$v^2 - v = \rho \text{ in } \mathfrak{F}. \tag{19}$$

Such a quantity is given in fact by

$$v = x + j^{-2}[\gamma\alpha + a + x(u + 1)] = x + j^{-2}w_0 . \tag{20}$$

For $v^2 = x^2 + xj^{-2}w_0 + j^{-2}w_0 x + j^{-4}w_0 w_0^{S^2}$. Now $xj^{-2} = j^{-2}(x + 1)$ since $j^{-2} = \gamma^{-1}j^2$. Hence $xj^{-2}w_0 + j^{-2}w_0 x = j^{-2}w_0(x + x + 1) = j^{-2}w_0$, $v^2 = x + a + j^{-2}w_0 + \gamma^{-1}w_0 w_0^{S^2} = v + a + \gamma^{-1}w_0 w_0^{S^2}$. Also $w_0^{S^2} = \gamma\alpha + a + (x + 1)(u + 1) = w_0 + (u + 1)$ since $x^{S^2} = x + 1$. Hence $w_0 w_0^{S^2} = w_0^2 + w_0(u + 1) = \gamma^2\alpha^2 + a^2 + x^2(u + 1)^2 + \gamma\alpha(u + 1) + a(u + 1) + x(u + 1)^2 = \gamma^2\alpha^2 + \gamma\alpha(u + 1) + a[a + (u + 1)^2 + (u + 1)] = \gamma^2\alpha^2 + \gamma\alpha(u + 1) + a(a + \alpha)$. But $a = \alpha_0 + \alpha u$, $a^2 = \alpha_0 + \alpha^2(u + \alpha)$, $a^2 + \alpha a = \alpha_0 + \alpha^3 + \alpha^2 u + \alpha_0\alpha + \alpha^2 u$ is in $\mathfrak{F}$. Also $a + \gamma^{-1}\gamma\alpha(u + 1) = \alpha_0 + \alpha u + \alpha u + \alpha = \alpha_0 + \alpha$ is in $\mathfrak{F}$, $v^2 - v$ is in $\mathfrak{F}$ as desired.

The proof just completed is essentially that given in [35] and we state the result as

LEMMA 3. *Every cyclic normal division algebra of degree four over any field* $\mathfrak{F}$ *contains a field* $\mathfrak{Z}$ *of* (13).

We next give what is, except for the complication of inseparability, the proof given in [26] of

LEMMA 4. *Let a normal division algebra* $\mathfrak{D}$ *of degree four over* $\mathfrak{F}$ *contain a quadratic subfield* $\mathfrak{U}$ *over* $\mathfrak{F}$. *Then* $\mathfrak{D}$ *has a quartic subfield* $\mathfrak{Z}$ *of* (13).

For if $\mathfrak{U}$ is inseparable over $\mathfrak{F}$ our result follows from Theorem 11.8 and Lemma 11.3. Hence let $\mathfrak{U}$ be separable over $\mathfrak{F}$ and $\mathfrak{V}$ be the $\mathfrak{D}$-commutator of $\mathfrak{U}$. Then $\mathfrak{V}$ has degree two over $\mathfrak{U}$, $\mathfrak{V}$ has order four over $\mathfrak{U}$, order eight over $\mathfrak{F}$. If S is the generating automorphism of $\mathfrak{U}$ over $\mathfrak{F}$ then S^2 is the identity automorphism and, by Theorem 4.5, there exists a regular quantity y in $\mathfrak{D}$ such that $uy = yu^S$ for every u of $\mathfrak{U}$. Then $uy^2 = y^2u$, $y^2 = v$ is in $\mathfrak{V}$. Now $\mathfrak{W} = \mathfrak{F}(v) \leqq \mathfrak{F}(y)$ and $\mathfrak{W} \leqq \mathfrak{V}$, $\mathfrak{F}(y)$ not in $\mathfrak{V}$ implies that $\mathfrak{W} < \mathfrak{F}(y)$. If $\mathfrak{W}$ has degree two over $\mathfrak{F}$ then either $\mathfrak{W}$ is inseparable, and our result follows again from Theorem 11.8 and Lemma 11.3, or $\mathfrak{W}$ is separable. If v is not in $\mathfrak{U}$ then $\mathfrak{Z} = \mathfrak{U}(v) = \mathfrak{U} \times \mathfrak{W}$ is the quartic field desired. Otherwise $\mathfrak{W} \leqq \mathfrak{U}$, $\mathfrak{W} = \mathfrak{U}$, $yv = vy$ implies that $yu = uy$ for every u of $\mathfrak{U}$ which is impossible since $yu^S = uy$, $u^S \neq u$ for some u of $\mathfrak{U}$. There remains the case $y^2 = \gamma$ in $\mathfrak{F}$. The $\mathfrak{D} > \mathfrak{Q} = (\mathfrak{U}, S, \gamma)$ over $\mathfrak{F}$, $\mathfrak{D} = \mathfrak{Q} \times \mathfrak{Q}_1$ by Theorem 4.6 where $\mathfrak{Q}_1$ is a normal division algebra of degree two over $\mathfrak{F}$, $\mathfrak{D} > \mathfrak{U} \times \mathfrak{W}$ as desired.

The remainder of our proof differs essentially from both that of the original proof of the author in [2] and that of [35]. It will be an application for the case where $\mathfrak{F}$ has characteristic not two of the theory of involutorial simple algebras, and will thus suggest a new tool for considerations of algebras of degree a power of two. We first prove the preliminary

LEMMA 5. *Let $\mathfrak{K}$ be a quadratic field over $\mathfrak{F}$ and $\mathfrak{D}_{\mathfrak{K}}$ contain a quadratic field over $\mathfrak{K}$. Then $\mathfrak{D}$ contains a quadratic field over $\mathfrak{F}$.*

For our result follows from Theorem 4.22 if $\mathfrak{D}_{\mathfrak{K}}$ is not a division algebra. Hence assume that $\mathfrak{D}_{\mathfrak{K}}$ is a division algebra and that $\mathfrak{D}$ contains no quadratic subfield so that if x is in $\mathfrak{D}$ and not in $\mathfrak{F}$ then $\mathfrak{F}(x)$ has degree four over $\mathfrak{F}$. Clearly $\mathfrak{F}(x) = \mathfrak{F}(x^2)$. Write $\mathfrak{K} = \mathfrak{F}(y)$ of degree two over $\mathfrak{F}$. Our hypotheses combined with Theorem 11.8 and Lemma 11.3 imply the existence of a separable quadratic subfield $\mathfrak{K}(d)$ of $\mathfrak{D}_{\mathfrak{K}}$, and we may assume that $d^2 = d + a$ with $a = \alpha_1 + \alpha_2 y$ in $\mathfrak{K}$, α_1 and α_2 in $\mathfrak{F}$. Then $d = d_1 + d_2 y$ for d_1 and d_2 in $\mathfrak{D}$ and not both in $\mathfrak{F}$,

$$(21) \qquad d_1^2 + d_2^2y^2 + (d_1d_2 + d_2d_1)y = (d_1 + \alpha_1) + (d_2 + \alpha_2)y.$$

If either $\mathfrak{K}$ is inseparable over $\mathfrak{F}$ of characteristic two or the characteristic of $\mathfrak{F}$ is not two we may take

$$(22) \qquad y^2 = \alpha \text{ in } \mathfrak{F}$$

and (21) is equivalent to

$$(23) \qquad d_1^2 + d_2^2\alpha = d_1 + \alpha_1, \qquad d_1d_2 + d_2d_1 = d_2 + \alpha_2.$$

Otherwise $\mathfrak{K}$ is separable over $\mathfrak{F}$ of characteristic two and we may take

$$(24) \qquad y^2 - y = \alpha \text{ in } \mathfrak{F},$$

so that (21) is equivalent to

$$(25) \qquad d_1^2 + d_2^2\alpha = d_1 + \alpha_1, \qquad d_1d_2 + d_2d_1 + d_2^2 = d_2 + \alpha_2.$$

In any case $d_2^2 = (d_1 + \alpha_1 - d_1^2)\alpha^{-1}$ in $\mathfrak{F}(d_1)$. If d_2 is in $\mathfrak{F}$ then d_1 not in $\mathfrak{F}$ implies that $d_1^2 - d_1$ is in $\mathfrak{F}$, $\mathfrak{F}(d_1)$ is a quadratic subfield of $\mathfrak{D}$, a contradiction. Hence $\mathfrak{F}(d_2)$ has degree four over $\mathfrak{F}$, $\mathfrak{F}(d_2) = \mathfrak{F}(d_2^2) \leqq \mathfrak{F}(d_1)$, $\mathfrak{F}(d_1) = \mathfrak{F}(d_2)$, $d_1d_2 = d_2d_1$. When $\mathfrak{F}$ has characteristic two and $\mathfrak{K}$ is inseparable we use (23) to obtain $2d_1d_2 = 0 = d_2 + \alpha_2$, d_2 is in $\mathfrak{F}$, contrary to proof. When $\mathfrak{K}$ is separable over $\mathfrak{F}$ of characteristic two we use (25) and $d_1d_2 + d_2d_1 = 0$ to obtain $d_2^2 = d_2 + \alpha_2$, $\mathfrak{F}(d_2)$ has degree two, contrary to proof. If, finally, $\mathfrak{F}$ has characteristic not two (23) implies that $2d_1d_2 = d_2 + \alpha_2$, $(2d_1 - 1)d_2 = \alpha_2$. Put $d_0 = d_1 - \frac{1}{2}$ and have $d_1^2 - d_1 = (d_0 + \frac{1}{2})^2 - (d_0 + \frac{1}{2}) = d_0^2 - \frac{1}{4} = \alpha_1 - \alpha d_2^2$, $2d_0d_2 = \alpha_2$, $4d_0^2d_2^2 = \alpha_2^2$, $4d_0^4 - d_0^2 = 4\alpha_1 d_0^2 - \alpha\alpha_2^2$. But $\mathfrak{F}(d_0) = \mathfrak{F}(d_1)$ of degree four over $\mathfrak{F}$, $\mathfrak{F}(d_0^2)$ clearly has degree two over $\mathfrak{F}$ contrary to our original hypothesis. This proves our lemma.

We assume now that $\mathfrak{D}^2 \sim 1$. If $\mathfrak{F}$ has characteristic two we apply Theorem 7.32 to obtain a pure splitting field $\mathfrak{K} = \mathfrak{K}_r$ of exponent two over $\mathfrak{F}$ of $\mathfrak{D}$ and thus, by Lemma 7.3, a sequence of fields $\mathfrak{K}_0 = \mathfrak{F} < \mathfrak{K}_1 < \mathfrak{K}_2 < \cdots < \mathfrak{K}_r$ where $\mathfrak{K}_i$ is a quadratic field over $\mathfrak{K}_{i-1}$. Then $\mathfrak{D}_i = \mathfrak{D}_{\mathfrak{K}_i} = (\mathfrak{D}_{i-1})_{\mathfrak{K}_i}$, $\mathfrak{D}_{r-1}$ is split by $\mathfrak{K}_r$ and hence $\mathfrak{D}_{r-1}$ contains a quadratic subfield by Theorem 4.27. By Lemma 11.5 so does $\mathfrak{D}_{r-2}$, and an immediate induction implies that $\mathfrak{D}$ has a quadratic subfield. The desired result follows by Lemma 11.4. Let $\mathfrak{F}$ then have characteristic not two. By Theorem 10.19 $\mathfrak{D}$ has an involution J over $\mathfrak{F}$. Since $\mathfrak{D}$ is not commutative we apply Theorem 10.5 to obtain a quantity $d \neq d^J$. Then $q = d - d^J \neq 0$, $q^J = -q \neq q$, q is not in $\mathfrak{F}$, $(q^2)^J = q^2$, $\mathfrak{F}(q^2) < \mathfrak{F}(q)$ since every a of $\mathfrak{F}(q^2)$ has the property $a = a^J$. Hence $\mathfrak{F}(q^2)$ is a quadratic subfield of $\mathfrak{D}$. By Lemma 11.4 we see that $\mathfrak{D}$ has the property of our theorem whenever $\mathfrak{D}^2 \sim 1$.

There remains the case where $\mathfrak{D}$ has exponent four so that, by Theorem 5.16, $\mathfrak{D}^2 \sim \mathfrak{Q}$ of degree and index two over $\mathfrak{F}$. Then there exists a quadratic splitting field $\mathfrak{K}$ over $\mathfrak{F}$ of $\mathfrak{Q}$ and $(\mathfrak{D}_{\mathfrak{K}})^2 \sim 1$. It follows by the proof above that $\mathfrak{D}_{\mathfrak{K}}$ has a quadratic subfield. Then Lemma 11.5 implies that $\mathfrak{D}$ has a quadratic subfield and the application of Lemma 11.4 completes the proof of Theorem 11.9.

We shall now pass to a discussion of the problem of constructing crossed products, in particular those of degree four.

7. The construction of crossed products. We saw in Chapter V that the multiplication table of a crossed product implies that multiplication is associative only if certain conditions are satisfied by the quantities of the defining factor set. These conditions (5.4) on a crossed product $(\mathfrak{Z}, a)$ of degree n over $\mathfrak{F}$ are always satisfied in the case of a cyclic algebra, but in general are n^3 conditions on n^2 quantities of $\mathfrak{Z}$ and hence involve n^3 coefficients in $\mathfrak{F}$. The number of redundant conditions is great even for $n = 4$ and it is desirable to eliminate these redundances in constructing particular algebras. This was accomplished in [97] for $n = 4$ with later generalizations to other cases in [141], [143], [145], [322], [25] and Chapter III of [142]. We shall proceed to a discussion of the general principles underlying the normalization by means of which the simplification is accomplished. Our first result in this direction is

Theorem 10. *Let $\mathfrak{C}$ be an algebra of order m over $\mathfrak{F}$ with a unity quantity e, S be an automorphism over $\mathfrak{F}$ of $\mathfrak{C}$, and $\mathfrak{A}$ be the supplementary sum*

$$\mathfrak{A} = \mathfrak{C} + y\mathfrak{C} + \cdots + y^{r-1}\mathfrak{C} \tag{26}$$

of linear sets $y^i\mathfrak{C}$ whose quantities are indicated formally by $y^i \cdot c$ for c in $\mathfrak{C}$. Assume that g is a quantity of $\mathfrak{C}$ such that $g = g^S \neq 0$, $cg = gc^{S^r}$ for every c of $\mathfrak{C}$, and define multiplication in $\mathfrak{A}$ by means of the distributive law and

$$(y^i \cdot c_1)(y^j \cdot c_2) = y^{i+j-\epsilon r} \cdot (g^\epsilon c_1^{S^j} c_2) \quad (i, j = 0, \cdots, r-1), \tag{27}$$

where $\epsilon = 0$ or 1 according as $i + j < r$ or $i + j \geqq r$. Then the operation of multiplication in $\mathfrak{A}$ is associative so that $\mathfrak{A}$ is an algebra of order mr over $\mathfrak{F}$ with e as unity quantity.

For the operation above is associative if and only if $a_1 = a_2$ for every

$$a_1 = (y^i \cdot c_1)[(y^j \cdot c_2)(y^k \cdot c_3)], \quad a_2 = [(y^i \cdot c_1)(y^j \cdot c_2)](y^k \cdot c) \quad (i, j, k = 0, \cdots, r-1), \tag{28}$$

and c_1, c_2, c_3 of $\mathfrak{C}$. Note that we are using $\mathfrak{C} = y^0\mathfrak{C}$, $c = y^0 \cdot c$ for every c of $\mathfrak{C}$. We put

$$i + j = A + \alpha r, \qquad j + k = B + \beta r, \qquad i + j + k = C + \gamma r \tag{29}$$

for ordinary integers A, B, C, α, β, γ such that

$$0 \leqq A < r, \qquad 0 \leqq B < r, \qquad 0 \leqq C < r. \tag{30}$$

Then clearly $\alpha = 0$ or 1, $\beta = 0$ or 1, $\gamma = 0$, 1, or 2, and

$$i + B = C + (\gamma - \beta)r, \qquad A + k = C + (\gamma - \alpha)r. \tag{31}$$

We now see that

$$\begin{aligned} a_2 &= y^A \cdot (g^\alpha c_1^{S^j} c_2)(y^k \cdot c_3) = y^C \cdot [g^{\gamma-\alpha}(g^\alpha c_1^{S^j} c_2)^{S^k} c_3] \\ &= y^C \cdot (g^\gamma c_1^{S^{j+k}} c_2^{S^k} c_3), \end{aligned}$$

since $(g^\alpha)^S = (g^S)^\alpha = g^\alpha$. Similarly

$$\begin{aligned} a_1 &= (y^i \cdot c_1)[y^B \cdot (g^\beta c_2^{S^k} c_3)] = y^C \cdot [g^{\gamma-\beta}(c_1^{S^B} g^\beta c_2^{S^k} c_3)] \\ &= y^C \cdot [g^\gamma (c_1^{S^{B+\beta r}} c_2^{S^k} c_3)] = a_2 \end{aligned}$$

since $B + \beta r = j + k$. Here we have used $cg = gc^{S^r}$, $cg^\beta = g^B c^{S^{\beta r}}$ for every c of $\mathfrak{C}$. This completes our proof of the associative law. That e is the unity quantity of $\mathfrak{A}$ is trivial.

Note that if we write y^i for $y^i \cdot e$ in $\mathfrak{A}$ the symbolic product $y^i \cdot c$ becomes the product $y^i c$, $g = y^r$, $\mathfrak{A}$ contains y^i for every i, and (27) may now be written as

$$(y^i c_1)(y^j c_2) = y^{i+j} c_1^{S^j} c_2 \tag{32}$$

for all integers i and j and all c_1 and c_2 of $\mathfrak{C}$.

We next prove the elementary

LEMMA 6. *Every automorphism S of a simple algebra $\mathfrak{C}$ over $\mathfrak{F}$ induces an automorphism in its centrum $\mathfrak{K}$ over $\mathfrak{F}$.*

For if k is in $\mathfrak{K}$ then $kc = ck$ for every c of $\mathfrak{C}$, S^{-1} is clearly an automorphism over $\mathfrak{F}$ of $\mathfrak{C}$, $kc^{S^{-1}} = c^{S^{-1}}k$, $k^S c = ck^S$ for every c of $\mathfrak{C}$, k^S is in $\mathfrak{K}$. Then $k \leftrightarrow k^S$ is clearly an automorphism of $\mathfrak{K}$.

The result just obtained will be used in the statement of

Theorem 11. *Let the centrum $\mathfrak{K}$ of a normal simple algebra $\mathfrak{C}$ of degree s over $\mathfrak{K}$ be a cyclic field of degree r over $\mathfrak{F}$, and let $\mathfrak{C}$ have an automorphism S over $\mathfrak{F}$ such that the induced automorphism*

$$k \leftrightarrow k^S \qquad (k \text{ in } \mathfrak{K}),$$

in $\mathfrak{K}$, is a generating automorphism of $\mathfrak{K}$ over $\mathfrak{F}$. Then the algebra $\mathfrak{A}$ of Theorem 11.10, *defined for this r and a regular quantity g, is normal simple of degree sr over $\mathfrak{F}$, and $\mathfrak{C}$ is the $\mathfrak{A}$-commutator of $\mathfrak{K}$.*

For let $a = c_1 + yc_2 + \cdots + y^{r-1}c_{r-1}$ with the c_i in $\mathfrak{C}$, a be in the $\mathfrak{A}$-commutator of $\mathfrak{K}$. By hypothesis there exists a quantity u in $\mathfrak{K}$ such that $\mathfrak{K} = \mathfrak{F}(u)$, $u = u^{S^t}$ if and only if t is divisible by r. Then $ua - au = \sum_{i=1}^{r} cj^{i-1}c_i(u^{S^{i-1}} - u) = 0$ if and only if $c_i(u^{S^i} - u) = 0$. Since $\mathfrak{K}$ is a field we have $c_i = 0$ for $i \neq 1$, a is in $\mathfrak{C}$, $\mathfrak{C} = \mathfrak{A}^{\mathfrak{K}}$. If a is in the centrum of $\mathfrak{A}$ it is in $\mathfrak{A}_{\mathfrak{K}}$ and hence is in $\mathfrak{C}$ and thus also in $\mathfrak{K}$. But $ya - ay = y(a - a^S) = 0$ if and only if $a = a^S$, a is in $\mathfrak{F}$. Hence $\mathfrak{A}$ is normal.

We let $\mathfrak{B}$ be a non-zero ideal of $\mathfrak{A}$ and among the non-zero quantities of $\mathfrak{B}$ we choose $b = c_0 + yc_1 + \cdots + y^t c_t$, with $c_t \neq 0$, such that t is the least possible. Since $y^r = g$ in $\mathfrak{C}$ we have $y^{-1} = y^{r-1}g^{-1}$ in $\mathfrak{A}$, and if $c_0 = 0$ we would have $y^{-1}b = c_1 + \cdots + y^{t-1}c_t$ in $\mathfrak{B}$ contrary to our definition of t. Hence $c_0 \neq 0$. We let $\mathfrak{K} = \mathfrak{F}(u)$ as above and, since $\mathfrak{B}$ is an ideal, the quantity $bu - u^{S^{-t}}b = yc_1(u - u^{S^{1-t}}) + \cdots + y^t c_t(u - u)$ in $\mathfrak{B}$. But $u^{S^{-t}} \neq u$ if $t > 0$ and $bu - u^{S^{-t}}b = c_{10} + \cdots + y^{t-1}c_{t0}$ contrary to our definition of t. Hence $t = 0$, $b \neq 0$ is in $\mathfrak{C}$. Since $\mathfrak{C}$ is simple we have $\mathfrak{C}b\mathfrak{C} = \mathfrak{C}$, the unity quantity e of $\mathfrak{A}$ is in $\mathfrak{C} = \mathfrak{C}b\mathfrak{C} \leqq \mathfrak{C}\mathfrak{B}\mathfrak{C} \leqq \mathfrak{B}$, $\mathfrak{A}e\mathfrak{A} = \mathfrak{A} \leqq \mathfrak{B}$, $\mathfrak{A} = \mathfrak{B}$ is simple. It is trivial to see that $\mathfrak{A}$ has order s^2r^2 and hence degree rs over $\mathfrak{F}$.

As a final result of a general nature we prove the *norm condition*

Theorem 12. *Let the algebra $\mathfrak{C}$ of Theorem* 11.11 *be a division algebra and r be a prime. Then the algebra $\mathfrak{A}$ is a division algebra if and only if there exists no quantity c in $\mathfrak{C}$ such that*

$$g = cc^S \cdots c^{S^{r-1}}. \tag{33}$$

For let $\mathfrak{A}$ be a division algebra and assume that (33) is satisfied for c in $\mathfrak{C}$. Then $c \neq 0$, $b = c^{-1}$ is in $\mathfrak{C}$. We put $y_0 = yb$ and since $c_0\mathfrak{C} = \mathfrak{C}$ for every $c_0 \neq 0$ of $\mathfrak{C}$ we have $y_0^i\mathfrak{C} = y^i(b^{S^{i-1}} \cdots b^S b)\mathfrak{C} = y^i\mathfrak{C}$. Then $\mathfrak{A} = \mathfrak{C} + y_0\mathfrak{C} + \cdots + y_0^{r-1}\mathfrak{C}$. Also $y_0^r = y^r b^{S^{r-1}} \cdots b^S b = cc^S \cdots c^{S^{r-1}}(c^{-1})^{S^{r-1}} \cdots (c^{-1})^S c^{-1} = 1$, $y_0^{r-1} + y_0^{r-2} + \cdots + y_0 + 1 = y_1 \neq 0$, $y_0 - 1 = y_2 \neq 0$, $y_1y_2 =$

$y_0^r - 1 = 0$ which is impossible in a division algebra. Conversely let (33) be not satisfied for any c of $\mathfrak{C}$. By Theorem 4.16 if $\mathfrak{K}_0$ over $\mathfrak{F}$ is equivalent to $\mathfrak{K}$ over $\mathfrak{F}$ then $\mathfrak{A}_{\mathfrak{K}_0} \sim \mathfrak{B}_0$ over $\mathfrak{K}_0$ of index s, a divisor of the index m of $\mathfrak{A}$. Now m divides sr and is divisible by s so that since r is a prime $m = sr$ or s. In the former case $\mathfrak{A}$ is a division algebra. In the latter case $\mathfrak{A} = \mathfrak{M} \times \mathfrak{D}$ where $\mathfrak{M} = (\mathfrak{K}, S, 1)$. Hence there exists a quantity y_0 in $\mathfrak{A}$ such that $y_0^r = 1$, $ky_0 = y_0k^S$ for every k in $\mathfrak{K}$. But $ky = yk^S$ for y as in (26) and having the property $y^r = g$. We see that $k^{S^{-1}}y = y(k^{S^{-1}})^S = yk$, $y^{-1}k^{S^{-1}} = ky^{-1}$, and that $y_0k = k^{S^{-1}}y_0$. Hence $y^{-1}y_0k = y^{-1}k^{S^{-1}}y_0 = ky^{-1}y_0$ so that, by Theorem 11.11, $b = y^{-1}y_0$ is in $\mathfrak{C}$, $y_0 = yb$, $y_0^r = 1 = g(b^{S^{r-1}} \cdots b^S b)$. Put $c = b^{-1}$ and have

$$b^{S^i}c^{S^i} = (bc)^{S^i} = 1,$$

$$cc^S \cdots c^{S^{r-1}} = g(b^{S^{r-1}} \cdots b^S b)(cc^S \cdots c^{S^{r-1}}) = g$$

contrary to our hypothesis.

The results we have obtained may be used as follows in the construction of crossed products. We let $\mathfrak{Z}$ be a normal field of degree n over $\mathfrak{F}$, $\mathfrak{G}$ be the automorphism group of $\mathfrak{Z}$ over $\mathfrak{F}$. Assume that $\mathfrak{G}$ has a normal divisor $\mathfrak{H}$ of order s, $n = rs$, and that the quotient group $\mathfrak{G}/\mathfrak{H}$ is cyclic of order r. Then

$$\mathfrak{G} = \mathfrak{H} + G\mathfrak{H} + G^2\mathfrak{H} + \cdots + G^{r-1}\mathfrak{H}, \tag{34}$$

where $G^r = H_1$ in $\mathfrak{H} = (H_1, H_2, \cdots, H_s)$. The set $\mathfrak{K}$ of all quantities k of $\mathfrak{Z}$ such that $k = k^{H_i}$ $(i = 2, \cdots, s)$ is a subfield of $\mathfrak{Z}$, and is cyclic of degree r over $\mathfrak{F}$ with generating automorphism

$$k \leftrightarrow k^G \qquad (k \text{ in } \mathfrak{K}). \tag{35}$$

We now let $\mathfrak{A}$ be a crossed product $(\mathfrak{Z}, a)$. The $\mathfrak{A}$-commutator $\mathfrak{C}$ of $\mathfrak{K}$ is the crossed product $\mathfrak{C} = (\mathfrak{Z}, a_1)$ over $\mathfrak{K}$ given by

$$\mathfrak{C} = u_{H_1}\mathfrak{Z} + u_{H_2}\mathfrak{Z} + \cdots + u_{H_s}\mathfrak{Z}, \tag{36}$$

and the problem of satisfying the associativity conditions for $\mathfrak{A}$ involves first the satisfying of these conditions for the algebra $\mathfrak{C}$ of degree s over $\mathfrak{K}$, a problem for an algebra of lower degree. Let us suppose this problem solved.

The algebra $\mathfrak{A}$ is now clearly an algebra $\mathfrak{A} = \mathfrak{C} + y\mathfrak{C} + \cdots + y^{r-1}\mathfrak{C}$ where $y = u_G$, G is the automorphism of (34). Now if we make the simple normalizations

$$y^r = u_D, \qquad u_{G^{(i)}H_j} = y^i u_{H_j} \qquad (i = 0, 1, \cdots, r-1; j = 1, \cdots, s), \tag{37}$$

and use the property that $\mathfrak{H}$ is a normal divisor of $\mathfrak{G}$, we obtain $HG = GH^{(1)}$ for every H of $\mathfrak{H}$, $u_H y = u_H u_G = u_{HG}a_{H,G} = u_{GH^{(1)}}a_{H,G} = yu_{H^{(1)}}a_{H,G}$. It follows that the automorphism S of Theorem 11.11 is the correspondence generated by

$$u_H^S = u_{H^{(1)}}a_{H,G}, \qquad z^S = z^G \qquad (z \text{ in } \mathfrak{Z}). \tag{38}$$

Thus we must have

$$(u_H u_T)^S = (u_{HT} a_{H,T})^S = u_{H^{(1)}T^{(1)}} a_{HT,G} a_{H,T}^G$$
$$= u_{H^{(1)}} a_{H,G} u_{T^{(1)}} a_{T,G} = u_{H^{(1)}T^{(1)}} a_{H^{(1)},T^{(1)}} a_{H,G}^{T^{(1)}} a_{T,G},$$

that is

$$a_{HT,G} a_{H,T}^G = a_{G^{-1}HG,G^{-1}TG} a_{H,G}^{G^{-1}TG} a_{T,G} \tag{39}$$

for every H and T of $\mathfrak{H}$. These are clearly only s^2 conditions. Moreover, we must have

$$u_H^{S^r} = y^{-r} u_H y^r = u_D^{-1} u_H u_D. \tag{40}$$

But $u_D u_{D^{-1}} = u_I a_{D,D^{-1}}$, and we saw in Section 5.5 that u_I is in $\mathfrak{Z}$,

$$(u_D)^{-1} = u_{D^{-1}} a \tag{41}$$

for a in $\mathfrak{Z}$. It follows that

$$u_D^{-1} u_H u_D = u_{D^{-1}H} a_{D^{-1},H} u_D a^{HD} = (u_{D^{-1}HD})(a_{D^{-1},H}^D)(a^{HD}).$$

However $u_H^{S^2} = [(u_{G^{-1}HG})(a_{H,G})]^S = (u_{G^{-2}HG^2})(a_{G^{-1}HG,G})(a_{H,G}^G)$, and we evidently obtain

$$(a_{D^{-1},H}^D)(a^{HD}) = (a_{G_1,G})(a_{G_2,G})^G \cdots (a_{G_r,G})^{G^{r-1}}, \qquad G_i = G^{i-r} H G^{r-i}, \tag{42}$$

for every H of $\mathfrak{H}$, a set of s additional conditions.

Conversely, when $\mathfrak{C}$ is associative and (39), (42) hold, the algebra $\mathfrak{A}$ defined by (26), with S defined by (38), and $g = y^r = u_D$ in $\mathfrak{C}$, has the properties of Theorem 11.11 and is the desired crossed product. Thus our method reduces the associativity conditions to those of $\mathfrak{C}$ together with $s^2 + s$ additional conditions, a set of at most $s^3 + s^2 + s < s^3 r^3$ conditions.

This method is of particular value in the case when $\mathfrak{G}$ is a solvable group so that $\mathfrak{A}$ may be constructed by a sequence of constructions of the type above. Moreover in this case the algebra of the initial step is a cyclic algebra and so is associative, our associativity conditions are then a finite sequence of conditions of the type given in (39), (42). In addition Theorem 11.12 then gives the conditions that the algebra we construct be a division algebra.

Further simplifications are possible when $\mathfrak{G}$ is an abelian group. For this and other normalizations see [322] and Chapter III of [142]. We now pass to a special case illustrating the theory.

Let $\mathfrak{Z} = \mathfrak{U} \times \mathfrak{B}$, where $\mathfrak{U}$ and $\mathfrak{B}$ are separable inequivalent quadratic fields over $\mathfrak{F}$. Then in Exercise 4 of Section A8.3 we saw that the algebra $\mathfrak{Z}$ is a normal quartic field over $\mathfrak{F}$ with automorphism group

$$\mathfrak{G} = (I, S_1, S_2, S_3), \qquad S_3 = S_1 S_2 = S_2 S_1, \qquad S_i^2 = I \ (i = 1, 2, 3). \tag{43}$$

The field $\mathfrak{U}$ is the set of all quantities $u = u^{S_2}$ of $\mathfrak{Z}$, the field $\mathfrak{B}$ is the set of all quantities $v = v^{S_1}$ of $\mathfrak{Z}$, and

$$u \leftrightarrow u^{S_1}, \qquad v \leftrightarrow v^{S_2} \qquad (u \text{ in } \mathfrak{U},\ v \text{ in } \mathfrak{B}) \tag{44}$$

are respective generating automorphisms of $\mathfrak{U}$ and $\mathfrak{V}$. Clearly $\mathfrak{Z} = \mathfrak{U}_{\mathfrak{V}}$ over $\mathfrak{V}$, $\mathfrak{Z} = \mathfrak{V}_{\mathfrak{U}}$ over $\mathfrak{U}$.

We let $\mathfrak{A}$ be a crossed product

$$\mathfrak{A} = \mathfrak{Z} + y_1\mathfrak{Z} + y_2\mathfrak{Z} + y_3\mathfrak{Z}, \tag{45}$$

where we have $u_{S_i} = y_i$ $(i = 1, 2, 3)$, and have made the normalization $u_I = 1$. Then $y_i^2 = a_i$ in $\mathfrak{Z}$ and $y_ia_iy_i^{-1} = a_i = a_i^{S_i}$. Hence a_2 is in $\mathfrak{U}$, a_1 is in $\mathfrak{V}$. The $\mathfrak{A}$-commutator of $\mathfrak{V}$ is the cyclic (quaternion) algebra $\mathfrak{C} = (\mathfrak{Z}, S_1, a_1) = \mathfrak{Z} + y_1\mathfrak{Z}$, and $\mathfrak{A} = \mathfrak{C} + y_2\mathfrak{C}$, y_2 is the quantity designated above as y. Now

$$y_1y_2 = y_2y_1a, \tag{46}$$

for a in $\mathfrak{Z}$, since $S_1S_2 = S_3 = S_2S_1$, $y_3 = y_2y_1$, $u_{S_2}u_{S_1} = u_{S_2S_1}a$.

The conditions (39) are obtained from $u_H^S = y_2^{-1}u_Hy_2$ for every H of $\mathfrak{H} = (I, S_1)$, and that given by $H = I$, $u_H = 1$ is clearly trivial in the present case. Thus we only need consider $y_1^S = y_2^{-1}y_1y_2 = y_1a$, $(y_1^S)^2 = a_1^S = a_1^{S_2} = (y_1a)^2 = a_1a^{S_1}a$. Hence the only significant condition of (39) is

$$a_1^{S_2}a_1^{-1} = a^{S_1}a. \tag{47}$$

The conditions (42) are the result of $y_1^{S_2} = (y_1a)^S = y_1aa^{S_2} = y_2^{-2}y_1y_2^2 = a_2^{-1}y_1a_2 = y_1(a_2^{S_1})^{-1}a_2$. Thus

$$aa^{S_2} = a_2(a_2^{S_1})^{-1} \tag{48}$$

is the only remaining associativity condition. We shall now use the simple structure of $\mathfrak{G}$ to obtain further simplifications of our conditions.

The group $\mathfrak{G}$ has a third subgroup (I, S_3) of index two and a corresponding (cyclic) quadratic subfield $\mathfrak{W}$ of all quantities w of $\mathfrak{Z}$ such that $w = w^{S_3}$. Evidently

$$w \leftrightarrow w^{S_1} = w^{S_2}$$

is a generating automorphism over $\mathfrak{F}$ of $\mathfrak{W}$. Now

$$y_3 = y_2y_1, \qquad y_3^2 = a_3 = a_3^{S_3} \tag{49}$$

so that a_3 is in $\mathfrak{W}$. But $y_3^2 = y_2y_1y_2y_1 = y_2^2y_1ay_1 = a_2a_1a^{S_1}$. It follows that $a^{S_1} = a_3(a_2a_1)^{-1}$ and, since $a_1 = a_1^{S_1}$ is in $\mathfrak{V}$ and $S_1^2 = I$, that

$$a = a_3^{S_1}(a_1a_2^{S_1})^{-1}. \tag{50}$$

Now (47) is equivalent to

$$a_3^{S_1}(a_1a_2^{S_1})^{-1}a_3(a_1a_2)^{-1} = (a_3a_3^{S_1})(a_2a_2^{S_1})^{-1}a_1^{-2} = a_1^{S_2}a_1^{-1}, \quad a_1a_1^{S_2}a_2a_2^{S_1} = a_3a_3^{S_1},$$

that is, to

$$N_{\mathfrak{W}|\mathfrak{F}}(a_3) = N_{\mathfrak{U}|\mathfrak{F}}(a_2)\cdot N_{\mathfrak{V}|\mathfrak{F}}(a_1). \tag{51}$$

Similarly (48) is equivalent to

$$a_3^{S_1}(a_1a_2^{S_1})^{-1}a_3(a_1^{S_2}a_2^{S_2S_1})^{-1} = a_3a_3^{S_1}(a_1a_1^{S_2})^{-1}(a_2^{S_1})^{-2} = a_2(a_2^{S_1})^{-1},$$

and hence to (51).

We have proved that (47) and (48) are equivalent to (51) and the condition that a_3 be in $\mathfrak{W}$. Thus our construction is completed by the formula (50) for a and the selection of a_1 in $\mathfrak{V}$, a_2 in $\mathfrak{U}$, a_3 in $\mathfrak{W}$ such that (51) holds. This completes our construction of all crossed products of degree four over $\mathfrak{F}$, a construction which includes that of all possible normal division algebras of degree four over $\mathfrak{F}$. For additional results on such algebras see the papers [12], [24], [25], and [48] referred to previously.

8. Literature on non-associative algebras. The structure of non-associative algebras is of considerable interest to algebraists. The most interesting of the earlier types are the Cayley-Dickson division algebras over the field of real numbers as discussed in [128] on page 72, on pages 14 and 15 of [129], and in [466]. These algebras also arise in the study of the non-associative algebras called the r-number systems of quantum mechanics. See [216], [217], [218], [219], [34]. The Cayley-Dickson systems were generalized in [281], [464] where algebras of similar type over a general field were studied.

Additional special types of non-associative algebras were studied in [125], [126], [127], [128], [271], [146]. However the examples most important at the present time are the Lie algebras arising in connection with the infinitesimal transformations connected with continuous groups. The modern theory of such algebras is connected with that of associative algebras as is indicated in [67], [86], [96], [206], [210], [435], [436], [437], [402], [403], [461], [465]. Simple Lie algebras are closely related to involutorial simple algebras, and their theory is given in [208], [213], [214], [233], [234]. For references to earlier work see the bibliographies in [435], [436], [437], [208]. See also [416] and the work on semi-linear transformations in [415], [61], [209], [295], [298], [317].

9. Riemann matrices. Let $\mathfrak{F}$ be a field of real numbers, ω be a matrix of r rows and $2r$ columns of complex elements, and $i = (-1)^{\frac{1}{2}}$. Then ω is called a *Riemann matrix* over $\mathfrak{F}$ if there exists a matrix $C = -C'$ with elements in $\mathfrak{F}$ such that $\omega C\omega' = 0$ and $i\omega C\bar{\omega}'$ is a positive definite Hermitian matrix. We call C a *principal matrix* of ω.

If β and B are non-singular square complex matrices of r and $2r$ rows respectively, and the elements of B are in $\mathfrak{F}$, the matrix $\beta\omega B$ is a Riemann matrix with principal matrix $B^{-1}C(B^{-1})'$ and $\beta\omega B$ is said to be *isomorphic* to ω. We call ω *impure* (over $\mathfrak{F}$) if it is isomorphic to

$$\begin{pmatrix} \omega_1 & 0 \\ \omega_3 & \omega_2 \end{pmatrix},$$

where ω_1 has r_1 rows and $2r_1$ columns, $0 < r_1 < r$; otherwise ω is called *pure.* If ω is impure it can actually be shown that ω_1 and ω_2 are Riemann matrices over $\mathfrak{F}$ and that ω is isomorphic to

$$\begin{pmatrix} \omega_1 & 0 \\ 0 & \omega_2 \end{pmatrix}.$$

For this see [41]. Thus the first problem of the theory of Riemann matrices is that of the proof of this result and the consequent theory of the reduction of an arbitrary Riemann matrix to pure components.

A matrix A with elements in $\mathfrak{F}$ is said to be a *multiplication* of ω if there exists a complex matrix α such that $\alpha\omega = \omega A$. The set $\mathfrak{A}$ of all multiplications of ω is an algebra over $\mathfrak{F}$ with the $2r$-rowed identity matrix I_{2r} as unity quantity. Moreover it may be shown that ω is pure if and only if $\mathfrak{A}$ is a division algebra.

The theory of Riemann matrices over the field of all rational numbers arose in connection with the study of algebraic correspondences of algebraic curves and their consequences on the period matrix ω of the abelian integrals of the first kind on the corresponding Riemann surface. Here ω can be shown to be a Riemann matrix. The subject is also connected with the period matrices of functions of r independent variables with $2r$ independent period systems. These connections are discussed in great detail in the report on such topics as given in [246]. A complete set of references is given, in particular, to the work of Lefschetz, Rosati, and Scorza, the principal writers on the subject up to the time of the report. See also the later [331], [14].

The more modern phase of the subject begins in the investigations of [8], [15], [21]. A complete exposition is then given in [41], [23], [38], where the final results are obtained. For proof of existence theorems see [33], [42]. These are of course investigations for the case of ω over a general $\mathfrak{F}$. See also [40].

In [440], [441], [442] H. Weyl connected the subject with that of generalized Riemann matrices of our Chapters VIII, X. This connection was amplified in [43] and an alternative treatment thereby provided. Let us investigate the connection briefly.

The equation $\omega C\omega' = 0$ implies that $\bar{\omega}C\bar{\omega}' = 0$. Now $\gamma = i\omega C\bar{\omega}'$ is positive definite and hence non-singular, $\omega C\bar{\omega}' = -i\gamma$, $\bar{\omega}C\omega' = i\bar{\gamma}$. Write

$$\Omega = \begin{pmatrix} \omega \\ \bar{\omega} \end{pmatrix},$$

so that Ω is a $2r$-rowed complex square matrix with the property

$$\Omega C\bar{\Omega}' = \begin{pmatrix} \omega C \\ \bar{\omega}C \end{pmatrix} (\bar{\omega}', \omega') = \begin{pmatrix} -i\gamma & 0 \\ 0 & i\bar{\gamma} \end{pmatrix}.$$

Then $\Omega C\bar{\Omega}'$ is non-singular, both Ω and C must be non-singular. We now may write

$$\Omega^{-1} = C\bar{\Omega}' \begin{pmatrix} i\gamma^{-1} & 0 \\ 0 & -i\bar{\gamma}^{-1} \end{pmatrix} = C(\bar{\omega}' i\gamma^{-1}, -\omega' i\bar{\gamma}^{-1}),$$

so that

$$\begin{aligned} R_\omega &= i\Omega^{-1} \begin{pmatrix} I_r & 0 \\ 0 & -I_r \end{pmatrix} \Omega = C(-\bar{\omega}'\gamma^{-1}, -\omega'\bar{\gamma}^{-1}) \cdot \begin{pmatrix} \omega \\ \bar{\omega} \end{pmatrix} \\ &= -C(\bar{\omega}'\gamma^{-1}\omega + \omega'\bar{\gamma}^{-1}\bar{\omega}) \end{aligned}$$

is the product of $-C$ by the sum of a complex matrix and its conjugate and is real. Evidently $R_\omega^2 = -I_{2r}$,

$$R_\omega C = i\Omega^{-1}\begin{pmatrix} I_r & 0 \\ 0 & -I_r \end{pmatrix}\begin{pmatrix} -i\gamma & 0 \\ 0 & i\bar{\gamma} \end{pmatrix}(\bar{\Omega}')^{-1}$$

$$= \Omega^{-1}\begin{pmatrix} \gamma & 0 \\ 0 & \bar{\gamma} \end{pmatrix}(\bar{\Omega}')^{-1}$$

is positive definite. It follows from our definitions of Section 8.13 that *R_ω is a generalized Riemann matrix with principal matrix C.*

Conversely, it is shown in [43] that if R is a real $2r$-rowed generalized Riemann matrix with a skew principal matrix C such that $R^2 = -I_{2r}$ then there exists a Riemann matrix ω such that $R = R_\omega$. The multiplication algebras of R_ω and of ω are the same algebra of matrices A, and R_ω is impure if and only if this is true of ω. This result reduces the whole question of the structure of the multiplication algebra of ω to that of R_ω, a problem completely solved by the results of Section 10.14. See also [114] for results on integral elements of the multiplication algebra of a Riemann matrix.

A Riemann matrix ω is called *real* if $\lambda\omega = \omega L$ where λ is a complex matrix, L has elements in $\mathfrak{F}$, $L^2 = I_{2r}$. For the geometric origin of this topic see [98], [99], [111], [112], [113], [243], [244]. The final solution of the principal questions on such matrices and connections with arbitrary generalized Riemann matrices of real elements was given in [43].

10. Supplementary reading. There are several references in the bibliography which are not properly a part of the theory of linear associative algebras. But they are sufficiently closely related to our subject to be of interest and we have listed them. We have already spoken in this connection of the papers on the theory of rings and ideals and shall not mention them further.

A second subject somewhat removed from that of these LECTURES is the theory of analytic functions of hypercomplex variables, that is, of variables in an algebra which is usually taken over the real or complex field. Similarly we have listed a number of other articles on subjects in which the number system is an algebra, for example those on matrices over an algebra, the theory of equations in an algebra, algebras over an algebra. Other general topics are connections of special algebras with the theory of numbers, the structure and representation of Boolean algebras and algebras over a Boolean algebra. We shall not list the particular references for these topics here but refer the reader to the bibliography where the titles of papers on these topics clearly indicate their content.

These are the major items on our list which are not precisely in our field. Let us then turn to our subject matter proper. We have already given sources for our first three chapters but would like to mention sources of the non-classical material on scalar extensions and inseparable fields of Sections 2.15 and 2.17. For this see [45], [119] as well as parts of [414]. The fundamental Lemma 3.3

was first proved in [78] but the proof we have given was suggested by that in [210].

The results of Chapters IV, V are due to R. Brauer, H. Hasse, E. Noether, and the author and are contained in [14], [15], [19], [27], [22], [20], [52], [74], [76], [77], [78], [80], [81], [82], [89], [179], [180], [181], [414]. Explicit references are indicated in [119]. However we will make some remarks about Chapter V. The exposition we have given has its origin in the theory given by H. Hasse in [179] which led to that of M. Deuring in [119]. Results of this type are contained in [5], [6], [15], [22], and the associativity conditions are essentially contained in the matrix representations of crossed products in [14], a paper which seems to have been overlooked by Deuring. We have already mentioned the work on crossed products of F. Cecioni and L. E. Dickson, but wish to mention also that the concept of a (non-cyclic) crossed product seems to have been considered first in [427]. We have also given references to papers using crossed product theory, for example [1], and wish finally to call attention to a result of R. Brauer proved in [81] as well as in [119] and [369]. This result states that every crossed product $(\mathfrak{N}, a)$ is similar to $(\mathfrak{N}_1, a_1)$ such that $\mathfrak{N}_1$ contains $\mathfrak{N}$ and the elements of the factor set a_1 are roots of unity.

We have already listed the numerous papers on the representation theory in Chapter VIII, but wish to make particular reference to the papers [74], [75], [76], [77], [78] of R. Brauer. References will be found in these papers to the fundamental work of G. Frobenius and I. Schur, and we leave the reader to locate these references in those papers. The material on cyclic algebras of Chapter VII has its origin in [22], [396], [397], [420], and [50] and was derived originally independently of the theory of crossed products.

The theory of factor sets has its origin partly in a generalization given in [78]. The author was led independently to similar results, never published, as a consequence of the representation theory for the multiplication algebras of Riemann matrices in [21]. It must be observed however that it is not yet known whether there exists any normal division algebra not a crossed product. Thus the interest in factor sets defined in [78] relative to an arbitrary maximal separable subfield of a normal division algebra is not yet quite as great as that of the factor sets occurring in crossed products. However the former subject may furnish an important tool for the investigation of this question of the existence of such division algebras. Brauer has recently used this method in the study of the structure of normal division algebras of degree five in [88], and has obtained a very interesting result for which we refer the reader to the paper in question.

We shall close with a brief mention of the theory of the arithmetics of algebras. The modern theory of this subject occupies approximately one half of the report on algebras of [119], and the author hopes that he may some day give an exposition of this subject as a part of a text on algebraic numbers, the theory of class fields, and the arithmetics of algebras. Fundamental results obtained since the report mentioned above are given in the numerous papers listed in our bibliography, in particular in the papers [150], [151], [152] of M. Eichler. Numerous papers in the tool subjects of class fields and valuation theory are also listed.

11. Bibliography. Our bibliographical list was compiled by the use of a number of sources and contains, in particular, a selection of the titles of nearly all papers on algebras and closely related topics appearing in the first eighteen volumes of the Zentralblatt. It includes relatively few titles of early papers on our subject but, in view of the supplementary lists of such titles already indicated, our bibliography should be quite completely adequate for a student of the modern theory of algebras.

The titles of the journals appearing in our list will naturally be abbreviated but we believe it worth while to spare the reader the very trying task of translating the abbreviations commonly used. Thus we shall begin our bibliography with a list of abbreviations and the corresponding full titles of journals as given in the Union List of Serials, a standard library reference.

BIBLIOGRAPHY

Acc. d'It.—Accademia d'Italia. Rome. Memorie.
Atti Nap.—R. Accademia delle Scienze. Naples. Atti.
Atti Tor.—R. Accademia delle Scienze di Torino. Atti.
Atti Lin.—Accademia Nazionale dei Lincei. Atti.
Acta Arith.—Acta Arithmetica.
Act. Sci. Ind. Paris—Actualités Scientifiques et Industrielles. Paris.
Ak. Berlin S. B.—K. Akademie der Wissenschaften. Berlin. Sitzungsberichte.
Ak. Vienna S. B.—K. Akademie der Wissenschaften. Vienna. Sitzungsberichte.
Am. J.—American Journal of Mathematics.
A. M. S. Bull.—American Mathematical Society. Bulletin.
A. M .S. Coll.—American Mathematical Society. Colloquium Publications.
A. M. S. Trans.—American Mathematical Society. Transactions.
Ann. di Mat.—Annali di Matematica Pura ed Applicata.
Ann. of Math.—Annals of Mathematics.
Bull. Aca. Sci. U. S. S. R.—Bulletin of the Academy of Sciences. U. S. S. R.
Calc. M. S. Bull.—Calcutta Mathematical Society. Bulletin.
Camb. Ph. S. Pro.—Cambridge Philosophical Society. Proceedings.
Cir. Mat. Pal.—Circolo Matematico di Palermo. Rendiconti.
Chi. M. S. J.—Chinese Mathematical Society. Journal.
Comm. Math. Helv.—Commentarii Mathematici Helvetici.
Comp. Math.—Compositio Mathematica.
C. R. Paris.—Institut de France. Académie des Sciences. Comptes Rendus.
Deut. Math.—Deutsche Mathematik.
D. M. V. Jahr.—Deutsche Mathematiker-Vereinigung. Jahresbericht.
Duke J.—Duke Mathematical Journal.
Edin. M. S. Pro.—Edinburgh Mathematical Society. Proceedings.
Erg. der Math.—Ergebnisse der Mathematik.
Fund. Math.—Fundamenta Mathematicae.
Giornale—Giornale di Matematiche.
Gött. Nach.—Gesellschaft der Wissenschaften zu Göttingen. Nachrichten.
Hamb. Abh.—Hamburg Mathematisches Seminar. Abhandlungen.
Hans. Abh.—Hansische Universität. Mathematisches Seminar. Abhandlungen.
Hiro. J.—Hiroshima Imperial University, Japan. Journal of the Faculty of Sciences.
Hokk. J.—Hokkaido Imperial University, Japan. Mathematical Society. Journal.
Imp. Aca. Tok. Pro.—Imperial Academy, Tokyo. Proceedings.
Ind. Aca. Pro.—Indian Academy of Sciences. Proceedings.
Ind. M. S. J.—Indian Mathematical Society. Journal.
Jap. J.—Japanese Journal of Mathematics.
Jassy Ann.—Jassy, Roumania. Universitatea. Annales Scientifiques.
J. de Math.—Journal de Mathématiques.
J. für Math.—Journal für die reine und angewandte Mathematik.
London M. S. J.—London Mathematical Society. Journal.
London M. S. Pro.—London Mathematical Society. Proceedings.
Math. Ann.—Mathematische Annalen.
Math. Zeit.—Mathematische Zeitschrift.
Mess. Math.—Messenger of Mathematics.
Monat.—Monatshefte für Mathematik und Physik.
P. N. A. S.—National Academy of Sciences. Proceedings.

N. R. C. Bull.—National Research Council. Bulletin.
Nat. Ges. Zur. V.—Naturforschende Gesellschaft, Zurich. Vierteljahreschrift.
P.-M. S. Jap.—Physico-Mathematical Society of Japan. Proceedings.
Rome Rend.—Rome Università. Seminario Matematico. Rendiconti.
R. S. Can. Trans.—Royal Society of Canada. Transactions.
R. S. Edin. Pro.—Royal Society of Edinburgh. Proceedings.
Toh. Imp. U. Sci. Rep.—Science Reports of the Tôhoku Imperial University. Sendai, Japan.
Schw. Nat. Ges. V.—Schweizerische Naturforschende Gesellschaft. Verhandlungen.
Soc. Pol. Mat. Ann.—Société Polonaise de Mathématiques. Annales.
Tôhoku J.—Tôhoku Mathematical Journal.
Tokyo J.—Tokyo. Faculty of Science. Journal.
Toronto St.—Toronto University. Studies.
Zeit. Phy.—Zeitschrift für Physik.
Zentralblatt—Zentralblatt für Mathematik und ihre Grenzgebiete.

Akizuki, Y. [**1**]: *Eine homomorphe Zuordnung der Elemente der galoisschen Gruppe zu den Elementen einer Untergruppe der Normklassengruppe*, Math. Ann., v. 112 (1936), 566–71.

Albert, A. A. [**2**]: *A determination of all normal division algebras in* 16 *units*, A. M. S. Trans., v. 31 (1929), 253–60; [**3**]: *On the rank equation of any normal division algebra*, A. M. S. Bull., v. 35 (1929), 335–8; [**4**]: *The rank function of any simple algebra*, P. N. A. S., v. 15 (1929), 272–6; [**5**]: *On the structure of normal division algebras*, Ann. of Math., v. 30 (1929), 322–38; [**6**]: *Normal division algebras in* $4p^2$ *units, p an odd prime*, Ann. of Math., v. 30 (1929), 583–90; [**7**]: *Direct products of rational quaternion algebras*, Ann. of Math., v. 30 (1929), 621–5; [**8**]: *The non-existence of pure Riemann matrices with normal multiplication algebras of order* 16, Ann. of Math., v. 31 (1930), 375–80; [**9**]: *A necessary and sufficient condition for the equivalence of rational generalized quaternion algebras*, A. M. S. Bull., v. 36 (1930), 535–40; [**10**]: *Determination of all normal division algebras in* 36 *units of type* R_2, Am. J., v. 52 (1930), 283–92; [**11**]: *A note on an important theorem on normal division algebras*, A. M. S. Bull., v. 36 (1930), 649–50; [**12**]: *New results in the theory of normal division algebras*, A. M. S. Trans., v. 32 (1930), 171–95; [**13**]: *A construction of all non-commutative rational division algebras of degree* 8, Ann. of Math., v. 31 (1930), 567–76; [**14**]: *The structure of pure Riemann matrices with non-commutative multiplication algebras*, Cir. Mat. Pal., v. 55 (1931), 57–115; [**15**]: *On direct products, cyclic division algebras, and pure Riemann matrices*, A. M. S. Trans., v. 33 (1931), 219–34; [**16**]: *On normal division algebras of type R in* 36 *units*, A. M. S. Trans., v. 33 (1931), 235–43; [**17**]: *On the Wedderburn norm condition for cyclic algebras*, A. M. S. Bull, v. 37 (1931), 301–12; [**18**]: *A note on cyclic algebras of order* 16, A. M. S. Bull., v. 37 (1931), 727–30; [**19**]: *On direct products*, A. M. S. Trans., v. 33 (1931), 620–5; [**20**]: *Division algebras over an algebraic field*, A. M. S. Bull., v. 37 (1931), 777–84; [**21**]: *The structure of matrices with any normal division algebra of multiplications*, Ann. of Math., v. 32 (1931), 131–48; [**22**]: *On the construction of cyclic algebras with a given exponent*, Am. J., v. 54 (1932), 1–13; [**23**]: *Algebras of degree* 2^e *and pure Riemann matrices*, Ann. of Math., v. 33 (1932), 311–8; [**24**]: *A construction of non-cyclic normal division algebras*, A. M. S. Bull., v. 38 (1932), 449–56; [**25**]: *Normal division algebras of degree four over an algebraic field*, A. M. S. Trans., v. 34 (1932), 363–72; [**26**]: *A note on normal division algebras of order sixteen*, A. M. S. Bull., v. 38 (1932), 703–6; [**27**]: *On normal simple algebras*, A. M. S. Trans., v. 34 (1932), 620–5; [**28**]: *A note on the equivalence of algebras of degree* 2, A. M. S. Bull., v. 39 (1933), 257–8; [**29**]: *On primary normal division algebras of degree eight*, A. M. S. Bull., v. 39 (1933), 265–72; [**30**]: *Non-cyclic algebras of degree and exponent four*, A.M.S. Trans., v. 35 (1933), 112–21; [**31**]: *A note on the Dickson theorem on universal ternaries*, A. M. S. Bull., v. 39 (1933), 585–8; [**32**]: *Normal division algebras over an algebraic number field not of finite degree*, A. M. S. Bull., v. 39 (1933), 746–9; [**33**]: *On the construction of Riemann matrices*, I, Ann. of Math., v. 35 (1934), 1–28; [**34**]: *On a certain algebra of *quantum mechanics*, Ann. of Math., v. 35 (1934), 65–73; [**35**]: *Normal division algebras of degree four over F of characteristic two*, Am. J., v. 56 (1934), 75–86; [**36**]: *Integral domains of generalized quaternion algebras*, A. M. S. Bull., v. 40 (1934), 164–76; [**37**]: *Normal division algebras over a modular field*, A. M. S. Trans., v. 36 (1934), 388–94; [**38**]: *A solution of the principal problem in the theory of Riemann matrices*, Ann. of Math., v. 35 (1934), 500–15; [**39**]: *On normal Kummer fields over a non-modular field*, A. M. S. Trans., v. 36 (1934), 885–92; [**40**]: *The principal matrices of a Riemann matrix*, A. M. S. Bull., v. 40 (1934), 843–6; [**41**]: *A note on the Poincaré theorem on impure Riemann matrices*, Ann. of Math., v. 36 (1935), 151–6; [**42**]: *On the construction of Riemann matrices*, II, Ann. of Math., v. 36 (1935), 376–94; [**43**]: *Involutorial simple algebras and real Riemann matrices*, Ann. of Math., v. 36 (1935), 886–964; [**44**]: *Normal division algebras of degree p^e over F of characteristic p*, A. M. S. Trans., v. 39 (1936), 183–8; [**45**]: *Simple algebras of degree p^e over a centrum of characteristic p*, A. M. S. Trans., v. 40 (1936), 112–26; [**46**]: *p-algebras over a field generated by one indeterminate*, A. M. S. Bull., v. 43 (1937), 733–6; [**47**]: *Modern Higher Algebra*, Chicago, 1937; [**48**]: *Non-cyclic algebras with pure maximal subfields*, A. M. S. Bull., v. 44 (1938), 576–9; [**49**]: *Quadratic null forms over a function field*, Ann. of Math., v. 39 (1938), 494–505; [**50**]: *On cyclic algebras*, Ann. of Math., v. 39 (1938), 669–82; [**51**]: *A note on normal division algebras of prime degree*, A. M. S. Bull., v. 44 (1938), 649–52.

ALBERT, A. A., and HASSE, H. [**52**]: *A determination of all normal division algebras over an algebraic number field*, A. M. S. Trans., v. 34 (1932), 722–6.

ARCHIBALD, R. G. [**53**]: *Diophantine equations in division algebras*, A. M. S. Trans., v. 30 (1928), 819–37.

ARTIN, E. [**54**]: *Über einen Satz von Herrn J. H. M. Wedderburn*, Hamb. Abh., v. 5 (1928), 245–50; [**55**]: *Zur Theorie der hyperkomplexen Zahlen*, ibid., 251–60; [**56**]: *Zur Arithmetik hypercomplexer Zahlen*, loc. cit., 261–89.

ARTIN, E., and SCHREIER, O. [**57**]: *Algebraische Konstruktion reeler Körper*, Hamb. Abh., v. 5 (1928), 83–115; [**58**]: *Eine Kennzeichnung der reell abgeschlossenen Körper*, ibid. 225–31.

ASANO, K. [**59**]: *Über die Darstellungen einer endlichen Gruppe durch reele Kollineationen*, Imp. Aca. Tok. Pro., v. 9 (1933), 574–6; [**60**]: *Zur Diskriminante einer Algebra*, Jap. J., v. 12 (1935), 51–8.

ASANO, K., and NAKAYAMA, T. [**61**]: *Über halblinear Transformationen*, Math. Zeit., v. 115 (1937), 87–114.

ASANO, K., and SHODA, K. [**62**]: *Zur Theorie der Darstellungen einer endlichen Gruppe durch Kollineationen*, Comp. Math., v. 2 (1935), 230–40.

BIRKHOFF, G. [**63**]: *On the combination of subalgebras*, Camb. Ph. S. Pro., v. 29 (1933), 441–64; [**64**]: *Note on the paper "On the combination of subalgebras,"* Camb. Ph. S. Pro., v. 30 (1934), 200; [**65**]: *Ideals in algebraic rings*, P. N. A. S., v. 20 (1934), 571–3; [**66**]: *On the structure of abstract algebras*, Camb. Ph. S. Pro., v. 31 (1935), 433–54; [**67**]: *Representability of Lie algebras and Lie groups by matrices*, Ann. of Math., v. 38 (1937), 526–32.

BOYCE, F. [**68**]: *Certain types of nilpotent algebras*, Chicago dissertation, 1938.

BRANDT, H. [**69**]: *Zur allgemeinen Idealtheorie*, Schw. Nat. Ges. V., v. 26 (1927), 318–19; [**70**]: *Idealtheorie in Quaternionenalgebren*, Math. Ann., v. 99 (1928), 1–29; [**71**]: *Idealtheorie in einer Dedekindschen Algebra*, D. M. V. Jahr., v. 37 (1928), 5–7; [**72**]: *Primidealzerlegung in einer Dedekindschen Algebra*, Schw. Nat. Ges. V., v. 28 (1929), 288–90; [**73**]: *Zur Idealtheorie Dedekindscher Algebren*, Comm. Math. Helv., v. 2 (1930), 13–17.

BRAUER, R. [**74**]: *Über Zusammenhänge zwischen arithmetischen und invariantentheoretischen Eigenschaften von Gruppen linearer Substitutionen*, Ak. Berlin S. B., v. 30 (1926), 410–6; [**75**]: *Über die Darstellung der Drehungsgruppe durch Gruppen linearer Substitutionen*, Berlin dissertation, 1925; [**76**]: *Untersuchungen über die arithmetischen Eigenschaften von Gruppen linearer Substitutionen*, Math. Zeit., v. 28 (1928), 677–96; [**77**]: ibid., II, v. 31 (1930), 733–47; [**78**]: *Über Systeme hyperkomplexer Zahlen*, Math. Zeit., v. 29 (1929),

79–107; [**79**]: *Die stetigen Darstellungen der complexen orthogonalen Gruppe*, Ak. Berlin S. B., v. 29 (1929), 626–38; [**80**]: *Über die algebraische Struktur von Schiefkörpern*, J. für Math., v. 166 (1932), 241–52; [**81**]: *Über die Konstruktion der Schiefkörper, die von endlichem Rang in bezug auf ein gegebenes Zentrum sind*, J. für Math., v. 168 (1932), 44–64; [**82**]: *Über den Index und den Exponenten von Divisionsalgebren*, Tôhoku J., v. 37 (1933), 77–87; [**83**]: *Über die Kleinsche Theorie der algebraischen Gleichungen*, Math. Ann., v. 110 (1934), 473–500; [**84**]: *Über die Darstellung von Gruppen in Galoisschen Feldern*, Act. Sci. Ind. Paris, 1935, 15 pp.; [**85**]: *Eine Bedingung für vollständige Reduzibilität von Darstellungen gewöhnlicher und infinitesimaler Gruppen*, Math. Zeit., v. 41 (1936), 330–9; [**86**]: *On algebras which are connected with the semi-simple continuous groups*, Ann. of Math., v. 38 (1937), 857–72; [**87**]: *Sur la multiplication des caractéristiques des groups continues et semi-simples*, C. R. Paris, v. 204 (1937), 1784–6; [**88**]: *On normal division algebras of index five*, P. N. A. S., v. 24 (1938), 243–6.

Brauer, R., Hasse, H., and Noether, E. [**89**]: *Beweis eines Hauptsatzes in der Theorie der Algebren*, J. für Math., v. 167 (1931), 399–404.

Brauer, R., and Nesbitt, C. [**90**]: *On the regular representations of algebras*, P. N. A. S., v. 23 (1937), 236–40; [**91**]: *On the modular representation of groups of finite order*, Toronto St., 1938, 21 pp.

Brauer, R., and Noether, E. [**92**]: *Über minimale Zerfällungskörper irreduzibler Darstellungen*, Ak. Berlin S. B., v. 32 (1927), 221–6.

Brauer, R., and Schur, I. [**93**]: *Zum Irreduzibilitätsbegriff in der Theorie der Gruppen linearer homogener Substitutionen*, Ak. Berlin S. B., v. 34 (1930), 209–26.

Bush, L. E. [**94**]: *Note on the discriminant matrix of an algebra*, A. M. S. Bull., v. 38 (1932), 49–51; [**95**]: *On Young's definition of an algebra*, A. M. S. Bull., v. 39 (1933), 142–8.

Casimir, H., and van der Waerden, B. L. [**96**]: *Algebraischer Beweis der vollständigen Reduzibilität der Darstellungen halbeinfacher Liescher Gruppen*, Math. Ann., v. 111 (1935), 1–12.

Cecioni, F. [**97**]: *Sopra un tipo di algebre prive di divisori dello zero*, Cir. Mat. Pal., v. 47 (1923), 209–54.

Cherubino, S. [**98**]: *Sulle varietà abeliane reale, e sulle matrici di Riemann reali*, I, Giornale, v. 60 (1922), 65–94; [**99**]: Ibid., II, v. 61 (1923), 47–68; [**100**]: *Sulle serie di potenze di una variabile, in un algebra*, Atti Lin., v. 22 (1935), 211–16.

Chevalley, C. [**101**]: *Sur un théorème de M. Hasse*, C. R. Paris, v. 191 (1930), 369–70; [**102**]: *Sur la théorie des restes normiques*, C. R. Paris, v. 191 (1930), 426–8; [**103**]: *Sur la structure de la théorie du corps de classes*, C. R. Paris, v. 194 (1932), 766–9; [**104**]: *Sur la théorie du corps de classes*, Paris dissertation, 1932; [**105**]: *La théorie du symbole de restes normiques*, J. für Math., v. 169 (1932), 141–57; [**106**]: *Sur la théorie du corps de classes dans les corps finis et les corps locaux*, Tokyo J., v. 2 (1933), 365–474; [**107**]: *Sur certains idéaux d'une algèbre simple*, Hamb. Abh., v. 10 (1934), 83–105; [**108**]: *Démonstration d'une hypothèse de M. Artin*, Hamb. Abh., v. 11 (1935), 73–5; [**109**]: *Sur la théorie du corps de classes*, C. R. Paris, v. 210 (1935), 632–4; [**110**]: *L'arithmétique dans les algèbres de matrices*, Act. Sci. Ind. Paris, 323, 1936, 35 pp.

Comessatti, A. [**111**]: *Fondamenti per la geometria sopra le superficie razionali dal punto di vista reale*, Math. Ann., v. 73 (1913), 1–72; [**112**]: *Sulle varietà abeliane reali* I, Ann. di Mat., v. 2 (1935), 67–106; [**113**]: Ibid. II, v. 3 (1926), 27–71; [**114**]: *Intorno ad un nuovo carattere delle matrici di Riemann*, Acc. d'It., v. 7 (1936), 81–129.

Conwell, H. H. [**115**]: *Linear associative algebras of infinite order whose elements satisfy finite algebraic equations*, A. M. S. Bull., v. 40 (1934), 95–102.

Darkow, M. D. [**116**]: *Determination of a basis for the integral elements of certain generalized quaternion algebras*, Ann. of Math., v. 26 (1926), 263–70.

Deuring, M. [**117**]: *Zur Theorie der Normen relativzyklischer Körper*, Gött. Nach., (1931), 199–200; [**118**]: *Galoissche Theorie und Darstellungstheorie*, Math. Ann., v. 107 (1932), 140–4; [**119**]: *Algebren*, Erg. der Math., v. 4, 1935, 143 pp.; [**120**]: *Über den Hauptsatz *der Algebrentheorie*, J. für Math., v. 175 (1936), 63–4; [**121**]: *Einbettung von Algebren in Algebren mit kleinerem Zentrum*, J. für Math., v. 175 (1936), 124–8.

Dickson, L. E. [**122**]: *Definitions of a linear associative algebra by independent postulates*, A. M. S. Trans., v. 4 (1903), 21–6; [**123**]: *On hypercomplex number systems*, A. M. S. Trans., v. 6 (1905), 344–8; [**124**]: *On finite algebras*, Gött. Nach., (1905), 358–93; [**125**]: *Linear algebras in which division is always uniquely possible*, A. M. S. Trans., v. 7 (1906), 370–90; [**126**]: *On commutative linear algebras in which division is always uniquely possible*, A. M. S. Trans., v. 7 (1906), 514–22; [**127**]: *On triple algebras and ternary cubic forms*, A. M. S. Bull., v. 14 (1908), 160–9; [**128**]: *Linear algebras*, A. M. S. Trans., v. 13 (1912), 59–73; [**129**]: *Linear algebras*, Cambridge, 1914, 73 pp.; [**130**]: *Linear associative algebras and abelian equations*, A. M. S. Trans., v. 15 (1914), 31–46; [**131**]: *On the relation between linear algebras and continuous groups*, A. M. S. Bull., v. 22 (1915), 53–61; [**132**]: *On quaternions and their generalization and the history of the eight square theorem*, Ann. of Math., v. 20 (1919), 155–71; [**133**]: *Arithmetic of quaternions*, London M. S. J., v. 20 (1931), 225–32; [**134**]: *Quaternions and their generalizations*, Mess. Math., v. 7 (1921), 109–14; [**135**]: *Impossibility of restoring unique factorization in a hypercomplex arithmetic*, A. M. S. Bull., v. 28 (1922), 438–42; [**136**]: *Algebras and their Arithmetics*, Chicago, 1923, 241 pp.; [**137**]: *The rational linear algebras of maximum and minimum ranks*, London M. S. Pro., v. 22 (1923), 145–62; [**138**]: *A new simple theory of hypercomplex integers*, J. de Math., v. 2 (1923), 281–326; [**139**]: *Algebras and their arithmetics*, A. M. S. Bull., v. 30 (1924), 247–57; [**140**]: *On the theory of numbers and generalized quaternions*, Am. J., v. 46 (1924), 1–16; [**141**]: *New division algebras*, A. M. S. Trans., v. 28 (1926), 207–34; [**142**]: *Algebren und ihre Zahlentheorie*, Zurich, 1927, 308 pp.; [**143**]: *New division algebras*, A. M. S. Bull., v. 34 (1928), 555–60; [**144**]: *Further development of the theory of arithmetics of algebras*, Toronto Congress Proceedings (1928), 173–84; [**145**]: *Construction of division algebras*, A. M. S. Trans., v. 32 (1930), 319–34; [**146**]: *Linear algebras with associativity not assumed*, Duke J., v. 1 (1935), 113–25.

Dorroh, J. L. [**147**]: *Concerning adjunctions to algebras*, A. M. S. Bull., v. 38 (1932), 85–8.

Eichler, M. [**148**]: *Untersuchungen in der Zahlentheorie der rationalen Quaternionenalgebren*, J. für Math., v. 174 (1936), 129–59; [**149**]: *Über die Idealklassenzahl total definiter Quaternionenalgebren*, Math. Zeit., v. 43 (1937), 102–9; [**150**]: *Bestimmung der Idealklassenzahl in gewissen normalen einfachen Algebren*, J. für Math., v. 176 (1937), 192–202; [**151**]: *Über die Einheiten der Divisionsalgebren*, Math. Ann., v. 114 (1937), 635–54; [**152**]: *Über die Idealklassenzahl hypercomplexer Systeme*, Math. Zeit., v. 43 (1938), 481–94.

Finan, E. J. [**153**]: *A determination of the domains of integrity of the complete matric algebra of order* 4, Am. J., v. 53 (1931), 920–8; [**154**]: *On the number theory of certain non-maximal domains of the total matric algebra of order* 4, Duke J., v. 1 (1935), 484–90.

Fitting, H. [**155**]: *Die Theorie der Automorphismenringe Abelscher Gruppen und ihr Analogon bei nicht kommutativen Gruppen*, Math. Ann., v. 107 (1932), 514–42; [**156**]: *Der Normenbegriff für die Ideale eines Ringes beliebiger Struktur*, J. für Math., v. 178 (1937), 107–22.

Fueter, R. [**157**]: *Reziprozitätsgesetze in quadratisch-imaginären Körpern*, Gött. Nach., (1927), 427–45; [**158**]: *Über eine speziale Algebra*, J. für. Math., v. 167 (1931), 52–61; [**159**]: *Analytische Funktionen einer Quaternionenvariablen*, Comm. Math. Helv., v. 4 (1932), 9–20; [**160**]: *Formes d'Hermite, groupe de Picard et théorie des ideaux de quaternions*, C. R. Paris, v. 194 (1932), 2009–11; [**161**]: *Quaternionenringe*, Comm. Math. Helv., v. 6 (1934), 199–222; [**162**]: *Zur Theorie der Brandtschen Quaternionenalgebren*, Math. Ann., v. 110 (1934), 650–61; [**163**]: *Über die analytische Darstellung der regulären Funktionen einer Quaternionenvariablen*, Comm. Math. Helv., v. 8 (1936), 371–8; [**164**]: *Die Theorie der regulären Funktionen einer Quaternionenvariablen*, Oslo Congress Proceedings, 1937, 75–91.

Furtwangler, P., and Taussky, O. [**165**]: *Über Schiefringe*, Ak. Vienna S. B., v. 7 (1936), 38.

GARVER, R. [166]: *Division algebras of order sixteen*, Ann. of Math., v. 28 (1927), 493–500.

GHENT, K. S. [167]: *A note on nilpotent algebras in four units*, A. M. S. Bull., v. 40 (1934), 331–8.

GRELL, H. [168]: *Zur Normentheorie in hyperkomplexen Systemen*, J. für Math., v. 162 (1930), 60–62.

GRIFFITHS, L. W. [169]: *Generalized quaternion algebras and the theory of numbers*, Am. J., v. 50 (1928), 303–14.

GRUNWALD, W. [170]: *Ein allgemeines Existenztheorem für algebraische Zahlkörper*, J. für Math., v. 169 (1933), 103–7.

HAANTJES, J. [171]: *Halblineare Transformationen*, Math. Ann., v. 114 (1937), 293–304.

HASSE, H. [172]: *Darstellbarkeit von Zahlen durch quadratische Formen in einem beliebigen algebraischen Zahlkörper*, J. für Math., v. 152 (1923), 113–30; [173]: *Äquivalenz quadratischer Formen in einem beliebigen algebraischen Zahlkörper*, J. für Math., v. 153 (1924), 158–62; [174]: *Bericht über neuere Untersuchungen und Probleme der algebraischen Zahlkörper*, D. M. V. Jahr. Supplementary v. 6 (1930), p. 38; [175]: *Neue Begründung und Verallgemeinerung der Theorie des Normenrestsymbols*, J. für Math., v. 162 (1930), 134–44; [176]: *Die Normenresttheorie relativ-Abelscher Zahlkörper als Klassenkörpertheorie im Kleinen*, J. für Math., v. 162 (1930), 145–54; [177]: *Über 𝔓-adische Schiefkörper und ihre Bedeutung für die Arithmetik hyperkomplexer Zahlkörper*, Math. Ann., v. 104 (1931), 495–534; [178]: *Beweis eines Satzes und Widerlegung einer Vermutung über das allgemeine Normenrestsymbol*, Gött. Nach., (1931), 64–9; [179]: *Theory of cyclic algebras over an algebraic number field*, A. M. S. Trans., v. 34 (1932), 171–214; [180]: *Additional note to the author's "Theory of cyclic algebras over an algebraic number field,"* ibid., 727–30; [181]: *Die Struktur der R. Brauerschen Algebrenklassengruppe über einem algebraischen Zahlkörper*, Math. Ann., v. 107 (1933), 731–760; [182]: *Über gewisse Ideale in einer einfachen algebra*, Act. Sci. Ind. Paris, 1934.

HASSE, H., and SCHILLING, O. [183]: *Die Normen aus einer normalen Divisionsalgebra über einem algebraischen Zahlkörper*, J. für Math., v. 174 (1936), 248–252.

HASSE, H., and SCHMIDT, F. K. [184]: *Die Struktur diskret bewerteter Körper*, J. für Math., v. 170 (1934), 4–63.

HAZLETT, O. [185]: *On the classification and invariantive characterization of nilpotent algebras*, Am. J., v. 38 (1916), 109–38; [186]: *On the theory of associative division algebras*, A. M. S. Trans., v. 18 (1917), 167–76; [187]: *On the arithmetic of a general associative algebra*, Toronto Congress Proceedings, v. 1 (1934), 185–91; [188]: *On division algebras*, A. M. S. Trans., v. 32 (1930), 912–25; [189]: *Integers as matrices*, Bologna Congress Proceedings, v. 2 (1928), 57–62.

HECKE, E. [190]: *Theorie der Algebraischen Zahlen*, Leipzig, 1923, 264 pp.

HENKE, K. [191]: *Zur arithmetischen Idealtheorie hyperkomplexer Zahlen*, Hamb. Abh., v. 11 (1935), 311–32.

HERTER, M. [192]: *Klassenzahl von Ringidealen ganzer Hurwitzscher Quaternionen und Hermitscher Formen*, Zurich dissertation, 1936, 59 pp.

HEY, K. [193]: *Analytische Zahlentheorie in Systemen hyperkomplexer Zahlen*, Hamburg dissertation, 1929.

HEYTING, A. [194]: *Die Theorie der linearen Gleichungen in einer Zahlenspezies mit nichtkommutativer Multiplikation*, Math. Ann., v. 98 (1927), 465–90.

HIRSCH, K. A. [195]: *A note on non-commutative polynomials*, London M. S. J., v. 12 (1937), 264–6.

HUA, L. K. [196]: *On a certain kind of operations connected with linear algebras*, Tôhoku J., v. 41 (1935), 222–46.

HULL, R. [197]: *Maximal orders in rational cyclic algebras of odd prime degree*, A. M. S. Trans., v. 38 (1934), 515–30; [198]: *The maximal orders of generalized quaternion division algebras*, A. M. S. Trans., v. 40 (1936), 1–11; [199]: *Note on the ideals of rational cyclic algebras of odd prime degree*, Am. J., v. 43 (1937), 384–87.

INGRAHAM, M. H. **[200]**: *Note on the reducibility of algebras without a finite base*, A. M. S. Bull., v. 38 (1932), 100–4; **[201]**: *A note on determinants*, A. M. S. Bull., v. 43 (1937), 579–80; **[202]**: *On a certain equation in matrices whose elements belong to a division algebra*, A. M. S. Bull., v. 44 (1938), 114–24.

INGRAHAM, M. H., and WOLF, M. C. **[203]**: *Relative linear sets and similarity of matrices whose elements belong to a division algebra*, A. M. S. Trans., v. 42 (1937), 16–31.

JACOBSON, N. **[204]**: *Non-commutative polynomials and cyclic algebras*, Ann. of Math., v. 35 (1934), 197–208; **[205]**: *A note on non-commutative polynomials*, ibid., 209–10; **[206]**: *Rational methods in the theory of Lie algebras*, Ann. of Math., v. 36 (1935), 987–91; **[207]**: *Totally disconnected locally compact rings*, Am. J., v. 58 (1936), 433–49; **[208]**: *A class of normal simple Lie algebras of characteristic zero*, Ann. of Math., v. 38 (1937), 508–14; **[209]**: *Pseudolinear transformations*, Ann. of Math., v. 38 (1937), 484–587; **[210]**: *A note on non-associative algebras*, Duke J., v. 3 (1937), 544–8; **[211]**: *A note on topological fields*, Am. J., v. 59 (1937), 889–894; **[212]**: *p-algebras of exponent p*, A. M. S. Bull., v. 43 (1937), 667–70; **[213]**: *Abstract derivation and Lie algebras*, A. M. S. Trans., v. 42 (1937), 206–24; **[214]**: *Simple Lie algebras of type A*, Ann. of Math., v. 39 (1938), 181–8.

JACOBSON, N., and TAUSSKY, O. **[215]**: *Locally compact rings*, P. N. A. S., v. 21 (1935), 106–8.

JORDAN, P. **[216]**: *Über eine Klasse nichtassoziativer hyperkomplexer Algebren*, Gött. Nach., (1932), 569–75; **[217]**: *Über Verallgemeinerungsmöglichkeiten der Formalismus der Quantenmechanik*, Gött. Nach., 1933, 209–14; **[218]**: *Über die Multiplikation quantenmechanischer Grössen*, Zeit. Phy., v. 80 (1933), 285–91.

JORDAN, P., VON NEUMANN, J., and WIGNER, E. **[219]**: *On an algebraic generalization of the quantum mechanical formalism*, Ann. of Math., v. 35 (1934), 29–64.

KETCHUM, P. W. **[220]**: *Analytic functions of hypercomplex variables*, Ann. of Math., v. 30 (1928), 641–67.

KODAIRA, K. **[221]**: *Über die Struktur des endlichen, vollständig primären Ringes mit verschwindendem Radikalquadrat*, Jap. J., v. 14 (1937), 15–21.

KORINEK, W. **[222]**: *Quadratische Körper in Quaternionenringen*, abstracted in the Zentralblatt, v. 3 (1932), 53; **[223]**: *Maximale kommutative Körper in einfachen Systemen hyperkomplexer Zahlen*, ibid., v. 7 (1933), 293; **[224]**: *Une remarque concernant l'arithmétique des nombres hyperkomplexes*, loc. cit.

KÖTHE, G. **[225]**: *Über maximale nilpotente Unterringe und Nilringe*, Math. Ann., v. 103 (1930), 359–63; **[226]**: *Die Struktur der Ringe deren Restklassenring nach dem Radikal vollständig reduzible ist*, Math. Zeit., v. 32 (1930), 161–86; **[227]**: *Abstrakte Theorie nichtkommutativer Ringe mit einer Anwendung auf die Darstellungstheorie kontinuierlicher Gruppen*, Math. Ann., v. 103 (1930), 545–72; **[228]**: *Ein Beitrag zur Theorie der kommutativen Ringe ohne Endlichkeitsvoraussetzung*, Gött. Nach., (1930), 195–207; **[229]**: *Schiefkörper unendlichen Ranges über dem Zentrum*, Math. Ann., v. 105 (1931), 15–39; **[230]**: *Über Schiefkörpern mit Unterkörpern zweiter Art über dem Zentrum*, J. für Math., v. 166 (1932), 182–84; **[231]**: *Erweiterung des Zentrums einfacher Algebren*, Math. Ann., v. 107 (1933), 761–6.

KRULL, W. **[232]**: *Über verallgemeinerte endliche Abelsche Gruppen*, Math. Zeit., v. 23 (1925), 161–96.

LANDHERR, W. **[233]**: *Über einfache Liesche Ringe*, Hamb. Abh., v. 11 (1937), 41–64; **[234]**: *Liesche Ringe von Typus A*, Hans. Abh., v. 12 (1938), 200–41.

LATIMER, C. G. **[235]**: *Arithmetics of generalized quaternion algebras*, Am. J., v. 48 (1926), 57–66; **[236]**: *On the finiteness of the class number in a semi-simple algebra*, A. M. S. Bull., v. 40 (1934), 433–5; **[237]**: *On ideals in generalized quaternion algebras, and Hermitian forms*, A. M. S. Trans., v. 38 (1935), 436–46; **[238]**: *On the fundamental number of a rational generalized quaternion algebra*, Duke J., v. 1 (1935), 433–5; **[239]**: *On the class number of a quaternion algebra with a negative fundamental number*, A. M. S. Trans., v. 40 (1936), 318–23; **[240]**: *On ideals in a quaternion algebra and representation* *of integers by Hermitian forms*, ibid., 439–49; [**241**]: *The quadratic subfields of a generalized quaternion algebra*, Duke J., v. 2 (1936), 681–4; [**242**]: *The classes of integral sets in a quaternion algebra*, Duke J., v. 3 (1937), 237–47.

LEFSCHETZ, S. [**243**]: *On the real folds of abelian varieties*, P. N. A. S., v. 5 (1919), 103–5; [**244**]: *Real hypersurfaces contained in abelian varieties*, ibid., 296–8; [**245**]: *On certain numerical invariants of algebraic varieties with application to abelian varieties*, A. M. S. Trans., v. 22 (1921), 327–482; [**246**]: *Selected topics on algebraic geometry*, N. R. C. Bull., v. 63 (1928), 310–95.

LEVITZKI, J. [**247**]: *Über vollständige reduzible Ringe und Unterringe*, Math. Zeit., v. 33 (1931), 663–91; [**248**]: *Über nilpotente Subringe*, Math. Ann., v. 105 (1931), 620–7; [**249**]: *On normal products of algebras*, Ann. of Math., v. 33 (1932), 377–402; [**250**]: *On automorphisms of certain rings*, Ann. of Math., v. 36 (1935), 984–92; [**251**]: *On the equivalence of the nilpotent elements of a semi-simple ring*, Comp. Math., v. 5 (1938), 392–402.

LITTLEWOOD, D. E. [**252**]: *The solution of linear congruences in quaternions*, London M. S. Pro., v. 32 (1931), 115–28; [**253**]: *Identical relations satisfied by an algebra*, ibid., 312–20; [**254**]: *On the classification of algebras*, loc. cit., v. 35 (1933), 200–40.

LITTLEWOOD, D. E., and RICHARDSON, A. R. [**255**]: *Fermat's equation in real quaternions*, London M. S. Pro., v. 32 (1931), 235–40; [**256**]: *Concomitants of polynomials in noncommutative algebra*, London M. S. J., v. 35 (1933), 325–79.

LO VOI, A. [**257**]: *Intorno alla construzione delle matrici di Riemann e alle loro moltiplicazioni complesse*, Cir. Mat. Pal., v. 55 (1931), 287–357; [**258**]: Ibid., 477–88.

MAASS, H. [**259**]: *Beweis des Normensatzes in einfachen hyperkomplexen Systemen*, Hans. Abh., v. 12 (1937), 64–9.

MACDUFFEE, C. C. [**260**]: *On the independence of the first and second matrices of an algebra*, A. M. S. Bull., v. 35 (1929), 344–9; [**261**]: *An introduction to the theory of ideals in linear associative algebras*, A. M. S. Trans., v. 31 (1929), 71–90; [**262**]: *The discriminant matrices of a linear associative algebra*, Ann. of Math., v. 32 (1931), 60–6; [**263**]: *The discriminant matrix of a semi-simple algebra*, A. M. S. Trans., v. 33 (1931), 425–32; [**264**]: *Ideals in linear algebras*, A. M. S. Bull., v. 37 (1931), 841–53; [**265**]: *The Theory of Matrices*, Erg. der Math., v. 2, 1933, 110 pp.; [**266**]: *Matrices with elements in a principal ideal ring*, A. M. S. Bull., v. 39 (1933), 564–84.

MAEDA, F. [**267**]: *Ring-decomposition without chain condition*, Hiro. J., v. 8 (1938), 145–67.

MALCEV, A. [**268**]: *On the immersion of an algebraic ring into a field*, Math. Ann., v. 113 (1937), 686–91.

McCOY, N. H. [**269**]: *On the characteristic roots of matrix polynomials*, A. M. S. Bull., v. 42 (1936), 592–600.

McCOY, N. H. and MONTGOMERY, D. [**270**]: *A representation of generalized Boolean rings*, Duke J., v. 3 (1937), 455–9.

MOISIL, G. C. [**271**]: *Remarques sur quelques types d'algèbres non-associatives*, Jassy Ann., v. 20 (1935), 10–36.

MOLIEN, T. [**272**]: *Zahlensysteme mit einer Haupteinheit*, abstracted in Zentralblatt, v. 17 (1937–8), 53.

MORI, S. [**273**]: *Über allgemeine Multiplikationsringe*, I, Hiro. J., v. 4 (1934), 1–26; [**274**]: Ibid., II, 99–109; [**275**]: *Über primäre Ringe*, loc. cit., v. 5 (1935), 131–9; [**276**]: *Über Totalnullteiler kommutativer Ringe mit abgeschwächtem U-satz*, I, loc. cit., v. 6 (1936), 139–46; [**277**]: Ibid., II, 257–69.

MORIYA, M. [**278**]: *Divisionsalgebren über einem $\mathfrak{p}$-adischen Zahlkörper eines unendlichen Zahlkörpers*, Imp. Aca. Tok. Pro., v. 12 (1936), 183–4.

MORIYA, M., and SCHILLING, O. F. G. [**279**]: *Zur Klassenkörpertheorie über unendlichen perfekten Körpern*, Hokk. J., v. 5 (1937), 189–205.

MOUFGANG, R. [**280**]: *Alternativkörper und der Satz vom vollständigen Vierseit* (D_9), Hamb. Abh., v. 9 (1933), 207–22; [**281**]: *Zur Struktur Alternativkörper*, Math. Ann., v. 110 (1934), 416–30; [**282**]: *Einige Untersuchungen über geordnete Schiefkörper*, J. für Math., v. 176 (1937), 203–23.

MURRAY, F. J., and VON NEUMANN, J. [**283**]: *On rings of operators*, Ann. of Math., v. 37 (1936), 116–229; [**284**]: Ibid., II, A. M. S. Trans., v. 41 (1937), 208–48.

NAGUMO, M. [**285**]: *Einige analytische Untersuchungen in linearen metrischen Ringen*, Jap. J., v. 13 (1936), 61–80.

NAKAMURA, T. [**286**]: *Ein Satz über p-adische Schiefkörper*, Imp. Aca. Tok. Pro., v. 10 (1934), 198–9.

NAKAYAMA, T. [**287**]: *Über die Definition der Shodaschen Discriminante eines normalen einfachen hyperkomplexen Systeme*, Imp. Aca. Tok. Pro., v. 10 (1934), 447–9; [**288**]: *Über das direkte Produkt zweier einfachen Algebren mit zueinander teilerfremden p-Indizen*, Jap. J., v. 12 (1935), 27–36; [**289**]: *Über das direkte Zerlegung einer Divisionsalgebra*, Jap. J., v. 12 (1935), 65–70; [**290**]: *Über die Beziehungen zwischen den Faktorensystemen und der Normklassengruppe eines galoisschen Erweiterungskörpers*, Math. Ann., v. 112 (1935), 85–91; [**291**]: *Eine Bemerkung über die Summe und den Durchschmitt von zwei Idealen in einer Algebra*, Imp. Aca. Tok. Pro., v. 12 (1936), 179–82; [**292**]: *Maximalordnungen und Erweiterung des Koeffizienten-Körpers eines hyperkomplexen Systems*, Jap. J., v. 13 (1937), 333–59; [**293**]: *Über die Algebren über einem Körper von der Primzahlcharakteristik*, Imp. Aca. Tok. Pro., v. 11 (1935), 305–6; [**294**]: Ibid., II, v. 12 (1936), 113–4; [**295**]: *Über die Klassifikation halblinearer Transformationen*, P.-M. S. Jap., v. 19 (1937), 99–107; [**296**]: *Divisionsalgebren über diskret bewerteten perfekten Körpern*, J. für Math., v. 178 (1937), 11–3; [**297**]: *Some studies on regular representations, induced representations, and modular representations*, Ann. of Math., v. 39 (1938), 361–9.

NAKAYAMA, T., and SHODA, K. [**298**]: *Über die Darstellung einer endlichen Gruppe durch halblinearen Transformationen*, Jap. J., v. 12 (1936), 109–22.

NEHRKORN, H. [**299**]: *Über absolute Idealklassengruppen und Einheiten in algebraischen Zahlkörpern*, Hamb. Abh., v. 9 (1933), 318–34.

VON NEUMANN, J. [**300**]: *On regular rings*, P. N. A. S., v. 22 (1936), 707–13; [**301**]: *Algebraic theory of continuous geometries*, P. N. A. S., v. 23 (1937), 16–22; [**302**]: *Continuous rings and their arithmetics*, ibid., v. 23 (1937), 341–9.

VON NEUMANN, J., and STONE, M. H. [**303**]: *The determination of representative elements in the residual classes of a Boolean algebra*, Fund. Math., v. 25 (1935), 353–78.

NOETHER, E. [**304**]: *Der Diskriminantensatz für die Ordnungen eines algebraischen Zahl oder Funktionenkörpers*, J. für Math., v. 157 (1927), 82–104; [**305**]: *Hyperkomplexe Grössen und Darstellungstheorie*, Math. Zeit., v. 30 (1929), 641–92; [**306**]: *Hypercomplexe Systeme in ihre Beziehungen zur kommutativen Algebra und Zahlentheorie*, Zurich Congress Proceedings (1932), 189–94; [**307**]: *Nichtkommutative Algebra*, Math. Zeit., v. 37 (1933), 514–41; [**308**]: *Der Hauptgeschlechtssatz für relative-galoissche Zahlkörper*, Math. Ann., v. 108 (1933), 411–19; [**309**]: *Zerfallende verschränkte Produkte und ihre Maximalordnungen*, Act. Sci. Ind. Paris, 1934.

NOWLAN, F. S. [**310**]: *On the direct product of a division algebra and a total matric algebra*, A. M. S. Bull., v. 36 (1930), 265–8.

NOWLAN, F. S., and HULL, R. [**311**]: *Sets of integral elements of certain rational division algebras*, R. S. Can. Trans., v. 31 (1937), 163–84.

OLSON, H. L. [**312**]: *Linear congruences in a general arithmetic*, Ann. of Math., v. 28 (1926), 237–9; [**313**]: *Doubly divisible quaternions*, Ann. of Math., v. 31 (1930), 371–4.

ORE, O. [**314**]: *Linear equations in non-commutative fields*, Ann. of Math., v. 32 (1931), 463–77; [**315**]: *Theory of noncommutative polynomials*, Ann. of Math., v. 34 (1933), 480–508; [**316**]: *On a special class of polynomials*, A. M. S. Trans., v. 35 (1933), 559–84.

OSIMA, M. [**317**]: *Über die Darstellung einer Gruppe durch halblineare Transformationen*, P.-M. S. Jap., v. 20 (1938), 1–5.

PALL, G. [**318**]: *The quaternionic congruence* $tat \equiv b \pmod{g}$, *and the equation* $h(gn + 1) = x_1^2 + x_2^2 + x^2$, Am. J., v. 59 (1937), 895–913.

PERLIS, S. [**319**]: *Maximal orders in rational cyclic algebras of composite degree*, A. M. S. Trans., v. 46 (1939), 82–96.

RANUM, A. [**320**]: *The groups belonging to a linear associative algebra*, Am. J., v. 49 (1927), 285–308.

RAUTER, H. [**321**]: *Quaternionen mit Komponenten aus einem Körper von Primzahlcharakteristik*, Math. Zeit., v. 29 (1929), 234–63.

REES, M. S. [**322**]: *Division algebras associated with an equation whose group has four generators*, Am. J., v. 54 (1932), 51–65.

REICHARDT, H. [**323**]: *Die Diskriminate einer normalen einfachen Algebra*, J. für Math., v. 173 (1935), 31–4.

RICHARDSON, A. R. [**324**]: *Hypercomplex determinants*, Mess. Math., v. 55 (1926), 145–52; [**325**]: *Concomitants relating to a division algebra*, ibid., 175–82; [**326**]: *Equations over a division algebra*, loc. cit., v. 57 (1928), 1–6; [**327**]: *Simultaneous linear equations over a division algebra*, London M. S. Pro., v. 28 (1928), 395–420.

RINEHART, R. F. [**328**]: *Some properties of the discriminant matrices of a linear associative algebra*, A. M. S. Bull., v. 42 (1936), 570–6.

RINGLEB, F. [**329**]: *Beiträge zur Funktionentheorie in hyperkomplexen Systemen* I, Cir. Mat. Pal., v. 57 (1933), 311–40; [**330**]: *Bemerkung zur Arbeit des Verfassers: "Beiträge zur Funktionentheorie in hyperkomplexen Systemen* I," ibid., 476–7.

ROSATI, C. [**331**]: *Sulle matrici di Riemann*, Cir. Mat. Pal., v. 53 (1929), 79–134.

ROSENBLUTH, E. [**332**]: *Die arithmetische Theorie und die Konstruktion der Quaternionenkörper auf klassenkörpertheoretische Grundlage*, Monat., v. 41 (1934), 85–125.

RUSAM, F. [**333**]: *Matrizenringe mit Koeffizienten aus endlichen Ringen ganzer Zahlen*, Erlangen dissertation, 1931, 71 pp.

SCHILLING, O. F. G. [**334**]: *Über gewisse Beziehungen zwischen der Arithmetik hyperkomplexer Zahlsysteme und algebraischer Zahlkörper*, Math. Ann., v. 111 (1935), 372–98; [**335**]: *Über die Darstellungen endlicher Gruppen*, J. für Math., v. 174 (1936), 188; [**336**]: *Zur algebraischen Theorie der Funktionenkörper mehrerer Variablen*, J. für Math., v. 175 (1936), 1–5; [**337**]: *Einheitentheorie in rationalen hyperkomplexen Systemen*, J. für Math., v. 175 (1936), 246–51; [**338**]: *Arithmetic in a special class of algebras*, Ann. of Math., v. 38 (1937), 116–19; [**339**]: *Some remarks on class field theory over infinite fields of algebraic numbers*, Am. J., v. 59 (1937), 414–22; [**340**]: *Class fields of infinite degree over p-adic number fields*, Ann. of Math., v. 38 (1937), 469–76; [**341**]: *The structure of certain rational infinite algebras*, Duke J., v. 3 (1937), 303–10; [**342**]: *Arithmetic in fields of formal power series in several variables*, Ann. of Math., v. 38 (1937), 551–76; [**343**]: *The structure of local class field theory*, Am. J., v. 40 (1938), 75–100; [**344**]: *A generalization of local class field theory*, ibid., 667–704.

SCHILLING, O. F. G., and HASSE, H. [**345**]: *Die Normen aus einer normalen Divisionsalgebra über einem algebraischen Zahlkörper*, J. für Math., v. 174 (1936), 248–52.

SCHILLING, O. F. G., and MORIYA, M. [**346**]: *Zur Klassenkörpertheorie über unendlichen perfekten Körpern*, Hokk. J., v. 5 (1937), 189–205; [**347**]: *Divisionsalgebren über unendlichen perfekten Körpern*, Hokk. J., v. 6 (1937), 103–11.

SCHMIDT, F. K. [**348**]: *Zur Klassenkörpertheorie im Kleinem*, J. für Math., v. 162 (1930), 155–68.

SCHOENEBERG, B. [**349**]: *Indefinite Quaternionen und Modulfunktionen*, Math. Zeit., v. 40 (1936), 94–109.

SCORZA, G. [**350**]: *Corpi Numerici e Algebra*, Messina, 1921; [**351**]: *Intorno alla teoria generale delle matrici di Riemann*, Cir. Mat. Pal., v. 41 (1916), 263–380; [**352**]: *Le algebre di ordine qualunque e le matrici di Riemann*, Cir. Mat. Pal., v. 45 (1921), 1–204; [**353**]: *Sopra un teorema fondamentale della teoria delle algebre*, Atti Lin., v. 20 (1934), 65–72; [**354**]: *Sulla struttura delle algebre pseudonulle*, ibid., 143–9; [**355**]: *Sulle algebre pseudonulle di ordine massimo*, Ann. di Mat., v. 14 (1935), 1–14; [**356**]: *Sopra una classe di algebre pseudonulle*, Atti Tor., v. 70 (1935), 196–211; [**357**]: *A propositio di un recente lavoro di A. A. Albert*, Atti Lin., v. 21 (1935), 727–32; [**358**]: *Le algebre del* 3.° *ordine*, Atti Nap., v. 20 (1938), no. 13, 1–14; [**359**]: *Le algebre del* 4.° *ordine*, Atti Nap., v. 20

(1935), no. 14, 1–83; [**360**]: *Nuovi contributi alla teoria generale delle algebre*, Atti Lin., v. 23 (1936), 915–20; [**361**]: *Nuovi contributi alla teoria generale delle algebra*, Rome Rend., v. 1 (1936), 59–82.

SHAW, J. B. [**362**]: *Synopsis of Linear Associative Algebras*, Washington, 1907.

SHODA, K. [**363**]: *Über die Automorphismen einer endlichen Abelschen Gruppe*, Math. Ann., v. 100 (1928), 674–86; [**364**]: *Über die Einheitengruppe eines endlichen Ringes*, I, Math. Ann., v. 102 (1929), 674–86; [**365**]: Ibid., II, Imp. Aca. Tok. Pro., v. 6 (1930), 93–6; [**366**]: *Über die mit einer Matrix vertauschbaren Matrizen*, Math. Zeit., v. 29 (1929), 696–712; [**367**]: *Über die Galoissche Theorie der halbeinfachen hyperkomplexen Systeme*, Math. Ann., v. 107 (1932), 252–8; [**368**]: *Bemerkungen über die Faktorensysteme einfacher hyperkomplexer Systeme*, Imp. Aca. Tok. Pro., v. 10 (1933), 57–70; [**369**]: *Über die endlichen Gruppen der Algebrenklassen mit einem Zerfällungskörper*, Imp. Aca. Tok. Pro., v. 10 (1934), 21–30; [**370**]: *Ein Kriterium für normale einfache hyperkomplexe Systeme*, Imp. Aca. Tok. Pro., v. 10 (1934), 195–7; [**371**]: *Diskriminantensatz für normale einfache hyperkomplexe Systeme*, Imp. Aca. Tok. Pro., v. 10 (1934), 315–7; [**372**]: *Diskriminantenformel für normale einfache hyperkomplexe Systeme*, Imp. Aca. Tok. Pro., v. 10 (1934), 318–21; [**373**]: *Hyperkomplexe Bedeutung der invarianten Idealklassen relativ-galoisscher Zahlkörper*, Imp. Aca. Tok. Pro., v. 12 (1935), 59–64; [**374**]: *Über die Maximalordnungen einer normalen einfachen Algebra mit einem unverzweiglen galoisschen Zerfällungskörper*, Jap. J., v. 13 (1937), 367–73.

SHODA, K., and NAKAYAMA, T. [**375**]: *Über das Produkt zweier Algebrenklassen mit zueinander primen Diskriminanten*, Imp. Aca. Tok. Pro., v. 10 (1934), 443–6.

SHOVER, G. [**376**]: *Class number in a linear associative algebra*, A. M. S. Bull., v. 39 (1933), 610–34.

SHOVER, G., and MACDUFFEE, C. C. [**377**]: *Ideal multiplication in a linear algebra*, A. M. S. Bull., v. 37 (1931), 434–8.

SKOLEM, T. [**378**]: *Zur Theorie der associativen Zahlensysteme*, Oslo, 1937.

SMITH, G. W. [**379**]: *Nilpotent algebras generated by two units i and j, such that i^2 is not an independent unit*, Am. J., v. 41 (1919), 143–64.

SPAMPINATO, N. [**380**]: *Sulle funzioni di una variabile, in un'algebra complessa, ad n unità, dotata di modulo*, Cir. mat. Pal., v. 57 (1933), 235–72; [**381**]: Ibid., II, v. 58 (1934), 105–43; [**382**]: Loc. cit., III, v. 59 (1935), 185–227; [**383**]: *Sulle funzioni totalmente derivabili in un'algebra reale e complessa dotato di modulo*, I, Atti Lin., v. 21 (1935), 621–5; [**384**]: Ibid., II, 683–7; [**385**]: *Una proprietà caratteristica delle funzioni totalmente derivabili*, I, Atti Lin., v. 21 (1935), 625–33; [**386**]: Ibid., II, 688–92.

SPEISER, A. [**387**]: *Gruppendeterminante und Körperdiskriminante*, Math. Ann., v. 77 (1916), 546–62; [**388**]: *Allgemeine Zahlentheorie*, Nat. Ges. Zur. V., v. 71 (1926), 8–48; [**389**]: *Idealtheorie in rationalen algebren*, Dickson's *Algebren*, Chapter 13; [**390**]: *Zahlentheorie in rationalen Algebren*, Comm. Math. Helv., v. 8 (1936), 391–406.

STAUFFER, R. [**391**]: *The construction of a normal basis in a separable normal extension field*, Am. J., v. 58 (1936), 585–97.

STONE, M. H. [**392**]: *The theory of representations for Boolean algebras*, A. M. S. Trans., v. 39 (1936), 37–111.

TANNAKA, T. [**393**]: *Zyklische Zerfällungskörper der einfachen Ringe über dem algebraischen Funktionenkörper*, Toh. Imp. U., Sci. Rep., v. 24 (1935), 165–73.

TAUSSKY, O. [**394**]: *Analytical methods in hypercomplex systems*, Comp. Math., v. 3 (1936), 399–407; [**395**]: *Rings with noncommutative addition*, Calc. M. S. Bull., v. 28 (1936), 245–6.

TEICHMÜLLER, O. [**396**]: *Verschränkte Produkte mit Normalringen*, Deut. Math., 1936, 92–102; [**397**]: *Multiplikation zyklischer Normalringe*, ibid., 197–238; [**398**]: *p-Algebren*, loc. cit., 362–88; [**399**]: *Diskret bewertete perfekte Körper mit unvollkommenem Restklassenkörper*, J. für Math., v. 176 (1937), 141–52; [**400**]: *Zerfällende zyklische p-algebren*, ibid., 157–60.

TELLER, J. H. D. [**401**]: *A class of quaternion algebras*, Duke J., v. 2 (1936), 280–6.

TOYODA, K. [**402**]: *On Casimir's theorem on semi-simple continuous groups*, Jap. J., v. 12 (1935), 17–20; [**403**]: *On the structure of Lie's continuous groups*, Toh. Imp. U. Sci. Rep., v. 25 (1936), 338–43.

TROST, E. [**404**]: *Über die Struktur der normalen Divisionsalgebren*, Zurich dissertation, 1933, 33 pp.

TSEN, C. C. [**405**]: *Divisionalgreben über Funktionenkörpern*, Gött. Nach. (1933), 335–9; [**406**]: *Algebren über Funktionenkörpern*, Göttingen dissertation, 1934; [**407**]: *Zur Stufentheorie der quasialgebraisch-Abgeschlossenheit kommutativer Körper*, Chi. M. S. J., v. 1 (1936), 81–92.

VANDIVER, H. S. [**408**]: *Note on an associative distributive algebra in which the commutative law of addition does not hold*, A. M. S. Bull., v. 42 (1936), 857–9.

VENKATARAYUDU, T. [**409**]: *On the linear algebra of classes of elements in a finite abelian group*, Ind. Aca. Pro., v. 5 (1937), 433–6; [**410**]: *Ideal theory of the abelian group algebra*, Ind. Aca. Pro., v. 7 (1938), 118–29.

VENKOV, B. [**411**]: *Über der Arithmetik der Quaternionen*, Bull. Aca. Sci. U. S. S. R., v. 16 (1922), 205–46; [**412**]: Ibid. (1929).

VAN DER WAERDEN, B. L. [**413**]: *Moderne Algebra*, vol. 1, Berlin, 1930; [**414**]: Ibid., vol. 2; [**415**]: *Gruppen von linearen Transformationen*, Erg. der Math., v. 4 (1935); [**416**]: *Die Klassifikation der einfachen Lieschen Gruppen*, Math. Zeit., v. 37 (1933), 446–62.

WAJNSTEJN, D. [**417**]: *Über die Clifford-Lipschitzen hyperkomplexen Zahlensysteme*, Soc. Pol. Mat. Ann., v. 16 (1938), 65–83; [**418**]: *α-Matrizen und Clifford-Zahlen*, Soc. Pol. Mat. Ann., v. 16 (1938), 162–75.

WARNING, E. [**419**]: *Bemerkung zur vorstehenden Arbeit von Herrn Chevalley*, Hamb. Abh., v. 11 (1935), 76–83.

WEBER, W. [**420**]: *Verschränkte Produkte mit Normalringen*, Deut. Math., v. 1 (1936), 90–2.

WEDDERBURN, J. H. M. [**421**]: *On a theorem in hypercomplex numbers*, R. S. Edin. Pro., v. 26 (1905), 1–3; [**422**]: *Note on hypercomplex numbers*, Edin. M. S. Pro., v. 25 (1906), 1–3; [**423**]: *On certain theorems in determinants*, Edin. M. S. Pro., v. 27 (1908), 1–3; [**424**]: *A theorem on finite algebras*, A. M. S. Trans., v. 6 (1905), 349–52; [**425**]: *On hypercomplex numbers*, London M. S. Pro., v. 6 (1907), 77–118; [**426**]: *A type of primitive algebra*, A. M. S. Trans., v. 15 (1914), 162–6; [**427**]: *On division algebras*, A. M. S. Trans., v. 22 (1921), 129–35; [**428**]: *Algebras which do not possess a finite basis*, A. M. S. Trans., v. 26 (1924), 395–426; [**429**]: *A theorem on simple algebras*, A. M. S. Bull., v. 31 (1925), 11–13; [**430**]: *Non-commutative domains of integrity*, J. für Math., v. 167 (1931), 129–41; [**431**]: *Boolean linear associative algebra*, Ann. of Math., v. 35 (1934), 185–94; [**432**]: *Lectures on Matrices*, A. M. S. Coll., v. 17, 1934, 200 pp.; [**433**]: *Note on algebras*, Ann. of Math., v. 38 (1937), 854–6; [**434**]: *A special linear associative algebra*, Edin. M. S. Pro., v. 5 (1938), 169–70.

WEYL, H. [**435**]: *Theorie der Darstellung kontinuierlicher halbeinfacher Gruppen*, Math. Zeit., v. 23 (1925), 211–309; [**436**]: Ibid., II, v. 24 (1926), 328–76; [**437**]: Loc. cit., III, 377–95; [**438**]: Loc. cit., *Appendix*, 789–91; [**439**]: *Idee der Riemannschen Fläche*, 2d edition, Leipzig, 1923; [**440**]: *On generalized Riemann matrices*, Ann. of Math., v. 35 (1934), 714–29; [**441**]: *Generalized Riemann matrices and factor sets*, Ann. of Math., v. 37 (1936), 709–45; [**442**]: *Note on matric algebras*, Ann. of Math., v. 38 (1937), 477–83; [**443**]: *Commutator algebra of a finite group of collineations*, Duke J., v. 3 (1937), 200–12; [**444**]: *The Classical Groups, Their Invariants, and Representations*, Princeton, 1939, 312 pp.

WICHMANN, W. [**445**]: *Anwendungen der p-adischen Theorie im Nichtkommutativen*, Monat., v. 44 (1936), 203–24.

WILLIAMSON, J. [**446**]: *The idempotent and nilpotent elements of a matrix*, Am. J., v. 59 (1936), 747–58.

WILSON, R. [**447**]: *The equation $px = xq$ in a linear associative algebra*, London M. S. Pro., v. 30 (1930), 357–66.

WITT, E. [**448**]: *Über die Kommutativität endlicher Schiefkörper*, Gött. Nach., 1931, 413; [**449**]: *Gegenbeispiel zum Normensatz*, Math. Zeit., v. 39 (1934), 462–7; [**450**]: *Riemann-Rochscher Satz und ζ-Funktion im Hyperkomplexen*, Math. Ann., v. 110 (1934), 12–28; [**451**]: *Zerlegung reeler algebraischer Funktionen im Quadrate; Schiefkörper in reellen Funktionen*, J. für Math., v. 171 (1934), 4–11; [**452**]: *Zwei Regeln über vershränkte Produkte*, J. für Math., v. 173 (1935), 191–2; [**453**]: *Konstruktion von galoisschen Körpern der Charakteristik p zu vorgegebener Gruppe der Ordnung p^f*, J. für Math., v. 174 (1936), 237–45; [**454**]: *Theorie der quadratischen Formen in beliebigen Körpern*, J. für Math., v. 176 (1937), 31–44; [**455**]: *Zyklische Körper und Algebren der Charakteristik p vom Grad, p^n*, ibid., 126–40; [**456**]: *Schiefkörper über diskret bewerteten Körpern*, ibid., 153–6; [**457**]: *Treue Darstellung Liescher Ringe*, J. für Math., v. 177 (1937), 152–60.

WOLF, L. A. [**458**]: *Similarity of matrices in which the elements are real quaternions*, A. M. S. Bull., v. 42 (1936), 737–43.

WYANT, E. K. [**459**]: *The ideals in the algebra of generalized quaternions over the rational number field*, abstract in A. M. S. Bull., v. 36 (1930), 204.

YOSIDA, K. [**460**]: *On the group imbedded in the metrical complete ring*, Jap. J., v. 13 (1936), 7–26; [**461**]: *A theorem concerning the semi-simple Lie group*, Tôhoku J., v. 44 (1937), 81–4.

YOUNG, J. W. [**462**]: *A new formulation for general algebra*, Ann. of Math., v. 29 (1927), 47–60.

ZASSENHAUS, H. [**463**]: *Lehrbuch der Gruppentheorie*, Leipzig, 1937.

ZORN, M. [**464**]: *Theorie der alternativen Ringe*, Hamb. Abh., v. 8 (1930), 123–47; [**465**]: *Note zur analytischen hyperkomplexen Zahlentheorie*, Hamb. Abh., v. 9 (1933), 197–201; [**466**]: *Alternativkörper und quadratische systeme*, ibid., 395–402; [**467**]: *The automorphisms of Cayley's non-associative algebra*, P. N. A. S., v. 21 (1935), 355–8; [**468**]: *On a theorem of Engel*, A. M. S. Bull., v. 43 (1937), 401–4.

INDEX